The Origins of Agriculture

AN EVOLUTIONARY PERSPECTIVE

The Origins of Agriculture

AN EVOLUTIONARY PERSPECTIVE

David Rindos

L. H. Bailey Hortorium
Cornell University
Ithaca, New York

With Foreword by Robert C. Dunnell

1984

ACADEMIC PRESS, INC.
(Harcourt Brace Jovanovich, Publishers)
Orlando   San Diego   San Francisco   New York   London
Toronto   Montreal   Sydney   Tokyo   São Paulo

ACADEMIC PRESS, INC.
Orlando, Florida 32887

United Kingdom Edition published by
ACADEMIC PRESS, INC. (LONDON) LTD.
24/28 Oval Road, London NW1 7DX

Library of Congress Cataloging in Publication Data

Rindos, David.
    The origins of agriculture.

    Bibliography: p.
    Includes index.
    1. Agriculture--Origin. I. Title.
GN799.A4R56   1983   630'.9'01   83-7165
ISBN 0-12-589280-2

PRINTED IN THE UNITED STATES OF AMERICA

84 85 86 87   9 8 7 6 5 4 3 2 1

"For Noah . . .
and all *his relations"*

Contents

5.
Feeding Behavior and Change in Diet

6.
Instability, Cultural Fecundity, and Dispersals

References

Index

Foreword

As an advocate of evolutionary theory as one scientific framework for the explanation of cultural variability, I am pleased to have been asked to write a brief foreword to David Rindos's *The Origins of Agriculture*. In this book Rindos has combined a subject of traditional anthropological and archaeological interest—the origins and dispersion of agriculture—with a new and powerful explanatory approach—evolutionary theory.

Probably no other concept in the development of anthropology has played a more central role than evolution or excited as much controversy and misunderstanding. Evolution was a key in the works of such founding fathers as Morgan and Tylor but was rejected as an overt conceptual approach in the early years of this century. It was revived in modified form by White and his students in mid-century, but in less than two decades receded to a position as only one of many interpretive themes. Stimulated in part by the influence of sociobiology and its incursions into traditionally anthropological areas, evolution is now returning to center stage.

Changes in political and social milieus have played their part in this love–hate relationship between anthropology and evolution. However, as a number of scholars have recently pointed out, the term *evolution* as used in anthropology refers to a multifarious group of ideas united only by name and a vague general resemblance to similarly denominated concepts elsewhere in science. From its scholarly beginnings in the work of Herbert Spencer, the anthropological and sociological tradition has pursued an independent course based in human–nature dichotomy. Although it benefited from the success of evolution in other areas, this analogous framework failed to internalize such key tenets of scientific evolution as random variation and natural selection. "Cultural evolution" developed as an interpretive algorithm that could be applied to traditional anthropological data in answering traditional anthropological questions. In contrast, evolutionary theory revolutionized biology by changing the kinds of questions as well as the way in which phenomena themselves were conceptualized. These major differences were obscured by the borrowing of terminologies.

Sociobiology has taken a different approach. It represents the application of modern biological evolutionary theory to cultural phenomena as conceptualized by anthropologists. Although insights into cultural variability have resulted, the effort has been plagued by the classic problems of uncritical borrowing. Biological evolution includes as theoretical elements empirical correlations that happen to be true, or at least dominantly true, only in nonhuman species. Consequently, the theory is proving to be insufficiently general to permit extension to human beings, for whom the mechanisms of trait transmission and stability of units of selection, to name but two parameters, are quite likely more various.

The sporadic love affair with evolution can be seen as a result of this history. On the one hand, anthropologists and archaeologists have been propelled toward evolutionary frameworks by their traditional interest in "why" questions and their desire to work more scientifically. On the other hand, none of the specific anthropological applications has been able to sustain detailed criticism. Without exception, the criticisms have been founded in the special characteristics of cultural evolution and sociobiology; yet they have typically been treated by their authors as a warrant to reject evolutionary theory in general. Thus, as Blute has put it, scientific evolution has remained an untried theory in anthropology.

The Origins of Agriculture fills this gap. By avoiding the historical traps of both cultural evolution and sociobiology, Rindos provides for the first time an evolutionary account of cultural phenomena that accommodates anthropological interests while remaining faithful to the scientific tenets of evolutionary theory. The contrast between Rindos's approach and that of cultural evolution in the anthropological tradition is marked. The subject matter and Rindos's particular interests in it do not produce as sharp a contrast with sociobiology, but the careful reader will differentiate the two with little difficulty.

One great strength of *The Origins of Agriculture* is the lucid exposition of the basic structure and concepts of evolutionary theory. Against this background, Rindos links the traditional interest in agriculture with evolutionary theory by reformulating the basic questions in terms that can be answered within that theoretical system. Thus domestication is not an event or invention; it is a mutualistic relation of varying degree between different species that arises under a specifiable set of conditions. As such, it is not unique to people, but its importance among human populations is not made less intelligible by denying its unique association. Similarly, agriculture is not an event or discovery. Rindos decouples it from domestication while still carefully showing the nature of the interaction between the two. He shows that agricultural systems are inherently unstable and identifies some of the consequences of that instability for human populations and for the history of such systems.

While Rindos's conclusions about agriculture and domestication are of interest in their own right, they are the more modest contribution of *The Origins of Agriculture*. This work is far more valuable as a research strategy demonstrating how evolutionary theory can be brought to bear on anthropological and archaeological questions of long-standing interest. In this respect *The Origins of Agriculture* is far from complete, but it is a solid start that can be generalized beyond agriculture. If by dint of detailed exposition and demonstration, Rindos's approach can avoid reinterpretation into the mold of cultural evolution, it will not only initiate a new direction in research on agriculture, but should stimulate far-reaching changes in anthropology and archaeology generally.

ROBERT C. DUNNELL

Preface

The study of human cultural change is an exercise in the explanation of variation, not speciation. Speciation requires significant genetic isolation and is usually associated with morphological change and genetic divergence. Human cultural change has occurred and proceeds independently of any significant genetic isolation or morphological change. This does not imply, however, that the variation in human lifeways resulting from cultural change has had no significant effects upon our species.

Most of the errors made in the name of evolution in the study of cultural change have arisen from a lack of appreciation for the distinction between variation and speciation. Sociobiologists have argued from variation (a precondition) to genetic change (a resultant state) on the basis of an analogy between cultural change and the speciation process. Cultural evolutionists have argued from a misplaced analogy between cultural change and phyletic evolution. They have confused the ''tree of life'' with the thicket of human behaviors. Cultural ecologists have attempted to describe the functional and adaptive significance of human behavior. Because speciation is frequently accompanied by the adaptive colonization of a new niche, cultural ecologists have projected a similar concept into the analysis of cultural change. And although these cultural physiologists have contributed much to our understanding of the variety of human experiences, they have improperly confused function (a ''how'' question) with evolutionary causation (a ''why'' question).

The purpose of this volume is to present an alternative approach to understanding cultural variation and change. I have one major aim: to demonstrate that domestication and the origin of agricultural systems are best understood by attempting to explicate the evolutionary forces that affected that development of domesticates and agricultural systems. By tradition the study of agricultural origins, as an example of cultural innovation, has been the privilege of social scientists. Yet, since the time of Darwin, domestication studies have been rigorously pursued by natural scientists also. In consequence the field has become somewhat of a chimera with alternative, often contradictory, types of explanations being used to describe, on the one hand, the origin of agricultural modes of subsistence and, on the other, the evolution of the plants that comprise agricultural systems. Resolution of the anomalies created by this bifurcated tradition requires not so much a balance in theory as a balanced treatment of the pertinent literatures. I hope that anthropologists and natural scientists, for whom this volume is intended, will appreciate the importance of information derived from fields other than the ones in which they were trained; the problem of agricultural origins, not academic fields, must determine the subjects to be reviewed.

The structure of this volume has turned out (somewhat, I must confess, to my surprise) to be tripartite. In the first two chapters I discuss cultural change, the domestication of plants, and the origin of agricultural systems in the most general of terms. I consider Darwinism in some depth, concentrating on the relationship between natural selection and cultural change. The fact that cultural change is not functionally a genetic process does not preclude a selectionist interpretation of cultural change. Behavioral modifications may affect the potential for survival and reproduction of humans and may thus be the basis for natural selection. In arguing that agricultural systems are amenable to analysis within this framework, I am attempting to combine within a single framework what I see as the best insights of two major schools of anthropological analysis— cultural ecology and historical particularism. I attempt to provide an explanation for the origin of agricultural systems that uses the methodology and insights of the environmentalists in conjunction with the historical perspective adopted by the historicists. The analysis is mechanistic in that one need assume only that the immediate responses of humans to their environment (which are, by and large, like those of any other animal) should enter into an analysis of change; it is selective in that change is understood in terms of differential fitness, and it is historical in that the process is seen as rooted in, and inseparable from, the history of humans as culturally evolving organisms.

Chapters 3 and 4 explore the world of domestication and agriculture and present a series of concepts that may permit a more natural explanation for these processes. Domestication is the result of coevolutionary interactions between humans and plants; it may be studied in numerous other animal–plant relationships and arises without recourse to either cultural adaptations or human intentionality. This relationship dictates not only the morphology of domesticated plants, but also the structure of the agroecology and the existence of plants such as weeds. The domestication relationship has three conceptually distinct aspects mediated by different types of human behavior and occurring in distinct environments. *Incidental domestication* is the result of human dispersal and protection of wild plants in the general environment. Over time this relationship will select for morphological changes in the plants, preadapting them for further domestication. *Specialized domestication* is mediated by the environmental impact of humans, especially in the local areas in which they reside. The most important outcome of specialized domestication is the development of a unique ecological niche—the agroecology. *Agricultural domestication,* the culmination of the other two processes, involves the further evolution of plants in response to the conditions existing with the agroecology; this last process is roughly equivalent to what has simply been termed *domestication* in the literature of agricultural origins.

The interactions of humans and plants have major effects on the areas in which people reside. After long periods of time, domesticates come to dominate regions of human habitation and new selective processes become important. The increasing predictability of the developing agroecology changes the direction of life-history selection for many cultivated plants bringing about a transition from an *r*-selected to a *K*-selected regime. One of the most important effects of specialized domestication upon humans has been an increase in "agrilocality." Analysis of the selective pressures existing under the three modes of domestication provides a rough approximation of the length of time each of them should be dominant: It is found that both incidental and early specialized domestication are characterized by negative feedback processes that reduce the rate of evolution of the symbiosis, whereas agricultural domestication is characterized by positive feedbacks that enhance the rate of its development.

The final two chapters of this volume present models for the origin and spread of agricultural systems based upon Darwinian evolutionary theory. Chapter 5 is a rather simple, albeit somewhat mathematical, treatment of how human feeding behavior may interact with the developing domestication symbiosis to produce first the domesticated plant and later the agroecology. This model demonstrates how a gradual process like coevolutionary domestication can account for radical changes in human subsistence patterns. It is shown that the type of relationship existing between humans and domesticates dictates an *exponential* relationship between abundance of domesticates and their relative contribution to the diet. This accounts for the apparent "sudden" appearance of agriculture subsistence in the archaeological record, predicts the existence of domestication symbioses at very early times, and accounts for the "broad spectrum revolution" that has been found to anticipate the appearance of developed agricultural systems in many areas of the world.

Although this model has utility in understanding the way in which agricultural systems may develop within a given region, it does not provide a complete view of the major role that agricultural subsistence played in the development of civilization. Thus in the final chapter I consider this question from a different perspective. Chapter 6 emphasizes the intrinsic properties of agricultural systems that were to bring about the spread of agricultural modes of subsistence. The spread of agricultural systems is rooted in the elaboration of evolutionary and ecological tendencies present in the specialized domestication relationship. Agricultural techniques transcend certain environmental limitations placed on the continued development of human–plant mutualism. Human behaviors that served to increase the carrying capacity of the environment for the domesticated plant were accompanied by increases in human population. At first, highly mutualistic societies, and later in time, agricultural societies came to dominate any given area. This was a result of the increases in carrying capacity created by mutualistic and agricultural domesticatory systems. However, the adoption of agricultural practices is correlated with changes in the subsistence base of populations and with changes in the morphology, distribution, and autecology of the domesticated plant; thus new instabilities are introduced into human subsistence patterns. The early evolution and subsequent dispersal of agricultural systems were directly tied to periodic drops in relative productivity that were the result of the interaction of changes in diet, the evolution of domesticates within the agroecology, and finally, the environmental sensitivity of the subsistence system. Agricultural practices maximizing instability in productivity have the highest rate of dispersal: Agricultural systems thus have not been successful because they were adaptations. Instead, the instabilities inherent in their evolution and functioning brought about their expansion and dispersal; a positive selection for instability has characterized agricultural systems from their very origins.

Perhaps the most important way in which this volume differs from others on the same subject is in the focus and type of information it attempts to provide. Two distinct approaches to understanding change have been used in the past. One approach is concerned with the evolution of behavior—what causes an animal species to change its behavior over time. Most previous models for the origin of agriculture have sought only to provide (and, I believe, in highly oversimplified terms) a direct *cause* for *change* in human behavior.

The approach taken in this volume seeks to understand the *history* of an organism or system in terms of the available evidence and general Darwinian principles. Questions of direct cause for particular changes in behavior must of necessity be left unanswered. Instead, a continuum of transitional forms is posited, and change is viewed as inherent to the system's functioning.

I would like to thank the L. H. Bailey Hortorium, Cornell University (which unintentionally subsidized much of the writing of this volume) for demonstrating the true depth of its conviction to the scientific study of the evolution of domesticated plants. Thanks are also due the Anthropology Department at Cornell (especially Tom Lynch, in whose course this whole business got started and who saw it through to the bitter end) for giving me the opportunity to be a practicing, rather than a theoretical, paleoethnobotanist, and for balancing in an awkward position with outstanding grace. Thanks also to the Graduate School for eventually proving to a cynic that Cornell is indeed "an institution where any person can find instruction in any study," and to the Miller Committee and the Department of Anthropology, University of Illinois–Urbana for allowing that study to become, again, instruction.

By longstanding tradition, here freely embraced, recognition must also be given to those who have criticized, at different times, segments of a veritable hurricane of drafts: Todd Cooke, Jim Bauml, John Henderson, Bruce Winterhalder, Natalie Uhl, Mark Cohen, Barbara Lynch, Tony Wonderly, Stan Green, Carl Bajema, Mike Whalen, Tom Lynch, Robert Dunnell, Don Lathrap, F. K. Lehman, John Lowe, and a pair of anonymous reviewers. While I must thank all of these scholars for pointing out embarrassing and sometimes absurd or silly statements in the text, I must admit I have not always been guided by their advice (as the reader may enjoy discovering). And although I would like to give them proper credit for those ideas originally theirs rather than mine, the strange thing about such culturally transmitted bits of information is that we frequently end up believing we invented them.

Finally, an inexpressible debt is owed Susan Straight just for being there.

CHAPTER 1

Agriculture, Evolution, and Paradigms

> There is no reason why the principles which have acted so efficiently under domestication should not have acted under nature.
>
> CHARLES DARWIN, *THE ORIGIN OF SPECIES*

In what may seem a bewildering transposition to most modern readers, Charles Darwin did not begin *The Origin of Species* with a discussion of such principles of evolution as variation in nature, the struggle for existence, natural selection, the geological record, or the modern distribution of species. Instead, these topics were delayed until he had presented his views on domestic gooseberries and pigeons. There is little doubt that Darwin had a profound interest in the changes that occur under cultivation[1]; his interest, however, was not that of the breeder or social scientist but that of the biologist. Domestication was far more than an analogy that might be used for the benefit of the uninitiated to show how natural selection winnows the variation existing in a species, thus bringing about fundamental evolutionary change; domestication was and is evolution. As Darwin notes in his introduction to *Origin:*

> It is, therefore, of the highest importance to gain a clear insight into the means of modification and coadaptation. At the commencement of my observations it seemed to me probable that a careful study of domesticated animals and cultivated plants would offer the best chance of making out this obscure problem. Nor have I been disappointed; in this and all other perplexing cases I have invariably found that our knowledge, imperfect though it be, of variation under domestication, afforded the best and safest clue. I may venture to express my conviction of the high value of such studies, although they have been very commonly neglected by naturalists. (1859:4)

The neglect of the great potential of studies of domestication that Darwin observed persists today, largely, I think, because of a lack of appreciation of his concept of unconscious selection.

---

[1]Several historians of science, including Ghiselin (1969), Vorzimmer (1970), Young (1971), Schweber (1977), and Ruse (1979), have stressed the primary role that domestication studies played in the creation of Darwin's theory of evolution by means of natural selection.

Unconscious selection was to Darwin the preeminent force behind the domestication of plants and animals.

> At the present time, eminent breeders try by methodical selection, with a distinct object in view, to make a new strain or sub-breed, superior to anything of the kind in the country. But for our purpose, a form of Selection, which may be called Unconscious, and which results from every one trying to possess and breed from the best individual animals, is more important. . . . Youatt gives an excellent illustration of a course of selection, which may be considered as unconscious, in so far that the breeders could never have expected, or even wished, to produce the result which ensued. (1859:34, 36)

Unconscious Selection (so conspicuously capitalized in this passage) differs from methodical selection in a most important way—that of long-range intent in breeding. Methodical selection is thus a systematic endeavor to modify a breed according to some predetermined standard, whereas unconscious selection is merely the result of man's immediate actions in effecting, over time, a change in a domesticate's gene pool.

The distinction between these two types of selection is considered at great length in Darwin's last and longest volume, *The Variation of Animals and Plants under Domestication:*

> With plants, from the earliest dawn of civilization, the best variety which was known would generally have been cultivated at each period and its seed occasionally sown; so that there will have been some selection from an extremely remote period, but without any prefixed standard of excellence or thought of the future. (1868:215)

> Unconscious selection graduates into methodical, and only extreme cases can be distinctly separated; for he who preserves a useful or perfect animal will generally breed from it with the hopes of getting offspring of the same character, but as long as he has not a predetermined purpose to improve the breed, he may be said to be selecting unconsciously. (1882:171)

> Unconscious selection in the strictest sense of the word, that is the saving of the most useful animals and the neglect or slaughter of the less useful, without any thought of the future, must have gone on occasionally from the remotest period and among the most barbarous nations. (1868:171)

> In fact, except that in the one case man acts intentionally, and in the other unintentionally, there is little difference [in the results] between methodical and unconscious selection. (1868:210)

The notion of intentionality has pervaded the study of domestication and the origin of agricultural systems. Natural scientists have ignored, in large part, the mechanistic processes that might underlie agricultural evolution because (as the unstated reasoning goes), if crops and agricultural systems are ultimately derived from individual or cultural choice or decision making, they are beyond the ken of the evolutionary biologist.

Yet, domesticates were originally wild and were thus adapted to conditions of life in nature; their traits were those of wild plants, not of cultivated ones, and as such they could not, initially, be distinguished from the other components of the flora. Are we to attribute a precognizance to Neolithic people that we would deny ourselves? And if not, how are we to account for the initial domestication of crop plants? One general model of the process calls for some sort of recognition by people of the advantages of changing their mode of subsistence. This notion extends the concept of methodical selection back in time to the earliest periods, but it misunderstands it and takes it to an improbable extreme: It demands a Lamarckian model for variation and selection—man's desire for a better crop or mode of subsistence must direct the variation to be found in nature. A Lamarckian viewpoint has guided most theories of the domestication of plants and of the origin of agricultural systems. When we claim that people chose to domesticate plants to provide a more stable and predictable source of food or that they became agricultural to solve an overpopulation problem, we are making the unconscious assumption that the plants involved were capable of responding "appropriately." The first task facing a Darwinian interpretation of the origin of agriculture is understanding why this viewpoint is so widely adopted.

Agriculture and the Paradigm of Consciousness

Gould points out that the essential difference between Lamarckism and Darwinism is in the types of variation they propose:

> Lamarckism is, fundamentally, a theory of *directed* variation. . . . variation is directed automatically toward adaptation. . . . Many people do not understand the essential role of directed variation in Lamarckism. They often argue: isn't Lamarckism true because environment does influence heredity—chemical and radioactive mutagens increase the mutation rate and enlarge a population's pool of genetic variation? The mechanism increases the *amount* of variation but does not propel it in favored directions. Lamarckism holds that genetic variation originates *preferentially* in adaptive directions.

> Darwinism, on the other hand, is a two-step process, with different forces responsible for variation and direction. Darwinians speak of genetic variation, the first step, as "random." This is an unfortunate term because we do not mean random in the mathematical sense of equally likely in all directions. We simply mean that variation occurs with no preferred orientation in adaptive directions. . . . Selection, the second step, works upon *unoriented* variation and changes a population by conferring greater reproductive success upon advantageous variants. (Gould 1980:79; emphasis original)

We may begin to appreciate the importance that Darwin placed upon unconscious selection. Methodical selection calls for selection with a "distinct object in view"; yet variation is random (i.e., unpredictable). Methodical selection, so important in the modern world, continually runs up against the problem of not encountering the necessary variation to breed for a desired end. Thus, most of the breeding that is now done seeks to *rearrange* existing variation. Although an occasional mutant form may be useful, it obviously cannot be created at will.

Darwin, a keen gardener and fancier, was well aware of this problem (1859:38–39,37)

> Man can hardly select, or only with much difficulty, any deviation of structure excepting such as is externally visible; and indeed he rarely cares for what is internal. He can never act by selection, excepting on variations which are first given to him in some slight degree by nature. . . . the gardeners of the classical period, who cultivated the best pears which they could procure, never thought what splendid fruit we should eat; though we owe our excellent fruit in some small degree, to their having naturally chosen and preserved the best varieties they could anywhere find.

Man selects, but his selection is similar to nature's—he selects the best, the most useful, the most desirable, the most vigorous, the most successful (in other words, the most "fit") plant present in the immediate environment at a given time. Splendid adaptations to man's desires occur, but they evolve over spans of time that preclude man's ever knowing what the fruits of his selection will be.

Man, like nature, is an unconscious agent selecting only for immediate benefit. Yet, most models proposed for the origin of agriculture have not stressed the unconscious aspects of man's interactions with plants. This in large part is because of a bias on the part of many who see in the origin of agriculture the beginnings of modern civilization. Forgetting that the concept of methodical selection has to do with the pursuit of goals by means of a sophisticated understanding of breeding systems, we apply it to man's earliest interactions with plants. We assume that modern man is in control of his destiny and forget that his options when dealing with the biological world are in large part dictated by processes over which he has no control. Man may indeed select, but he cannot direct the variation from which he must select.

We tend to set man outside of nature: we see culture and civilization estranging man from the natural order. We view civilization as a form of organization based on control both of nature and of human beings, and having properly identified agriculture as the foundation of civilization, we proceed to read this "paradigm of consciousness"[2] back into time to account for the very origin of agriculture, and thus of civilization itself.

> Elemental men are simply important animals in the ecological community; advanced hunters and gatherers have an important effect on the ecological community, but do not control it; domesticators of plants and animals, however, exert a great deal of influence on the physical environment and often actually control its ecological balance. Domestication has revolutionary importance because it . . . makes possible a vast increase in the quantity and degree of stabilization of the food supply. This . . . allows for a resultant, correlative increase in population, and since large populations are necessary

---

[2]Although certain workers (e.g., Wagener 1977) have used the word *mentalism* here, I prefer the term *paradigm of consciousness* because *mentalism* has a well-established and quite different meaning in philosophy (viz., that objects have no reality save in the mind of the observer); analogous logic demands rejection of *idealism*. *Paradigm* is here used in the sense of Kuhn (1962).

for the development of the civilized way of life, domestication is at the foundation of civilization. (Watson and Watson 1969b:93–94)

Theory should account not only for the "invention" of agriculture but also for its acceptance and the widespread economic transformation of human society which resulted. It is the latter rather than the invention per se which is the important historical event. (Cohen 1977a:6)

Domestication was the invention that made populous and complex human societies viable. It has thus proved to be the single most important intervention man has ever made in his environment. (Isaac 1970:1)

As cultivated plants are a prerequisite and an integral part of every advanced civilization, so are they the creation of man, and, considering the important role that their development played in our cultural evolution, we might well say that their creation has been one of the greatest achievements of the human mind. (Schwanitz 1966:2)

The history of mankind is a long and diverse series of steps by which he has achieved ecologic dominance . . . largely he has prospered by disturbing the natural order. (Sauer 1969:3–4)

The most fateful and portentous development in the whole story of man was his learning how to produce food by intention, instead of harvesting it from natural productions. The rise and fall of empires, emergence of religious leaders, and variations in social and political structures are relatively trivial subjects compared to the domestication of plants and animals. (Harlan and de Wet 1973:51)

It seems obvious that in some sense agriculture is the basis of civilization, at least *our* Western civilization. This assumption is rather self-serving, however, because it places our mode of subsistence at the pinnacle of human development. Nevertheless, it is difficult for us to identify with people of the Paleolithic but easy and automatic to identify with the early agriculturalists. What we are is, in part, understandable through the comprehension of how we came to be this way. Thus, it is not surprising that some workers have considered agricultural origins in terms of such seemingly modern concerns as ecological and population problems.

The ways of the hunters are beginning to show us how we are failing as human beings and as organisms in a world beset by a "success" that hunters never wanted. (Shepard, quoted in Reed 1977a:4)

As food production became more efficient, villages arose and in time cities came into existence, and civilization was on its way. . . . We might argue that it was neither leisure time nor a sedentary existence but the more rigorous demands associated with an agricultural way of life that led to great changes in man's culture. (Heiser 1973:2).

Agriculture generally means larger groups, larger settlements, and . . . the greater the size of the group, the greater the level of tension and conflict. . . . Also, agriculture, at least in its large-scale and more intensive forms, generally means widening difference in social status. (Pfeiffer 1976:25)

If agriculture is the basis of civilization and civilization is characterized by conscious control and production, then agriculture must have originated in the same consciousness and control. Our actions are seen as *motivated* actions;

understanding of the origins of agriculture must therefore include reflections on the motives and cultural choices involved in, or consequent on, such actions. Understanding the origin of agriculture is seen as a step toward understanding the "unnatural" state of civilization. This, I believe, is the fundamental motivation behind the application of the paradigm of consciousness to agricultural origins.

It is, of course, logically possible that agriculture did arise with the recognition that culture and civilization could be the means to manipulate the environment for man's immediate benefit. I am not going to dispute the paradigm of consciousness in its most general form; I am only interested in trying to point out that it is not required for the creation of a model for the origin of agriculture. Moreover, at least in this case, it is not only unnecessary, but is also a source of confusion: it distracts us from the fundamental processes underlying the origin of cultivated plants.

The idea that we as a culture, a nation, or a species are in conscious control of our environment and thus of our destiny is one part truth, one part rhetoric, and two parts wishful thinking. The various controls of society are often inward-directed and are generally more efficient at rationalizing the status quo than at bringing about directed long-term change. It is all very well to speak of social or cultural goals if these are restricted to the realizable or if this is simply a shorthand way of describing change that has occurred over time; to use the goals or problem-solving abilities of people as an *explanation* for long-term historical change is another matter entirely. The danger of this approach is that we may attribute powers to people or to culture that they do not have. People could not create the variation that would permit domestication, they could only select; and they could not have known how important the products of their selection would become. Darwin understood the variation occurring in domesticates far better than most modern scholars and was therefore reluctant to attribute to humans any great deal of awareness of the *ultimate* effects of their selection from among this variation. This difference in approach is essential to a Darwinian view of variation, selection, and evolution. Thus, Darwin ends Chapter 1 of *Origin* (1859:43) by stating: "Over all these causes of Change, the accumulative action of Selection, whether applied methodically and quickly, or unconsciously and slowly but more efficiently, seems to have been the predominant Power."

The success of agriculture in the modern world has made it tempting to create models for its origin that are little more than descriptions of the benefits that would have accrued to a society embarking upon a pristine[3] agricultural mode of subsistence. In attempting this sort of modeling, however, we make four major errors. First, we assume that the ultimate benefits of agriculture for a society would have been observable in its early stages. As we shall see, there are many reasons why they would not have been. Second, we assume that a society aware of these benefits would act to take advantage of them. This argument fails

---

[3]By *pristine* I mean arising by internal processes rather than by diffusion.

for numerous reasons; for the present, we might only ask why the decision to take advantage of these benefits was not made earlier. Third, we remove a historical event from its context. The transition to a pristine agricultural way of life had to be relatively gradual; the mechanisms of plant evolution and most of what we know about cultural change both demand this assumption. Thus, any change in subsistence had to be integrated into earlier modes of subsistence to produce a viable system in every phase of its development. Finally, we do not provide any explanation for the development of agriculture—any process, force, or event that can be identified as causing the transition to the new mode of subsistence.

That we humans are now agricultural animals is indisputable; that at some point we or our progenitors were otherwise seems self-evident. Yet, despite all of the recent advances in our understanding of our own prehistory, there has been little change in our description of this profoundly important change in subsistence. This has not been for lack of effort—the interacting concepts of domestication and agricultural origins have occupied the attention of both natural and social scientists for well over a century. Most workers have approached these problems without making a distinction between them; the origin of agricultural systems and the origin of the plants that are used in these systems have been seen as the results of one process.

The assumption that agriculture and the domesticated plant are results of one and the same process is probably rooted in long-standing traditions concerning mankind and culture. From the classical Greeks to nineteenth-century historians, it was assumed that humans evolved through three distinct, sequential, and progressive cultural stages: hunting–gathering, pastoral-nomadic, and agricultural-settled. Agriculture was seen as the natural and, indeed, inevitable outcome of cultural development; because every society went through the same series of stages, agriculture could be held to be the same in all societies.

Theories of change based on a concept of progressive evolution have a long and prestigious pedigree and have been exceedingly slow to die. This is true both in the natural and in the social sciences. Perhaps the greatest misconception commonly held about Darwin is that he was responsible for the concept of evolution. Although the identification of the earliest evolutionist is of little importance to us here, we might note that evolution as the *transmutation of form through time* has been around for a long time. The Ionian philosopher, Anaximandros, wrote in the sixth century B.C. of an origin of life in simple forms, inhabiting the waters, which gradually became adapted to life on land. He believed that people, too, were part of this progressive, evolutionary chain. It was up to Aristotle, however, to express the concept in a form that would survive well past the time of Darwin (Sarton 1970:I,497):

> Plato tended to assimilate change with corruption; Aristotle, on the contrary, conceives change as a motion toward an ideal. Plato rejected . . . progress, while Aristotle accepted it. Things change because of the potentialities inherent in them; they change in order

to attain or to approximate their perfection. The Idea of Form is *in* the thing (like the adult in the embryo). . . . Evolution proceeds as it does, not because of material causes producing natural consequences, pushing them on . . . but by final causes pulling them ahead. . . . The world is gradually realized because of a transcendental Design, or call it Divine Providence.

Darwin was well aware that the significant difference in the interpretation of nature that he was providing was not in its *result* (evolution and adaptation) but in its *means* (the natural selection of individual variation). As he notes in the Historical Sketch prefaced to the second edition of *Origin* (1860:xii):

Passing over allusions to [evolution] in the classical writers, the first author who in modern times has treated it in a scientific spirit was Buffon. But as his opinions fluctuated greatly at different periods, and as he does not enter on the causes or means of the transformation of species, I need not here enter on details.

Lamarck was the first man whose conclusions on the subject excited much attention. This justly celebrated naturalist . . . did the eminent service of arousing attention to the probability of all change in the organic, as well as in the inorganic world, being the result of law and not of miraculous interposition. . . . With respect to the means of modification, he attributed something to the direct action of the physical conditions of life, something to the crossing of already existing forms, and much to the use and disuse, that is, to the effects of habit. . . . But he likewise believed in a law of progressive development; and as all the forms of life thus tend to progress, in order to account for the existence at this time of simple productions, he maintains that such forms are now spontaneously generated.

But no matter how hard Darwin might have fought for the idea of evolution as descent modified to changing conditions of life, evolutionary theory was accepted with a bias that went against the most fundamental ideas of Darwin (Bernal 1971; © 1965, 1969 by J. D. Bernal).

Darwin did more than assert evolution; he provided a mechanism—*natural selection*—that destroyed the justification for the Aristotelian category of final causes. No wonder the theologians, whose world-picture was finalistic, repudiated it. Even more shocking was the idea that man himself—that unique end to creation—was nothing more than a remarkably successful ape. This seemed not only to shatter the doctrine of religion, but also the eternal values of rational philosophy. Both were to recover from the blow only too easily. . . . For the doctrine found supporters as well as enemies. It was a weapon in the hands of materially minded industrialists. . . . It seemed to give a scientific blessing to the exercise of unfettered competition and to justify the wealth of the successful by the doctrine of the *survival of the fittest*. (p. 558)

The *Origin of Species* arrived at a time when its message was badly needed. It was taken up by the radical, anti-clerical wing in economics and politics, made as it was very largely in the image of its own theories of *laisser-faire* and self-help. It made possible the justification of everything that was going on in the capitalist world, the ruthless exploitation of man by man, the conquest of the inferior by superior peoples. Even war itself could be justified by comparison with Nature "red in tooth and claw." The old excuse for the dominance of classes or races, that they were chosen people or the sons of Gods, had faded, and new excuses were needed to justify their continuation in a rational and scientific world. Darwinism provided it, although this was the last thing Darwin himself wanted. (p. 662)

The concept of progressive evolution, so sharply denied by Darwin, was too useful a concept to die—at least for the race, class, culture, and sex both occupying the highest position and promulgating this new doctrine of "Social Darwinism" (a euphonious although incongruous label).

Cultural Evolution

Progressive evolutionism, even shorn of its racist and elitist connotations, is tempting largely because it makes life easy for the scholar who invokes it as explanation. It functions in both the natural and the social sciences to simplify the process of explaining historical change: If evolution is progressive, we need only demonstrate that a phenomenon is an "improvement" on conditions existing at an earlier time. Though it seems harsh to reduce a rather large body of scholarship to these terms, the error is so common and takes such a multitude of forms that in this case oversimplification is beneficial because, as we shall see, progressive evolution has a way of cropping up in the most unseemly settings. In its most subtle form, progressive evolutionism holds that cultural evolution is guided by goals and desires. In this context, the presumption is that agriculture was adopted as the necessary way to increase productivity.

Even while Darwin was writing, L. H. Morgan was preparing his seminal volume, *Ancient Society* (1877). In this work, we may see an early anthropological view of progressive evolution applied to changes in the human subsistence pattern:

> The important fact that mankind commenced at the bottom of the scale and worked up, is revealed in an expressive manner by their successive arts of subsistence. Upon their skill in this direction the whole question of human supremacy on the earth depended. Mankind are the only beings who may be said to have gained an absolute control over the production of food. . . . It is accordingly probable that the great epochs of human progress have been identified more or less directly with the enlargement of the sources of subsistence. (Morgan 1964:24)

Here we see not only the progress of man to a higher form of civilization but also the previously mentioned identification of agricultural subsistence with civilization itself. Little however is said about the means of transition.

Although vague ideas such as cultural progress may underlie the theories of academicians, these theories seldom remain unadorned for long. Anthropologists holding to the tenets of the Boasian historical particularist school, although accepting progressive evolution (Harris 1968), generally believe that environmental and biological processes are irrelevant to the *origin* of cultural traits (e.g.,

Herskovitz 1940). Arising as a reaction to the excesses of an unfettered and misunderstood Darwinism, this school has generally held that, although they may have a role as modifying variables, biological phenomena are subservient to endogenous cultural events. Thus, many ethnographers have sought to find the genesis of agricultural practices in "universal" cultural patterns. A fertile field of conjecture, has been the effect upon subsistence of the "natural" division of labor in society.

From the early days of serious ethnography, sex-role specialization in human societies was claimed to be both consistent and significant: The task of men was hunting and war whereas women took care of gardens and children. Bachofen (1861) read the sexual roles of his society back into time to explain the origin of agricultural systems, envisioning "matriarchal stage" in the development of civilization that provided the initiating cause for agriculture. In this stage, the important female gatherers in a hunting–gathering society first protected and then learned to cultivate food plants—and they had the power necessary to impose their discovery upon the men. Frazer, in his influential *Golden Bough,* developed the feminine motif for the origin of agriculture in substantial detail, backed by the appropriate mythological references. He held that the digging of root crops would have acted to enrich the soil and thus to increase yields, which in turn would have permitted society the opportunity for permanent settlement. "On the whole then, it appears highly likely that as a consequence of a certain natural division of labor between the sexes, women have contributed more than men towards the greatest advance in economic history, namely the transition from a nomadic to a settled way of life, from a natural to an artificial basis of subsistence" (1912:129). Although Frazer did not have to explain how women ever convinced men of the utility of their invention (perhaps a reflection of suffrage agitation?), we might note in passing that they also did not have much to do with establishing the situation in which they were to make their discovery. But, putting aside their social and political implications, many of the themes advanced by Frazer have remained with us: The role of digging in enriching the soils and the relationship between agriculture and settlement have continued to occupy many workers' minds.

Charles Reed (1977b:563) has laid the foundation of animal domestication on the doorstep of the hormone estrogen:

> Little girls, increasingly as they grow, have estrogens coursing in their bloodstreams; little girls play with dolls, have maternal instincts. They are not yet, as their mothers may be, inured to killing and the necessities of killing; a little girl might well adopt, protect and tend a weaned lamb, kid or baby pig, thus establishing the one-to-one social relationship necessary for abolition of the flight reaction.

Thus, arguments based on differences between the sexes continue to be advanced. However, I must take this opportunity to complain, since I am a man who played with dolls as a child, cried when his guinea pig died, and has recently discovered his paternal "instincts." Nevertheless, I expect such arguments to persist.

Another approach to the problem of agricultural origins that grew out of the notion of cultural stages is what I call the *religious model*. Hahn (1909), one of its early proponents, maintained that keeping cattle for sacrifice allowed men to learn the techniques of feeding and care that were eventually to yield true domestication. Isaac develops Hahn's basic approach in remarkable depth, seeing, for example, the origin of vegeculture in the "ritual experiment in which plants were 'slain,' i.e., cut up and buried" (1970:110) and arguing that the "initial domestication of plants and animals constituted such a break with the past that an intellectual revolution must have preceded the economic development" (1970:115). Allen (1897), pondering the necessity of cleared soil for effective agriculture, suggested that agriculture probably began with offerings of food grains on the graves of the newly departed. As corroborating evidence he used modern examples of death rites and their connection with agriculture. He was impressed by the widespread association of ritual sacrifices with planting ceremonies and saw it as a survival of the idea that it was necessary to have a corpse to make crops grow.

I must confess a certain weakness for these religious theories of the origin of agriculture. As intentionalistic systems, they have the advantage of maintaining the primacy of consciousness in man. They also have the advantage of explaining directly, and in one simple step, the invention of agricultural techniques such as planting or the saving of the best seed. Theories that postulate agricultural behaviors evolving as a slow process cannot make as clear and readily understandable a case. Evolutionary theories may only grope at the reconstruction of a probable history for the development of agricultural behavior, based on what evidence we have available at present. Unfortunately, however, religious theories must be rejected when confronted by the evidence that agricultural systems have been developed by other animals (unless we are willing to attribute religion to ants or termites). In the end, we must come to terms with the fact that the evolution of the plants that comprise agricultural systems is controlled by the same forces both in nature and within the agroecology—that plants are incapable of responding directly to the human need for changes in morphology or productivity.

The best-known modern description of pure cultural evolution as the cause of agriculture occurs in the earlier papers of Braidwood, who thought agriculture developed when cultures located in a suitable environment reached the appropriate stage to accomplish this feat (Braidwood 1967). The progression of social and technological events is a product of society's increasing ability to control its own destiny. Agriculture arises when there is a "suitable cultural level," and, having arisen, changes both culture and plants. "Why did incipient food production not come earlier? Our only answer at the moment is that culture was not yet ready to achieve it'' (Braidwood and Willey 1962:342). The underlying problem here is obvious: No mechanism, nor even hypothesis, is advanced to make sense out of the change. Cultural evolution is culture evolving, and it is culture that creates the conditions for its own further growth. We should not be too quick to judge Braidwood, however, for we shall see that he had reasons for adopting this view.

Binford took advantage of the straw man unintentionally created by Braidwood and unleashed an attack upon the obvious weakness inherent in simple progressive evolutionary schemes (1968:322):

> In his statements Braidwood proposes that cultivation is the expected, natural outcome of a long, directional evolutionary trend limited only by the presence in the environment of domesticatable plants and animals. This is clearly an orthogenetic argument. . . . The vital element responsible for the directional series of events appears to be inherent in human nature; it is expressed . . . in such phrases as ''increased experimentation'' and ''increased receptiveness.''
>
> It is argued here that vitalism, whether expressed in terms of inherent forces originating the direction of organic evolution or in its more anthropocentric forms of emergent human properties which direct cultural evolution, is unacceptable as an explanation. Trends which are observed in cultural evolution require explanation; they are certainly not explained by postulating emergent human traits which are said to account for these trends.

Orthogenesis and ''emergent human properties'' are a bit too much for anyone to swallow—unless they are the most palatable ideas available—and it appears that, for Braidwood, they were.

Environmental Determinism and Orthogenesis

The economic interpretation of prehistoric Europe presented by Clark (1952) was, as noted by Green (1980b:314), a ''benchmark in the development of archaeological thought on cultural change.'' But just as the surveyor improperly measuring the altitude of a benchmark errs not only for himself, but for all subsequent workers utilizing his work, so the equilibiurm model of Clark was to mislead a generation of scholars seeking to understand the origin of agricultural practices.

The core concept in the work of Clark was that culture and environment form a gestalt that tends toward equilibrium. Thus, change grows not out of progressive tendencies inherent in the culture, but out of the society's efforts to reestablish an equilibrium disrupted by change either in the culture or in the environment. As a research strategy, this model is not without its uses, emphasizing, as it does, the dynamic that exists between people and the environment. Further, the central role it assigns to economic behavior as the basis for the interpretation of artifactual diversity does much to rescue the description of prehistoric cultural differences from a vague, idealistic world of undefinable typological difference. At the same time, however, the emphasis upon the cultural endowment of the society as the mechanism of adaptation, the way in which the equilibrium may be maintained, yields a less-than-convincing explanation for cultural change. In practice, interaction gives way to reaction, and cultural change comes to be merely the adaptive response of culture to environmental change: cultural "adaptation" and teleology come to occupy the place of cultural "progression" and orthogenesis—at best, a very small step for man in explaining a great leap for mankind.

Although Clark's work was important for casting environmentalism in a modern mold (a task that was formalized and brought into contemporary settings in the work of Julian Steward [1955] who stressed the importance of the economic *cultural core*), the concept of culture–environment interaction had been floating around in various guises for a long time. As Geertz puts it (1963:1),

> the recent burst of efforts to adapt the biological discipline of ecology . . . to the study of man is not simply one more expression of the common ambition of social scientists to disguise themselves as "real scientists," nor is it a mere fad. The necessity of seeing man against the well-outlined background of his habitat is an old, ineradicable theme in anthropology, a fundamental premise.

Indeed, it is not only in anthropology that environmentalism has been a guiding paradigm. As Stocking (1968:225) has pointed out, "The men who established the social sciences as academic disciplines in the United States around the turn of the century were for the most part environmentalists, and many of them were in fact reacting against the biological determinist and conservative implication of Spencerial Social Darwinism." One possible reason for the prominence of environmentalism in its various forms in the social sciences is the role it plays as a "liberal" alternative not only to the racist and elitist theories of the Social Darwinists, but also to the revolutionary theories of the Marxists. Hardesty (1977:2) has said, "the rise of 'technological determinism' as espoused by Marxist social philosophy also contributed to its resurgence. Environmental determinism was a rebuttal to the antienvironmental position of Marxist writers." We may even go so far as to wonder if the "equilibrium" approach so favored in the 1950s and 1960s was in part a reading back into time of bourgeois concepts of government as "rationalizing" a capitalist economy in an effort to maintain economic stability.

Environmentalism in its simplest form is environmental determinism, a doctrine claiming a strict causal link between environmental conditions and cultural action. It is in this rather primitive form that environmentalism was to enter the literature of agricultural origins in the writings of V. Gordon Childe.

Childe (1925) has been justly credited (and, more recently, damned) for popularizing the concept of the *Neolithic Revolution* for the transition from food procurement to food production systems and the numerous developments in cultural history correlated with this shift. Childe's well-known *riverine-oasis hypothesis* holds that increasing aridity occurring in the early Neolithic placed people in intimate association with domesticatable plants and animals that were also forced to concentrate in limited areas in which water was available. "Enforced concentration by the banks of streams and shrinking springs would entail an intensive search for means of nourishment. Animals and men would be herded together. . . . Such enforced juxtaposition might promote that sort of symbiosis between man and beast implied by the word domestication" (Childe 1951:25). Of course, Childe was not the first to advance a climatic stimulus for domestication; in fact, earlier workers anticipated him in most particulars. As early as 1908, Pumpelly was considering the influence of climate on culture. He believed that desiccation had forced the herding cultures of the ancient Near East into oases in which society had to adopt an agricultural way of life in order to survive. Toynbee adopted a similar point of view in his widely read *Study of History* (1935:304–305). It was Childe's work, however, that was to prove most influential for archaeology, because, in contrast to previous workers, he "presented a series of propositions specific enough to be tested through the collection of paleoenvironmental and paleoanthropological data" (Binford 1968:322). And it is in this setting that we return to the work of Braidwood, for as Binford has remarked, "If it was Childe who first provided a set of testable propositions as to the conditions under which food-production was achieved, it was Braidwood who actively sought the field data to test Childe's propositions" (Binford 1968:322).

Although Braidwood (1951) evidenced a certain laudable scepticism concerning the hypothesis he was testing, he was nevertheless a committed environmentalist (Braidwood and Reed 1957):

> We . . . suggest . . . that the archaeologist (by the very nature of his training in the social sciences or humanities) is unprepared to treat much of the evidence with the sophistication it deserves. . . . understanding of how [domestication] came about, and of the exact nature of its consequences will require knowledge of the plants and animals and their paleo-environments as well as of the biological and cultural natures of the men who first achieved food-production. (p. 19)

> It is our thesis that detailed understanding of the subsistence levels are basic [and] that such detailed understandings cannot be achieved by the archaeologists and anthropologists working alone, but that they can be achieved . . . by archaeologists, anthropologists and natural scientists working together. Our first goal would thus be understanding of man in the succession and variety of eco-systems of which he has been a part. (p. 30)

However, when the data came in, the hypothesis went out.

> We and our natural science colleagues reviewed the evidence for possible pertinent fluctuations of climate and of plant and animal distributions . . . and convinced ourselves that there is no such evidence . . . for changes in the natural environment . . . as

might be of sufficient impact to have predetermined the shift to food production. (Braidwood and Howe 1960:142)

Lacking firm evidence of climatic change, Childe's simple hypothesis of
climate determination for the origin of agriculture could not stand. Yet, if one
particular incidence of environmental determinism is disproven, why should
Braidwood retreat into positing orthogenetic cultural evolution? I believe the
answer lies in the assumption of intent that underlies models based on environmental determinism.

Numerous environmental stimuli (including climatic, demographic, and
geographical ones) have been posited as initiating the transition to food production. Yet, if one reflects upon the process by which these stimuli become translated into cultural change, it will be seen that the stimuli themselves are not held
directly responsible for the transition; rather, mankind is held to *respond* to
changed conditions by inventing agriculture, with the environmental stimulus
providing only the reason for the invention. Intent, cultural invention, and a
Lamarckian mode of variation underlie environmental determinism. As has been
noted by Philip Wagener (1977:70–72):

> The defect of environmental determinist reasoning, even at its best, concerns its basic
> logic. . . . The environmental determinist did not promise to discover what conditions
> would always insure the invention . . . nor did they ever describe in detail the processes
> supposedly sufficient to implement the influence they claimed. In fact . . . [they] fell
> back on mentalism—either proposing that certain climatic conditions . . . stimulate
> mental activity, to which they attributed all differential progress; or that particular
> conditions revealed themselves at crucial moments in a way that forced insight on men
> and so led them to inventions.

Braidwood may have fallen from grace, but his fall was a short one: orthogenetic
cultural evolution is environmental determinism lacking the initiating environmental stimulus. Rather than pointing out the gulf separating orthogenetic cultural evolution and environmental determinism, Braidwood's work calls our attention to their close relationship.

Environmental determinism still has some strong adherents; foremost
among them is Wright (1968, 1970, 1977), who notes that "the chronological
coincidence of important environmental and cultural changes [in the Near East]
during the initial phases of domestication is now well enough documented that it
cannot be ignored" (1968:338). However, much the worse for Childe, pal-
eoecological studies have shown that at the putative time for the origin of agricultural systems, the "climate in this region went from dry to moist rather than
the reverse, thus vitiating the essence of Childe's hypothesis" (1977:281). This
need not, however, cause problems for the environmental determinist: One need
only show a correlation with a change in climate for the determinism to function— the specifics may be taken care of by the appropriate cultural response.
And if change is lacking, cultural response may still be used to explain away

many an awkward noncorrelation: "[In the New World] the evidence for climatic control is weak, and the case for domestication may instead rest with cultural factors" (Wright, 1977:282).

Climatic influence may play a role in almost any consideration of the origin of agriculture in a specific region, for no reasonable person could deny at least a limiting role to environmental conditions. Because it is only in studies relating to a specific region that climate can be held to play the decisive role in the origin of agriculture, it is not surprising that, on the whole, climatic determinists are frequently also diffusionists. If intention is to account for the origins of agriculture, even if initiated by a climatic event, we must posit that this interaction has been sufficiently rare as to explain why agriculture did not begin earlier in human history. It is not surprising that most who believe in a climatically induced agricultural genesis also accept the notion of one or a limited number of hearths in which agriculture originated and from which it spread.

Braidwood, in the final analysis, was responding to the inability of environmentalism to predict cultural change and fell into the historical particularist camp—a school that might be described as cultural evolution without the evolution. Yet, despite all of the problems inherent in this school of thought, we must admit it capable of advancing a telling critique of poorly reasoned environmentalist arguments. Carter, perhaps the preeminent historical particularist currently writing on the origin of agriculture, argues that agriculture was an invention— not an invention conditioned by this or that environmental or demographic or geographic particular, but invention "dependent upon individual genius" (1977:90). And Carter sticks to his guns, yielding no more ground than absolutely necessary: "Granted that genius functions in a particular cultural setting complete with antecendents and so forth, still there is a difference between emphasis on processes which will produce the end result by some inexorable functioning versus the flash of creative genius" (1977:90); "To argue that climatic change stimulated man to invent agriculture would be suspect from the beginning for this places the initiative outside of man" (1977:97).

The gist of Carter's antienvironmental arguments is as follows: (1) People were predominantly gatherers of plant foods for a very long time before the advent of sophisticated agricultural systems. (2) If agriculture is the result of a natural process and man is more or less genetically uniform, agriculture must have come early in human history and must have been discovered independently by most cultures. (3) If the process culminating in agriculture was the result of environmental or demographic stress, all evidence (save agriculture, in this context) indicates that these stresses would have occurred frequently during the history of our species. (4) Agriculture is neither an obvious nor necessarily the "best" subsistence system. (5) Agriculture could scarcely be a response to needs because the needs it satisfies would not arise until after it came into existence. (6)

We have consistently confused the improvement of plants under cultivation with the initiation of domestication—and domestication is the real issue. These are sound arguments, requiring answers. Environmentalists would do well to take them seriously, for they point to several of the major errors that have been made in the description of the interaction of man and his environment: adaptationism, Lamarckism, and cultural teleology.

Cultural Ecology

The cultural ecologists of the 1950s and 1960s created a hypothetical (and often Panglossian) world of adapted cultural systems, in which it was becoming difficult to understand how change could ever occur. Cultural equilibrium- and systems-models had become the main armaments of environmentalism, and they were leading the environmentalists into a dead end by doing such a fine job of explaining peoples' relationships with nature that change itself was becoming incomprehensible. As Flannery put it (1969:75), "The basic problem in human ecology is why cultures change their mode of subsistence at all." But before I point to the fundamental theoretical and tactical errors that were to lead Flannery along with most other environmentalists into this sad situation, it would be good to review briefly the major contributions of this school if for no other reason than to be spared the friendship of those who would otherwise read my rather strong attack against the models used by the cultural ecologists as a repudiation by me of their basic approach.

Binford's "Post-Pleistocene adaptations" (1968) not only ably demolished earlier approaches to the origin of agricultural systems, but helped initiate a fruitful research strategy into basic changes in human subsistence. Building upon the earlier model of Childe, Binford sought to rescue a mechanistic view of cultural history from what he correctly perceived as the dead end of vitalism and orthogenesis. Nevertheless, he still had to account for the inability of Childe's model to stand up to Braidwood's empirical test. He thus chose, wisely I believe, to cast his arguments in terms of a more inclusive and wide-ranging theoretical framework, attempting to apply the insights of scholars such as Clark and Steward to the question of cause for changes in subsistence patterns (Binford 1968:323):

> If our aim is the explanation of cultural differences and similarities in different places and at different times, we must first isolate the phenomena we designate "cultural." Culture is all those means whose forms are not under direct genetic control . . . which serve to adjust individuals and groups within their ecological communities. If we seek understanding of the origins of agriculture . . . we must analyze these cultural means as

adaptive adjustments. . . . Adaptation is always a local problem, and selective pressures favoring new cultural forms result from non-equilibrium conditions in the local ecosystem. Our task, then, becomes the isolation of the variables initiating directional change in the internal structuring of ecological systems.

To undertake this type of analysis, Binford had to make what seemed, at the time, to be an innocuous assumption—that equilibrium was a normal state for a culture's relationship with its environment. In this he followed and quoted Leslie White's influential volume, *The Evolution of Culture* (White 1959b:284): "Cultural systems relate man to habitat, and an equilibrium can be established in the relationship as in others. When an equilibrium has been established culturally between man and habitat, it may be continued indefinitely until it is upset by the intrusion of a new factor." Although I will attack this assumption in some depth throughout this volume, at the time it was probably a rather reasonable one. Two long-standing assumptions about "primitive" man were being reconsidered: Increasing evidence was accumulating that our image of "man the hunter" was a less-than-accurate one for most preagricultural societies, and we were beginning to reject the rather Victorian notion of primitive people fighting off imminent starvation and struggling continuously for survival in an environment they could not control. The late 1950s and early 1960s were also a rather optimistic and economically stable time and this may well have been reflected in the views of the world being preached at that time. Not only in the social sciences did we find equilibrium and peace reigning, but this attitude was widespread throughout academia: Odum's influential *Fundamentals of Ecology* (1959) was finding a receptive audience for its systems theory models of homeostasis and stability; Wynne-Edwards (1962) found few dissenters from his pleasant idea that species as species have evolved marvelous adaptations to prevent overpopulation and its distasteful consequences; only the incompetent were denied tenure, and mortgages were cheap and easy to get. In this environment, the concept of an evolved equilibrium between culture and environment probably appeared to most scholars all but self-evident.

The specific way in which Binford sought to rescue Childe's theories in the face of denial by the archaeological record was by pointing out that, although Childe's independent variable, climate, was not applicable in this case, his basic approach was correct. The fundamental contribution of Childe was thus the recognition that agriculture is a "structurally new adaptive means in an ecological niche not previously occupied by cultural systems" (Binford 1968:325). Braidwood had therefore erred not only in subscribing to orthogenesis and teleology, but also in failing to appreciate that agriculture was far more than the "natural outcome of a long, directional evolutionary trend limited only by the presence in the environment of domesticatable plants and animals" (1968:322); Braidwood had missed the very essence of agriculture: the fact that it was a new system by means of which people related to their immediate environment. Braidwood, like many others before and after him, assumed that all one had to do was "plug in" the appropriate plants and animals to invent a functioning agricultural system.

The variable that Binford advanced to take the place of Childe's climate was *population pressure*. For Binford (unlike many modern theorists to be discussed later), this pressure existed as a disequilibrium between populations and resources, and was due to a breakdown in control mechanisms (Binford 1968:328):

> As long as one could assume that man was continually trying to increase his food supply, understanding the "origins of agriculture" simply involved pinpointing those geographic areas where the potential resources were and postulating that man would inevitably take advantage of them. With the recognition that equilibrium systems regulate population density below the carrying capacity of an environment, we are forced to look for those conditions, which might bring about disequilibrium and bring about selective advantage for increased productivity.

Binford used his equilibrium model to show not only that Childe was correct in assuming that climate might be an important variable, but also that if no evidence of climatic change can be found, a disturbance in population densities will be responsible for important adaptive shifts in subsistence strategies (Binford 1968:328):

> According to the arguments developed here, there could be only two . . . sets of conditions that would bring about disequilibrium and selection for increased productivity: 1) A change in the physical environment which . . . would decrease the amounts of available food. . . . This is essentially the basis for Childe's propinquity theory. 2) Change in the demographic structure of a region which brings about the impingement of one group on the territory of another would also upset an established equilibrium system, and might serve to increase the population density of a region beyond the carrying capacity of the natural environment. Under these conditions manipulation of the natural environment in order to increase its productivity would be highly advantageous.

Binford argued that population density, even for hunting and gathering peoples, will not be uniform: Areas favorable to human habitation will be separated by zones that are less favorable. When population builds up within the "optimal" habitats, excess population will be released to the less favorable ones because the equilibrium between population and resources must be maintained. However, the very act of releasing these "daughter" groups places stress upon those areas least likely to be able to absorb it—the suboptimal zones. It is therefore in the suboptimal zones that we expect to find the greatest pressures for the adoption of new modes of subsistence, for it is only through the utilization of a new type of relationship with the environment that the system can create an equilibrium in the face of long-standing and frequent disturbances brought on by the influx of immigrant groups. Although one may question the assumptions upon which this scenerio is based, it must be admitted that the argument is mechanistic as far as it goes.

Binford must receive credit for providing many of the ingredients necessary to the formation of a science of cultural ecology. Rather than viewing history as determined by a particular external force or by the mere functioning of culture itself, he emphasized the development of culture as interaction with the environment. Although environmental changes—for example, in sea level, precipitation or temperature—are important, people are seen as interacting with, rather than merely reacting to, these changes. Demographic processes and change in settlement patterns are seen as the results of cultural responses, which, in turn, create new possibilities within an altered environment. Also, once these processes are initiated they proceed at least some distance without having to rely upon vitalism in any form. Finally, Binford recognized that a multitude of routes may be taken as a response to the same initial conditions. Thus he showed admirable restraint in claiming a general inevitability of the processes he describes (Binford 1968:331): "such advantage [as flows from the adoption of agriculture] does not insure that these developments will inevitably occur. In many cases these problems are met by changes which might be called regressive in that the changes in adaptation which occur may be in the direction of less complex cultural forms." Binford was one of the first scholars to recognize that the question "Why don't all cultures evolve agriculture?" is an empty query.

Another important contributor to the development of an ecological approach to the understanding of cultural change was Flannery. Perhaps even more than Binford, he appreciated the importance of the ecological and evolutionary parameters governing the development of agricultural systems. This is clear even in his early papers:

> With the wisdom of hindsight we can see that, when the first seeds had been planted, the trend from "food collecting" to "food producing" was underway. But from an ecological standpoint the important point is not that man *planted* wheat but that he (i) moved it to niches to which it was not adapted, (ii) removed certain pressures of natural selection, which allowed more deviants from the normal phenotype to survive, and (iii) eventually selected for character not beneficial under conditions of natural selection. (Flannery 1965:1251. Copyright 1965 by the American Association for the Advancement of Science.)

Flannery also stressed at an early date the importance of viewing domestication as a mechanistic process of long duration that was inseparable from the total environment in which the process was occurring. Speaking of the origin of agriculture in the Near East he wrote

> The food-producing revolution . . . is here viewed not as the brilliant invention of one group or the product of a single environmental zone, but as the result of a long process of changing ecological relationships between groups of men . . . and the locally available plants and animals which they had been exploiting on a shifting, seasonal basis. In the course of making available to all groups the natural resources of every environmental zone, man had to remove from their natural contexts a number of hard-grained grasses and several species of ungulates. . . . Shielded from natural selection by man, these small breeding populations underwent genetic change in the environment to which they

had been transplanted, and favorable changes were emphasized by the practices of the early planter or herder. (Flannery 1965:1255. Copyright 1965 by the American Association for the Advancement of Science.)

In this same paper, Flannery began to stress the importance to the eventual development of agricultural systems of a "flexible 'broad-spectrum' collecting pattern keyed to the seasonal aspects of the world resources [in various] . . . environmental zones with perhaps a certain amount of seasonal migration from zone to zone" (1965:1250). It was in this context that it first became possible to realize that the development of agricultural systems could be related to processes of extreme antiquity (a line of reasoning that unfortunately was never given the attention it merited): "From 40,000 to 10,000 B.C., man worked out a pattern for exploiting the natural resources of this part of the world, and I suspect that this pre-agricultural pattern had more to do with the beginning of food production than any climatic 'shock stimulus.'" (Flannery 1965:1251)

This concern with the interaction of a changing diet and patterns of exploitation of the biome was developed in some detail in another paper in which Flannery attempted to generalize his model by applying it to the New World. Here we find seasonality and scheduling integrated into the basic equilibrium model to explain the origin of agriculture and its timing. Important in this exegesis is the recognition that the subsistence strategy that preceded agriculture was *functionally* related to the origin of agriculture itself.

> Seasonality and scheduling . . . prevented intensification of any one procurement systems to the point where the wild genus was threatened; at the same time, they maintained a sufficiently high level of procurement efficiency so there was little pressure for change. . . . Under conditions of fully-achieved and permanently-maintained equilibrium, prehistoric cultures might never have changed. That they did change was due at least in part to . . . a series of genetic changes which took place in one or two species of Mesoamerican plants which were of use to man. (Flannery 1968:79)

Thus we may see that the new system of environmental relationships develops as a mechanistic consequence of the relationship itself (given the existence of the "necessary" genetic change in beans and maize). Flannery then went on to explicate the effects of these changes in morphology of plants upon the scheduling and subsistence pattern. This gradual change was to have major effects upon the demographic structure of the people participating in the developing system.

> Starting with what may have been (initially) accidental deviations in the system, a positive feedback network was established which eventually made maize cultivation the most profitable single subsistence activity in Mesoamerica. The more widespread maize cultivation, the more opportunities for favorable crosses and back-crosses; the more favorable genetic changes, the greater the yield; the greater the yield the higher the population, and hence the more intensive cultivation. . . . What this meant initially was that [the system of] Wild Grass Procurement grew steadily at the expense of, and in competition with, all other procurement systems . . . the system increased in complexity by necessitation of a *planting* period (in the spring) as well as the usual *harvesting* period (in the fall). It therefore competed with both the spring-ripening wild plants

> . . . and the fall-ripening [ones]. . . . . It competed with rainy-season hunting of deer and
> peccary. And it was a nicely self-perpetuating system. . . . Because of the functional
> association between band size and resource, human demography was changed . . . an
> amplification of the rainy-season planting and harvesting also mean an amplification of
> the time of microband coalescence . . . in fact, increased permanence of the microband
> may have been *required* by the amplified planting and harvesting pattern. (Flannery
> 1968:80–81)

Flannery had developed a description for changes in human subsistence patterns
in which the major variables were related to a general model. Here, the process,
once initiated, gradually proceeded in a mechanistic fashion towards the known
outcome—agriculture.

As often happens to a model, Flannery's model had created a new group of
"origins" to be explained: if we can explain agriculture as an outgrowth of the
broad-spectrum revolution, what is the origin of the broad-spectrum revolution?
Problems like this breed like fleas and make authors uncomfortable. Flannery
found his solution in Binford's demographic-pressure model. Thus the first
"revolution"—the broad-spectrum one—was the result of pressures that
"would have been felt most strongly in the more marginal areas which would
have received overflow from the expanding populations of the prime hunting
zones, raising their densities to the limit of the land's carrying capacity" (Flannery 1969:78). The same process could then be extended to produce a unified
theory of the origin of agriculture. Unfortunately, however, as we shall see,
Binford's model required a certain amount of intent on man's part to make it
work, and Flannery (1969:81) adopted, at least in part, this same narrow focus,
leading to a position that smacked of demographic and/or climatic determinism
(an association he was later to regret [1973:284]). Also, for reasons not entirely
clear to me, Flannery in his 1969 paper abandoned the notion of an interaction of
scheduling and resources. Although the archaeological evidence from the Near
East did not indicate the clear transhumance practiced in the New World, scheduling problems are on the whole as important to sedentary societies as they are to
mobile ones.

In any case, Flannery's basic commitment to ecological principles permitted
him to offer a far more reasonable picture of the effects of agriculture upon
human subsistence patterns than had been previously presented. In part, this was
due to his careful observation of contemporary agricultural systems: If one
adopts a gradualistic and mechanistic view of the origin of agriculture it is not
necessary to assume that early agricultural systems were qualitatively different
from modern ones. Thus, if Binford's approach hurt Flannery's models in one
way, it helped in another by reinforcing the recognition that agriculture is a
"niche" inhabitable by man (Flannery 1969:95):

> Recent ethnographic studies do not indicate that hunters and gatherers were "starv-
> ing". . . . Nor does the archaeological record in the Near East suggest that the average

Neolithic farmer was any better nourished than the average Palaeolithic hunter. More-over, over much of the Near East, farming does not necessarily constitute a "more stable" subsistence base or a "more reliable" food supply. The real consequence of domestication was to (1) change the means of production in society, (2) make possible divisions of labor . . . and (3) lay the foundations for social stratification. . . . It also (4) increased man's potential for environmental destruction, so that eventually it would have been impossible for him to return to his former means of subsistence, had he wanted to.

The specific model Flannery advanced to organize his concepts on equilibrium and its disturbance was a cybernetic one. According to Flannery, the use of a cybernetics model has certain advantages:

> It does not attribute cultural evolution to "inventions," "experiments," or "genius," but instead enables us to treat prehistoric cultures as systems. It stimulates inquiry into the mechanism that counteract change or amplify it, which ultimately tells us something about the nature of adaptation. Most importantly it allows us to view change not as something arising *de novo*, but in terms of quite minor deviations in one small part of a previously existing system, which, once set into motion, can expand greatly because of positive feedback. (Flannery, 1968:65)

This model, based upon a series of incorrect assumptions linking change to adaptation, equilibrium, and systems models, tends to obscure change in the very process of attempting to explicate it. Despite their strong commitment to a materialistic view of cultural change, the cultural ecologists were, eventually, forced to argue in terms of intent and teleology. Yet, as early as 1968, Flannery[4] anticipated the argument I will advance here: that the development of agricultural systems cannot be understood without appreciating the evolutionary consequences of the symbiosis that came to be established between man and plant.

> Preliminary studies of the food debris from these caves indicates that certain plant and animal genera were always more important than others, regardless of local environment. These plants and animals were the focal points of a series of procurement systems, each of which may be considered one component of the total ecosystem of the food collecting era. They were heavily utilized—"exploited" is the term usually employed—*but such utilization was not a one-way system. Man was not simply extracting energy from his environment, but participating in it; and his use of each genus was part of a system*

[4]Of course, Flannery and Binford were not working in a vacuum, nor were they the only cultural ecologists publishing during this period. Many other writers contributed to the development of a mechanistic view of man's relationship to his environment, including forerunners such as Ames (1939), Anderson (1956), Sauer (1947, 1956), Vavilov (1926, 1951) and Shokovskii (1962). Ideas advanced by these people were amplified, tested and developed in various directions by the careful work of scholars such as H. G. Baker (1965, 1970), Birdsell (1972), Bray (1974, 1976, 1977), Bronson (1972, 1977), Brothwell (1975), de Wet (1975), Harlan (1965, 1967, 1970, 1971, 1976, 1977), Harlan and de Wet (1965, 1973), Harlan, de Wet, and Stempler (1976), Harlan and Zohary (1966), D. R. Harris (1967, 1969, 1972, 1973, 1977a, 1977b), Hawkes (1969), Higgs (1976), Higgs and Jarman (1969, 1972), Janzen (1973), Lewis (1972), Lynch (1973), MacNeish (1964a, 1964b, 1967), Reed (1977a, 1977c) and Zohary (1969, 1970).

which allowed the latter to survive, even flourish, in spite of heavy utilization. Many of these patterns have survived to the present day, among Indian groups like the Paiute and Shoshone . . . or the Tarahumara of Northern Mexico . . . thus allowing us to postulate some of the mechanisms built into the system, which allowed the wild genera to survive. (Flannery 1968:69; emphasis mine)

Adaptation, Equilibrium, and Systems Theory

The chief rhetorical device employed by cultural ecologists of all persuasions is the concept of equilibrium and its disturbance. In the application of this device, they have used (and misused) theories drawn from the ecological literature that are largely concerned with the *description* of existent systems. When used for the purposes of description, little criticism can be made of the working assumptions of equilibrium and adaptation; in studying a system it is reasonable to assume that it is, in fact, functioning. However, we must clearly recognize that the utilization of this device is *predicated* upon these assumptions and it therefore does *not* follow that application of these assumptions is presumptive proof of either adaptation or stability. Furthermore, the description of systems is an entirely different matter from the explanation of the *origin* or *evolution* of systems and applying the equilibrium–stability–adaptation model here inevitably leads to Lamarckism.

The assumption that culture maintains, or tends to maintain, an equilibrium with its environment is seductive if for no other reason than it allows us to posit a comprehensible order in the relationship between people and their environment: we need only know that if change does not occur the equilibrium is functioning, whereas if change occurs the equilibrium has been disturbed. It also saves us the difficulty of trying to discover possible mechanisms for change. We need only name the feedback processes: Negative feedbacks will prevent movement from a given stage, and positive feedbacks will accentuate the process of movement away from a given situation. But hypothetical feedbacks provide no more than the illusion of explanation because we have, in fact, specified nothing we did not know before.

Cultural ecology has relied, in large part, upon systems theory for its arsenal of arguments, and this reliance has placed severe constraints upon the rigor of its explanations (Athens 1977). We have here a fine example of the tail of modeling wagging the dog of explanation.

Fundamentally, systems theory is predicated upon the preexistence of a specifiable set of equations that can be used to describe the functioning of the system. It does no good to connect a series of black boxes with arrows, produce a flow chart, and thereby claim that one has captured the workings of a system. We have, of course, defined a system, but without specifying the rules governing interrelationships within this system we have created little more than a meaningless construct. Unless we are willing to claim that our model system is only *analogous* to the real system we are modeling, the arrows must stand for *known* functions describing the relationships between our boxes. Without specifying *how* components are interelated we have created little more than an intuitive guess at a system. Berlinski, in a scathing attack on the sloppy application of systems analysis in the social sciences, notes:

> In the classical tradition, there is no engineering without some prior specification of a dynamical model; Professor Kalman writes that "the model is obtained from physical measurement or physical laws; its actual determination is outside of the scope of control theory or even systems theory". . . . *That the models meet the criteria of control theory, or systems engineering generally, is an assumption and not a consequence of the systems-analytic technique.* And it is an assumption of some strength; in the classical tradition, at any rate, the models are severely specialized: differential equations first, single input second, constant coefficients third, linearity fourth. The social scientist prepared to make the case for control theory or even for the vastly more flabby thesis that [social] systems are somehow input–output devices, must *logically* stand ready with a suitable dynamic theory: there is no looking to control theory itself for such a structure. (Berlinski 1976:154–155, original emphasis; © by the MIT Press.)

Systems theory may be applied only after the dynamic properties governing a system have been discovered. Furthermore, the system amenable to description in terms of systems theory must meet certain mathematical preconditions in its functioning—namely, single inputs, constant coefficients, and linearity. Attempting to squeeze social systems into this straight jacket creates, at best, discomfort; attempting to describe the evolution of any system in these terms is downright irresponsible. Natural systems, including cultural systems, are simply too complex and unpredictable (at least for the present and the foreseeable future) to be modeled in these terms.

Instead of a description of the dynamics of social systems, we have been given an article of faith—the doctrine of homeostasis and adaptation—that is supposed to make the mechanistic description of cultural change unnecessary. As Little and Morren put it,

> Biological systems, whether viewed as individual organisms, populations, or communities, must maintain some degree of homeostasis . . . such a dynamic equilibrium is sustained by various *feedback mechanisms* that control the steady state and maintain it at a given level. . . . Restoration of homeostasis and the consequent reduction of strain is an *adaptation* or adjustment to the original conditions of environmental stress. . . . such adjustments are a basic property of all living systems, including human biological, behavioral and life-support systems. (Little and Morrel 1976:29, emphasis original).

Now equilibrium and feedbacks may be useful if we are describing thermostats and their functioning, but their applicability to the evolution of ecological communities, much less to social and cultural systems, is far from proven. More to the point, this approach proves a harsh taskmaster: If emphasis is placed upon the equilibrium, we find change difficult to understand; if, instead, we place emphasis upon change, our "negative feedback" is not inherently specifiable—thus, in practice we are free to choose the one that best fits our prejudices. It is all, at best, a rather sloppy sort of science.

The biggest problem, however, is the goal-direction inherent in systems theory. Systems theory was designed to describe systems intentionally created for specified goals. Thus, if we describe a nonchanging system, equilibrium must be seen as the goal of its functioning, and, if describing change, the establishment of a new equilibrium becomes, again, the goal. And how are social–environmental systems to have goals? This problem is rather neatly handled by means of the doctrine of adaptation. Because natural and social systems are "inherently adaptive," we are free to posit that (whether by means of intentional striving or by means of some vague sort of group selection) adaptation is a goal; and because adaptive systems exist in the equilibrium state, the ultimate effect of adaptation is maintenance or reestablishment of an equilibrium. *Quod erat demonstrandum.* As we have seen, Binford (1968:323; emphasis mine) argues that culture is "all those means whose forms are not under direct genetic control . . . which serve to *adjust* individuals and groups within their ecological communities."

Again, according to Flannery (1969:75):

> Starvation is not the principal factor regulating mammal population . . . Instead, evidence suggests that other mechanisms, including their own social behavior, homostatically maintain . . . populations *below* the point at which they would begin to deplete their own food supply. This is not to say that paleolithic populations were not limited by their food supply, obviously they were. But in *addition* they engaged in behavior patterns *designed* [my emphasis] to maintain their density below the starvation level.

Attributing goals to systems leads to sloppy thinking and, ultimately, to teleology. It is necessary to give some meaning to the arrows on the flow chart that we have constructed. To say that equilibrium or adaptation (they get confused in the literature) is the goal of cultural change tells us less than we knew before, for now we must explain *how* culture recognizes perturbations and *how* it reacts to correct them.

In practice, the issue is resolved by recourse to the paradigm of consciousness. For example, Binford, in considering the effects of population stress in the suboptimal areas, notes: "it is in the context of such situations of stress in environments, with plant and animal forms *amenable to manipulation,* that we would expect to find conditions favoring the development of plant and animal domestication" (1968:332; emphasis mine). Flannery says, "It is possible, therefore, that cultivation began as *an attempt to produce artificially,* around the margins of the 'optimum' zone stands of cereals as dense as those in the heart of the zone" (1969:89; emphasis mine). Thus, despite their strong repudiation of the vitalism inherent in Braidwood's work, both Binford and Flannery ultimately resort to arguments based on "emergent" human properties, such as intent. Intentionality in this context is the *deus ex machina* that permits us to reconcile adaptation with equilibrium, and goals with means. Change is putative evidence of a disturbance to an earlier equilibrium, and because the goal of society is to reestablish an adaptive state (that is, an equilibrium state), society must *choose* an *appropriate* solution to the disturbance.

So it is that we return to our discussion of Lamarckism. As we have noted, Lamarckism is characterized by an evolutionary one-step: change in environment → appropriate adaptive response. The whole structure the cultural ecologists created falls apart if we reject the doctrine that the ultimate result of a cultural system's functioning will be adaptation (and the equilibrium that characterizes it). Thus we see, as a final irony, that both Flannery and Binford, despite their attack on Braidwood's vitalism, fall into teleological arguments, which are merely orthogenetic ones with the time dimension put into reverse. To claim that an evolutionary change occurred *because* it was adaptive is to misunderstand evolution.

Adaptationism (Gould and Lewontin 1979; Lewontin 1979b) is the belief that adaptation is the *goal* or even the *inevitable result* of evolutionary change. It is usually justified on the basis that Darwin's theory of evolution was advanced to explain adaptation. But adaptation per se simply was not a problem to those who denied evolution or to those who saw it controlled by the hand of Deity— indeed, "arguments from design," teleological though they may be, are intellecturally more economical in explaining adaptation than is natural selection. Likewise, Lamarckian theories of directed variation that accept the direction action of environment as causative for evolutionary change (and we should remember that this idea was accepted, at least in part, by Darwin) were a relatively efficient way to explain adaptation. Darwin's concept of natural selection of existing variation is much less so.

Darwin's contribution to our view of nature was the recognition that *fitness* is the result of natural selection. Adaptation is not inconsistent with natural selection; indeed, much of *The Origin of Species* is devoted to proving that adaptations may be explained by means of his two-stage model; however, we must realize that this was necessary to deflate arguments from design or theories of directed variation and to prove that an entirely mechanistic process of selection may give us the same results as vitalistic, orthogenetic, or teleological ones. Moreover, implicit in Darwin's concept of variability is his recognition that not all traits are adaptive—on the simplest level, where could the raw material for natural selection come from if not from "random" (that is, not predictively adaptive) variation?

But fitness should not be seen as proceeding inevitably toward adaptation and adaptation only. Fitness may be conferred in many other ways: for example, by means of processes such as sexual selection (a topic Darwin did much to elucidate), kin selection, and changes in developmental rates. Although we may claim that the result of all of these processes is *adaptation,* this paints the term with such a broad brush that precious distinctions are easily lost. Darwin also understood that his model of natural selection could as easily explain extinction as speciation, something rival theories could not do. Vitalistic or orthogenetic evolutionary theories could only explain "progress" in evolution, whereas natural selection could also make "degradation" and extinction comprehensible.

The fundamental error of adaptationism is arguing from the end of adaptation. But in this case the argument is couched in an evolutionary terminology that tends to throw a smokescreen around the issue at stake. It distracts us from the question of *how* change occurs—fitness may yield adaptation, but evolution only proceeds on the basis of differential fitness. Darwin's description of *evolution* (a term he assiduously avoided because of its contemporaneous orthogenetic implications) as "descent with modification" is not only salutary but broader and more inclusive than most characterizations of evolution seen these days. It is also, in its very materialism, far more stirring:

> [As my theory is explored and investigated] the other and more general departments of natural history will rise greatly in interest. The terms used by naturalists, of affinity, relationship, community of type, paternity, morphology, adaptive characters, rudimentary and aborted organs, etc., will cease to be metaphorical and will have a plain significance. When we no longer look at an organic being as a savage looks at a ship, as something wholly beyond his comprehension; when we regard every production of nature as one which has had a long history; when we contemplate every complex structure and instinct as the summing up of many contrivances, each useful to the possessor, in the same way as any great mechanical invention is the *summing up of the labour, the experience, the reason and even the blunders of numerous workmen;* when we thus view each organic being, how far more interesting—I speak from experience— does the study of natural history become! (Darwin 1859:485–486; emphasis mine)

Equilibrium Lost

The systems modeling adopted by cultural ecologists was eventually bound to get them into trouble. The credibility of a materialistic argument suffers when emergent human properties must be invoked to make the system function. Carter was probably speaking for many scholars unwilling to express their doubts when he pointed out the uncounscious teleology inherent in most environmentalists' arguments (Carter 1977:97):

> Except in modern times when we have a class of inventors, WHAT invention arose through necessity? Cows were not "invented" for milk, nor chicken for flesh or eggs, nor did the Eskimo invent clothes, oil lamps and skin boats so he could live in the arctic.

Even in our time, the auto and airplane were not invented to fill needs. They were invented more in the spirit of play and the need FOLLOWED the invention.

These are general truths. The physical environment does not compel, suggest, hint, encourage—it is simply there. Men perceive their environment through cultural filters and behave accordingly. Man did not invent in response to needs. Neither did he invent in response to opportunity.

If the ultimate cause of agriculture is consciousness, and if humans ultimately are conscious animals, it is indeed more elegant to make consciousness the proximate, initiating, and *sole* cause. If we are to rely upon emergent human properties, let us at least have the invention emerge.

Of course, Carter's problem is his conviction that agriculture is an *idea,* and it is from this Platonic source that most of his problems flow. As we have noted, one of the major contributions of the cultural ecologists was the recognition that agriculture is a *system.* But their description of the system's functioning was to get them into as much trouble as the mentalist device they used to solve the problem their assumptions had created: "Since all parts of the system are in a perpetual state of flux, there can be no single 'cause' of change. Within the framework of the equilibrium model, . . . the question 'What caused agriculture?' is as meaningless as 'What caused elephants?' " (Bray 1974:78). Although there is a great deal of truth in this statement, it is nevertheless a cop-out. "What caused agriculture?" is, in fact, a meaningful question that *must* be addressed.

The uncritical adoption of the equilibrium model made cultural ecology a self-defeating enterprise. Yet, the downfall of this model left a void in the theoretical description and analysis of the relationship of people to their environment. As Green (1980b:314) has observed:

The equilibrium perceived by many anthropologists has been called into question as an observation bias, or an artifact of short-term study. Moreover the increased complexity or cultural endowment of so-called civilized societies has certainly not allowed for the domination of the ecological framework. More important, perhaps, than either of these problems is that the assumption of equilibrium masks all questions of stability and change. Equilibrium models have therefore recently yielded to models of change which rely on perturbation in the form of pressures and/or cultural innovation as driving variables.

Unfortunately, most of these recent hypotheses for the origin of agriculture have, in effect, called for an abandonment of the insights gained by the earlier cultural ecologists (whose problem, we should remember, was bad modeling, not an improper theoretical approach).

Most of these "new" models are retreats to a simple-minded determinism or to eclectic stews that throw together every variable thus far advanced to explain the origin of agriculture—as if a multiplication of causes might compensate for a lack of theory. Thus, from the standpoint of theoretical rigor, the determinists are pleasanter company than the eclectics, although the latter are less likely to offend one's sensibilities. And despite strong disavowals, both groups are indebted to the cultural ecologists for many of their arguments. Unfortunately, it has been, by and large, the errors of cultural ecology that have survived; we have not thrown out the baby with the bath water, but saved the bath water and thrown out the baby.

The cultural ecologist's greatest contribution was the recognition that agriculture was a new system by which people interacted with the environment. Their methodology was their second great contribution, for they at least attempted to construct a mechanistic model for the development of new systems out of existing modes of environmental interaction. But, as we have seen, they ended up trapped between the errors of intent and equilibrium. The recently developed approach of "demographic determinism" has attempted to take the basic approach of the cultural ecologists and rephrase it so as to avoid the problems inherent in cultural ecology. In this I believe they have failed; in fact, they have done worse than their predecessors.

Several scholars have attempted to interpret the origin and development of agricultural systems by focusing on the influence of population growth upon technological innovation and land-utilization patterns (Abernathy 1976; Cohen 1975, 1977a, 1977b; Cohen et al. 1980; Grigg 1976; Harner 1970; Hassan 1978, 1981; Simon 1978; Smith 1972a, 1972b; Spooner 1972; and Zubrow 1975). In this they have stressed the importance of the work of Boserup (1965) in pointing out the heuristic utility of population growth as an independent variable in the analysis of cultural change. However, insofar as any of these authors represent population growth as an endogenous function of human demographics, they misrepresent Boserup's views (1965:14; emphasis mine):

> The neo-Malthusian school has resuscitated the old idea that population growth must be regarded as a variable dependent mainly on agricultural output. I have reached the conclusion . . . that in many cases the output from a given area of land responds far more generously to an additional input of labor than assumed by neo-Malthusian authors. If this is true, the low rates of population growth found (until recently) in pre-industrial communities cannot be explained as the result of insufficient food supplies due to overpopulation, and *we must leave more room for other factors in the explanation of demographic trends. It is outside the scope of the present study, however, to discuss these other factors—medical, biological, political, etc.*—which may help to explain why the rate of growth of population in primitive communities was what it was. Throughout, our inquiry is concerned with the *effects* of population changes on agriculture and *not with the causes* of these population changes.

Boserup is not a "demographic determinist," and she has made clear that, although population is to be treated as an independent variable, it is *not* to be treated as an ultimate cause. I am in agreement with her on the importance of the ultimate responsiveness of the agroecosystem, although, unlike her, I am concerned with the earliest phases of the development of this system and have placed a corresponding emphasis on the biological aspects of this interaction.

Cohen has applied a demographic perspective to the problem of the early history of agricultural subsistence. Rather than even attempt to explain the mechanisms underlying the origin of agricultural practices, he takes evasive action (1977a:22–23).

> Taken as a whole [the] evidence indicates two very important things about the "concept" of domestication: first, that any human group dependent in some degree on plant materials, possessing the rudiments of human intelligence, and having any sort of home-base camp structure (of the type associated with all men probably since *Homo erectus*) will be almost bound to observe the basic processes by which a seed or shoot becomes a plant. Ignorance of this basic principle is almost inconceivable.

In this attitude, Cohen is among a large company of scholars who have seen the "idea" of agriculture as self-evident. Thus his argument is similar to that of White 20 years earlier (1959b:283): "We are not to think of the origin of agriculture as due to the chance discovery that seeds thrown away sprouted. Mankind knew all this and more for tens of thousands of years before cultivation of plants began." Although this argument is a bit too idealistic for my taste, it has, at least in some sense, a real meaning—a meaning that is not lost on Cohen (1977a:23):

> This suggests that the conceptual break, or for that matter the operational distinction, between agricultural and nonagricultural practices is not very great. . . . there is a continuum in the degree of assistance different human populations offer to the species on which they depend . . . Thus agriculture should be conceived more as a de facto accumulation of new habits than as a conceptual breakthrough.

Cohen correctly identifies domestication and the origin of agricultural systems as a gradual process; in a similar manner he recognizes that agriculture does not inevitably bring with it an increase in the labor-efficiency quality or reliability of the subsistence base. The crunch comes, however, in his treatment of these insights:

> In brief, I intend to argue that human population has been growing throughout its history, and that such growth is the *cause,* rather than the result, of much human "progress" or technological change. . . . While hunting and gathering is an extremely successful mode of adaptation for small hunting groups it is not well adapted to the support of large or dense human populations. I suggest therefore that the development of agriculture was an *adjustment* which human populations were *forced to make in response to their own increasing numbers.* . . . *agriculture . . . represent[s] no great conceptual break with traditional subsistence patterns; and . . . it is therefore not ignorance, but rather lack of need* that prevents some groups of people from becoming agriculturists . . . agriculture is not easier than hunting and gathering and does not provide a higher quality, more palatable or more secure food base. . . . [It] has only one advantage . . . that of providing more calories . . . *it will thus be practiced only when necessitated by population pressure.* (Cohen 1977a:14–15; emphasis mine)

The important problem requiring an answer for Cohen is not the ultimate cause of the origin of agricultural systems, but rather the proximate cause of the *adoption* of agricultural systems. He treats agriculture as a self-evident idea that merely awaited the appropriate situation for its application and attempts to explain the conditions that created this situation—none other than our old friend, population pressure: "The development of agriculture was an adjustment which human populations were forced to make in response to their own increasing numbers" (Cohen 1977a:14–15). Cohen turns the systems theory of the cultural ecologists inward and, like the cultural evolutionists, finds the impetus for cultural change in an endogenous process, here identified as population growth and its concomitant pressures. What Cohen proposes is thus, in effect, an evolutionary process that is mediated by population growth and the response of culture to this growth. Cultural evolutionism and demographic determinism are alike in proposing endogenous development rather than an interactive process. The cultural ecologists, in contrast, created a system mediated by a series of interactions between the population and its immediate environment. Although complexity is no indication of validity, simplistic and eufunctional arguments have provided ample ammunition for recent critics of demographic determinism (e.g., Bray 1976, 1977; Bronson 1972, 1977; Hassan 1978; Polgar 1975) just as they occasioned fierce attacks upon cultural evolutionism at an earlier time.

Cohen turns the equilibrium model of the cultural ecologists upside-down. Probably reflecting our times as much as the believers in homeostasis reflected theirs, he advances a model in which "human societies have in fact grown throughout their history and have encroached progressively upon their resources to the extent that the continuous development of new adaptive strategies and the continuous redefinition of ecological relationships were necessary" (1977a:16). Thus, rather than seeing cultural change as arising from the reestablishment of an equilibrium lost to outside disturbance, Cohen argues that human cultures have been continually attempting to establish some approximation of an equilibrium. Cultural change grows out of readjustments *necessitated* by an endogenous phenomenon—that of population growth. In this way, like many liberal economists, Cohen is an optimistic Malthusian: Although not denying the reality of population growth, he believes that technological solutions may result from the very "need" to find a solution.

Of course, scrutinizing needs and finding solutions to them is a highly cognitive business; thus it is not surprising to find consciousness and intent advanced in no uncertain terms throughout Cohen's work. Like cultural ecology, demographic determinism requires an adaptive and appropriate response to a given situation, and like cultural evolution the response must be openly goal-directed.

Cohen's rejection of equilibrium-related processes is not total, however. Although denying that cultures maintain a homeostatic relationship with the environment, he asserts that cultures do have "highly effective mechanisms by which population pressure is balanced from region to region" (1977a:16). He argues, in essence, that it was not until the niches available to hunters and gatherers throughout the world were filled that population pressure could become effective in initiating the adoption of agricultural means of subsistence. Therefore, "it is not unreasonable to find a roughly synchronous building up of population pressure over very large portions of the globe, with the result that agriculture was 'invented' or adopted by most of the world's population within the same fairly short time span" (1977a:16).

This emphasis upon synchronism in the adoption of agricultural modes of subsistence is the most original part of Cohen's presentation. Although some scholars have been troubled by the apparent simultaneity of the origin of agriculture in various parts of the world, most have ignored the issue; others have used it as proof of the importance of diffusion in the history of agricultural subsistence patterns. Cohen sets himself apart from both of these groups and shows his commitment to the creation of a general theory by confronting the evidence head-on and by attempting to provide at least a quasi-mechanistic resolution to the problem:

> Probably in response to the excessive zeal of early diffusionists who explained the arrival of agriculture in one region largely on the basis of the dissemination of ideas from another, modern archaeologists have stubbornly insisted on studying the causes of agriculture within the limits of local archaeological sequences and local variables. *The most striking fact about early agriculture, however, is precisely that it is such a universal event.* Slightly more than 10,000 years ago, virtually all men lived on wild foods. By 2,000 years ago the overwhelming majority of men lived by farming. In the four million year history of *Homo sapiens,* the spread of agriculture was accomplished in about 8,000 years . . . the problem is [thus] not just to account for the beginnings of agriculture, but to account for the fact that so many human beings made this economic transition in so short a time. (Cohen 1977a:5, original emphasis)

This aspect of Mr. Cohen's position is not only provocative, but also, at least for the foreseeable future, unassailable: *Something* does appear to have happened during the period of time he points to, though its precise nature is never clearly explained. Thus, his would-be critics find themselves in a no-win situation. Because what Cohen is addressing is not the origin of agriculture but its adoption, any evidence of domestication from earlier periods of time is incapable of casting doubt upon this hypothesis. If, for example, we were to find domesticates 20,000 or even 50,000 years old, he could respond that he recognizes that the "idea" of domestication was around for a very long time, and then go on to argue that population pressure simply had not yet reached the point where *reliance* upon agriculture was called for. Likewise, "synchronism", is here defined rather broadly as some 8,000 years—(at what point would the event no longer be synchronous—10,000 years? 100,000?) Further, because Cohen does not reject diffusion for at least *some* cases of the adoption of agriculture, evidence of diffusion may not be used to invalidate his hypothesis. Again, neither Cohen nor anyone else employing this argument has considered it threatened by the existence of *modern* nonagricultural societies: These are merely examples of "reversions" or (for the demographic determinists in particular) the consequence of the impracticality of agriculture, the absence of population pressure, or the reversal of the direction of population pressures. Finally, even if we were capable of proving, in terms acceptable to the demographic determinists, that the adoption of agriculture worldwide was *not* a synchronous event, they could simply say that they weren't really talking about the origin or adoption of agriculture at all, but about the origin of village and city life. (Recalling our earier discussion of civilization and agriculture, this is what I believe Cohen actually had in mind throughout his discussion.)

Although unassailable theories ("nonrejectable" in today's jargon) are always suspect, they may of course still be accurate. And because any theory inevitably contains assumptions not subject to disproof ("primitive terms") it would be most unfair to reject them solely on this basis. Thus we arrive at a standoff: I am not particularly impressed by the "simultaneity" argument; it seems to me too vague and too difficult to test. Every phenomenon has a beginning and it is a misunderstanding of historical phenomona to ask, why not earlier?—for any time this question is answered, it may simply be repeated in the new context. Therefore, the goal of science is to advance mechanistic models of causation that explain, not so much the initial appearance of a phenomenon, but its functioning, mechanics, and consequences. I also cannot help but see an analogy with organic evolution: The origin of any species presents the same appearance of "sudden" and "unprecedented" evolution and it is frequently found over a great geographic range at the time of its first appearance in the paleontological record. These phenomena, however, are usually interpreted in part as an artifact of preservation; it is not until the species is fairly common that we are likely to find it preserved. Also, they are seen in part as a product of the nature of speciation itself: successful species—that is, ones that become abundant—have frequently found an unfilled niche, a new tactic for survival, that creates the appearance of rapid evolutionary development and early dissemination over wide areas. Thus, the final reckoning for the simultaneity hypothesis will only come when we have far more data than we do now on early subsistence and settlement habits. Resolution of this problem is likely to be a pleasure denied our generation.

Demographic determinism, just like its predecessor and close relative, cultural ecology, is concerned with an interaction of demography and technology. Both posit that population increase preceded and was the ultimate cause for technological change and both agree that adaptation mediated by consciousness was the proximate cause; both have tried to provide a more or less mechanistic model for the wide distribution and success of agricultural modes of subsistence. Their most important difference, however, lies in their view of the *nature* of agriculture. Whereas the cultural ecologists stressed the systemic nature of agriculture and tried to produce a functional and integrated view of its development out of preexisting modes of subsistence, the demographic determinists have tended to see it as an ''idea awaiting its time.'' In this, they must share a bed with the atheoretical cultural evolutionists and historical particularists, an arrangement upsetting to all concerned. And it is not surprising that, given this basic prejudice, all three schools have received their share of grief from members of what I call the *eclectic school.*

The eclectics are not bound by any particular set of theoretical convictions—indeed, as in their writings, it is the lack of theory that here binds. Representing the mainstream of Western anthropological and botanical thought on the origin of agriculture, they have, nevertheless, maintained fidelity to the systemic description of human interactions with nature. Despite the failure of equilibrium modeling, they have carried on the work of cultural ecology and have managed major contributions to our understanding of agricultural systems and their function and origins. The remainder of this volume is, if anything, an attempt to summarize and unify their insights and researches within what I hope is a simpler and more inclusive theoretical framework. The general approach of this group has been summarized by Bronson and it is one with which I can, in large part, agree.

> The population-centered model of subsistence evolution may be pedagogically useful but it is of doubtful value as a research guide. Much more satisfactory is the rather subliminal model that seems actually to guide much of the research on post-Pleistocene adaptations, whatever the explicit theoretical orientation of the individual researcher may be. The leading characteristics of this model are complexity, factor feedback, and instability. A great many agencies—sedentariness, epidemiology, genetics, environmental structure, technologies of subsistence and non-subsistence, political evolution, economic development warfare, the density and distribution in space of populations—are recognized as potential influences, without seriously contending that any necessarily have priority. The relationship between each pair of these is visualized as one of feedback; the chicken-and-egg quality of interactions between adaptational factors has long been recognized by most specialists. And the rather Augustinian notion that all recent (i.e., post-Pleistocene) adaptive patterns are intrinsically unstable is gaining ground again after a brief setback during the heyday of functionalism. Change, driven by the sheer impossibility of keeping so many interacting factors out of disequilibrium, is a normal condition. What requires explaining is stability not change. (Bronson 1977:45)

I am not sure, however, that a polarization of stability and change, or of equilibrium and disequilibrium, is a good way of understanding either chickens or eggs. Both are parts of one process, and it is the process that requires our attention. Furthermore, careful study of processes may give us new insights into old problems. The chicken and egg problem, itself, is amenable to an evolutionary solution: We know that birds evolved from reptilelike ancestors, all of which laid eggs. Therefore, we may confidently claim that, *in evolutionary terms,* the egg did come before the chicken.

CHAPTER 2

Darwinism and Culture

Most writers on the emotions and on human nature seem to be treating rather of matters outside nature than of natural phenomena following nature's general laws. They appear to conceive man to be situated in nature as a kingdom within a kingdom: for they believe that he disturbs rather than follows nature's order, that he has absolute control over his actions, and that he is determined solely by himself.

BARUCH SPINOZA (1632), *ETHICS*

The fundamental observation of Darwinian evolution[1] is that evolution occurs through the interaction of the environment and the individual organism. No two individuals are exactly alike, but all face basically the same problems of survival. The differences between individuals influence their survival and ultimate reproductive success. This emphasis upon the differential survival and reproduction of individuals distinguishes Darwinian evolution from all other dynamic theories of change. As Lewontin (1974:4–5) has put it,

> The essential nature of the Darwinian revolution was neither the introduction of evolutionism as a world view (since historically this is not the case) nor the emphasis on natural selection as the main motive force in evolution (since empirically that may not be the case), but rather the replacement of the metaphysical view of variation among organisms by a materialistic one. . . . [Darwin] called attention to the *actual* variation among *actual* organisms as the most essential and illuminating fact of nature. Rather than regarding the variation among members of the same species as an annoying distraction, as a shimmering of the air that distorts our view of the essential object, he made that variation the cornerstone of his theory.

Thus the great contribution of Darwinian evolution is the understanding that the differences in survival and reproduction between related organisms are largely a

---

[1]More accurately, *neo-Darwinian* evolution. The model proposed by Darwin has undergone extensive revision and amplification. The approach, however, has remained unchanged in its stress upon a two-step evolutionary process involving the natural selection of existing, undirected variation. Throughout this discussion it will be important to remember that although genetics, including population genetics, has been of particular importance in elucidating the basis of variation and the population effects of selection, natural selection acts upon the total phenotype—including behaviors, even though these may *not* be genetically determined.

function of the differences in traits individuals present to the environment. Organisms having particular traits tend to survive and reproduce in greater numbers.

One of the most important scholarly debates of our generation concerns the applicability of this specific Darwinian insight for the explanation of cultural change. The debate has often waxed nasty, yet the several sides are dealing with an issue of fundamental concern to both the natural and social sciences. And although the debate itself has moved simultaneously in several directions, the fundamental question has remained a simple one: Is cultural change the result of a natural evolutionary process? I believe it is, but my position is one that must be presented at some length if misunderstandings are to be minimized.

In this chapter I review several interrelated literatures bearing upon cultural change. First, I abstract the fundamentals of modern Darwinian evolutionary theory, especially as related to the central concepts of fitness and adaptation. One of the purposes of this discussion will be to stress the need for a clear distinction between *functional* or proximate, and *evolutionary* or ultimate modes of explanation. The first is concerned with *how* organisms function in the environment, whereas the second is concerned with *why* they are as they are. If the distinctive aims and methods of these two approaches are not appreciated, confusion will be the likely result. Second, I present evidence that learned behaviors have been acted upon by natural selection and that this process has had major evolutionary effects. In this context, it will be necessary to point out the types of confusions that may arise when considering the intimate and often interactive relationship between variation and natural selection. Third, I consider in some depth the school of thought traditionally known as *cultural evolution*. It will be shown that cultural evolutionism bears no relationship whatsoever to Darwinian evolution and that, in fact, it grew out of an entirely different scientific and philosophical tradition. Major errors in evolutionary thought in both the natural and social sciences may be traced to fundamental disagreements with Darwinism as advanced and defended by members of this school. Finally, I introduce some of the modern literature dealing with cultural change in what I believe to be a new and exciting manner. Cultural selectionism is a mechanistic and materialistic approach to cultural change that seeks an evolutionary explanation for *why* cultural variation exhibits the forms it does. It places major emphasis upon variability, heredibility, and differential fitness in explaining changes in the *learned* behavior of animals (especially our own species).

Natural Selection, Fitness, and Adaptation

All organisms that survive to reproduce have offspring that resemble them. Thus, the selective pressures occurring at any given time will have effects on the composition of future populations. The fittest individual is the one with the greatest relative reproductive success—the one that has more offspring than other, related organisms. Thus the total population of offspring inevitably resembles the more fit. Over time, the force underlying evolution—natural selection— brings about a change in the representation of traits in a population.

Natural selection, however, is not a metaphysical force interposed between the organism and the environment, but is rather the summation of all of the relationships between the organism and the environment that affect the organism's reproductive success. In a sense, natural selection is part of the organism (Simpson 1953, 1967): Natural selection in this sense does not *cause* the changes that occur in populations over time. Rather it *is* the disparities present in organisms, which, over time, accumulate and allow us to describe the result as evolution. In other words, natural selection is not an external force determining which individuals are to survive, but a concept that (rather artificially) lumps together any and all events based on differential reproduction that cause change in the characteristics of a population over time. This change is what we call *evolution.*

Darwinian evolution is predicated upon variability, differential fitness, and hereditability. It may occur only if three conditions are met: (1) individuals in a population differ in their traits, (2) a correlation exists between the possession of certain traits and the probability of an individual's survival and reproductive success, and (3) the success of the offspring reflects the success of the parents. In its broadest formulation, Darwinism is silent upon the *source* of variation, the *reasons* for differential fitness, and even the *method* of transmission of traits (Lewontin 1970). Instead, it only claims that natural selection is mediated by differential fitness.

Fitness has been given numerous meanings and interpretations in the biological literature (Burian, in press; Mayr and Provine 1980) and, somewhat surprisingly, we lack an extended discussion of the topic. Nevertheless, two extreme conceptions may be recognized. The first, *Darwinian fitness,* is pure differential reproductive success over time: the reproduction that has, in fact, occurred.[2] This concept, appealing to those with inductivist and historicist tendencies, advances no specific reason to account for general evolutionary change. In practice, it forms the foundation for all evolutionary analyses: We are sure of little else except that certain organisms have had greater differential reproductive success, that this success was related to traits they possessed, and that by means of their enhanced Darwinian fitness they have come to represent the mode of variation within their taxon. Darwinian fitness, defined in terms of the conse-

---

[2]The recently advanced concept of propensity fitness (Mills and Beatty 1979) is a fine attempt at modifying Darwinian fitness to account for several logical and philosophical, but few practical, problems (Rosenberg 1982).

quences rather than the causes of natural selection, makes no a priori assumptions concerning the nature of the traits that enhance fitness, and is thus incapable of providing a general reason for changes in relative fitness values.

The second type of fitness we will recognize is *engineering fitness*, which in the way it functions in argument, is the polar opposite of Darwinian fitness; an "argument from design" is invoked to explain differential reproductive success. We assume that any absolute or relative advantage in the design or function of traits will account for the observed historical changes in fitness values; that an advantage accrues to individuals with traits that are better "solutions" to environmental "problems." Engineering fitness imposes a major constraint upon the types of traits presumed to increase differential reproductive success. It adds a fourth condition to Darwinian evolution: Those traits that increase the efficiency of the organisms in the environment tend to increase reproductive success and will be selected for.

Whereas Darwinian fitness may be criticized for providing no basis for the identification of the reasons for survival and increased reproductive success, engineering fitness must be faulted for seriously limiting the types of traits presumed to be of evolutionary importance so that fitness and adaptation become synonymized. Thus, several authors, including Thoday (1958b) and Dobzhansky (1968), have called for a clear conceptual distinction between fitness and adaptation, a position that I also endorse. Engineering fitness frequently may be appropriate for the analysis of evolutionary change, but it is certain to mislead if used as the exclusive measure of changes in fitness values.

Engineering fitness also must be criticized for its frequently invoked corollary that evolution exhibits directionality. If fitness is enhanced by increased adaptiveness, increasing adaptation is the only possible, or most likely, result of evolution. Proponents of such an adaptationist view of evolution have long cited Darwin in defense of their claim. Indeed, Darwin was primarily concerned with the evolution of adaptation: but this was both a tactical and conceptual issue. Darwin, we must remember, was confronting two opposing theories that sought to explain the biological world. The first, creationism, used the fit of organism to the environment as proof of the existence of a master plan. Therefore, Darwin had to demonstrate that the marvelous adaptations found in nature could result from a mechanistic and blind process. The second and perhaps more successful theory opposing natural selection, however, was promulgated by the many scientists who accepted orthogenetic, frequently teleological, evolutionary models for organic change. Provine notes that:

> Contrary to popular belief, the reaction among most biologists to Darwin's idea of the evolution of species was not strongly adverse, especially after the initial impact of the *Origin*. The idea of evolution was not new, having appeared prominently in the works of Buffon, Erasmus Darwin, Lamarck, and many others. . . . Darwin's idea of natural selection, however, did arouse a very strong reaction. The random production of variation, with the relentless elimination of the less fit variants, ran entirely against the

prevalent view of "design" in nature. . . . That the beauty and harmony of living creatures was the result of chance rather than design was abhorrent to most minds. . . . Design in nature was not inconsistent with evolution since the unfolding of new organic forms could be seen as resulting from a higher order, from a master plan. But Darwin's idea of natural selection denied this possibility. . . . Most of the serious attacks upon Darwinism centered upon the idea of natural selection. (Provine, W. B., *The origins of theoretical population genetics*, pp. 10–11. Published by The University of Chicago Press. Copyright © 1971 by The University of Chicago Press.)

Proponents of natural selection theory have long been wary of attributing any inherent directionality to evolutionary processes. It is widely held that adaptationism implies a directionality prohibited by natural selective processes. Thus, the view of evolution that holds that natural selection optimizes design of organisms or increases their absolute fit to the environment has come in for more than its share of discussion and criticism (Caplan 1981; Gould and Lewontin 1979; Lewontin 1979b; Maynard Smith 1978; Stern 1970; Williams 1966). First, under an adaptationist interpretation, engineering fitness becomes an intellectually sterile approach in that it provides a tailor-made explanation for any conceivable change (save, of course, extinction). "Why did Y evolve out of X?" "Because Y was more adapted than X." Assuming that increasing adaptation is the only result, or even the predominant result, of evolution does little to elucidate the means of evolutionary change and, in fact, it may not be literally true (Beatty 1980; Cody 1974; Lewontin 1979a; Maynard Smith 1978). Second, it is necessary to be able to show that an adaptation was actually that which was selected rather than a post hoc effect of other traits that were, in fact, the object of selection (Gould and Vrba 1982, Williams 1966). We must demonstrate that the adaptation was an end of a means, rather than merely the fortuitous effect of more general causation (Falk 1981). Burian (in press) has pointed out that numerous philosophical, linguistic, and conceptual problems stand in the way of properly applying the concept of adaptation to Darwinian evolution, and that until these issues are clarified it will be difficult to discuss, no less resolve, the underlying biological problems.

Given the manifold difficulties that stand in the way of accepting an adaptation-centered view of natural selection, it would seem wise to take particular pains to avoid explaining evolutionary change solely in terms of adaptive or optimizing scenarios. Instead, we should attempt to keep in mind the fact that seemingly maladaptive or adaptation-neutral processes may have been of major significance in any evolutionary history. By placing our emphasis upon the *reasons* for increases in Darwinian fitness, it is possible to combine the insights provided by both interpretations of fitness. But in applying this type of an approach we must be open to myriad possibilities. As Williams has put it,

Natural selection commonly produces fitness in the vernacular sense. We ordinarily expect it to favor mechanisms leading to an increase in health and comfort and a decrease in danger to life and limb, but the theoretically important kind of fitness is that which promotes ultimate reproductive survival. Reproduction always requires some sacrifice in well-being, and some jeopardy of physiological well-being, and such sacrifices may be favorably selected, even though they may reduce fitness in the vernacular sense of the term.

We ordinarily expect selection to produce only "favorable" characters, but here again there are exceptions. . . . A given gene substitution may have one favorable effect and another unfavorable one in the same individual. . . . The same gene may produce mainly favorable effects in one individual but mainly unfavorable effects in another, because of difference in environment or genetic background. . . . Another frequent outcome of natural selection is the promotion of long-term survival of the population. . . . Here again, however, there are exceptions. The constant maximization of mean fitness in some population might bring about an increasing ecological specialization, and this might mean reduced numbers, restricted range and vulnerability to changed conditions. . . . Probably most evolutionary increases in body size cause a decrease in numbers. (1966:26–28)

Understanding why the maximization of fitness does not inevitably mean an increase in population over time depends, in part, upon the recognition that organisms cannot change without also affecting the environment in which further change will occur. Emlen makes this point with the

hypothetical case of a horse population which is capable of cropping the grass to a height of two inches and in doing so gains sufficient energy to maintain a stable population. Suppose that a new genetic innovation suddenly appears which enables its owners to crop the grass to one inch. Momentarily, the energy available to the population increases, until enough forage is eaten to one inch that the old genotypes are incapable of surviving. A new equilibrium population density will be reached which may or may not be higher than the original one. As adaptations of this sort continue to appear, there will eventually be a point reached at which the grass can no longer replace itself so efficiently and its productivity goes down. At this point less food is available to the population which must consequently decline. Finally, when the shoots of grass are cropped short enough, no grass reproduction occurs, the food supply disappears, and the horses go extinct. Selection responds to the *relative* fitness values of different genotypes and a genotype may increase in relative fitness by dragging down the fitness of other genotypes. (Emlen, *Ecology,* © 1973. Addison-Wesley, Reading, MA., pp. 89, 371–372. Reprinted with permission.)

Because neither organism nor environment is static, evolution must be understood as a historical phenomenon mediated by a materialistic process. We seek to understand *why* and *how* certain organisms came to be adapted to certain environmental conditions or, conversely, why they failed to do so. The theory of natural selection, with its demand for understanding changes in relative fitness in organisms, places rather severe limitations on hypotheses advanced to elucidate any particular historical sequence. By making our analysis consistent with this model, we can establish standards for the acceptance of evidence. We must be able to give good reasons for believing that the appearance of new traits would correlate with greater fitness for certain organisms. We are also invoking a biological "law" which, until proven inapplicable to a given situation, must be assumed to be in force. The major value of the concepts of natural selection and fitness is that they point out to us that organisms survive differentially because of differences in traits. Focusing on the organism and its traits, and the local environment in which it lived, we are challenged to try to discover just what traits were responsible for the organism's survival.

An understanding of the evolutionary results of the interaction of organism and environment helps us to avoid certain problems that arise if we consider adaptation alone. Prime among these is the fact that, if all organisms are by definition adapted, it is difficult to see how evolutionary change can be characterized by adaptation. One popular way out of this dilemma is to posit that the environment is constantly decaying relative to existing organisms; thus selection acts merely to maintain "levels" of adaptation. This notion has been whimsically named the *Red Queen hypothesis* (Van Valen 1973), after the character in Lewis Carroll's *Through the Looking Glass* who had to keep running just to stay in the same place. The relationships that develop between evolving organisms and changing environments allow us to reconcile descriptive studies based on adaptation with historical evolutionary studies based on fitness. However, we must recognize that this is only a general theoretical resolution of the problem and no substitute for a careful analysis of specific cases.

The rigorous definition of natural selection, in terms of the reasons underlying the maximization of fitness, permits a more subtle and insightful understanding of evolutionary processes. It opens our eyes to possibilities that are obscured if we see natural selection merely as the means by which new adaptations are created or as the result of an incomprehensible historical process. *Fitness* explains differential reproductive success among a group of related organisms; *adaptation* describes the myriad relationships between these organisms and their environment (including internal processes and other organisms). When organisms survive they are by definition "adapted"; what we seek to understand when we consider adaptation is the relationship between the organism and its total environment. In an evolutionary analysis of fitness, however, we seek to understand why traits changed over time; that is, what the relationship was between relative fitness and particular traits. Thus the study of evolutionary phenomena involves two complementary approaches: the study of adaptations (description) and the study of evolutionary change (historical analysis). Both are valid, but they have different goals. The error we must avoid is mistaking description for explanation.

Mayr (1961, 1982) has pointed out that two major types of investigation, corresponding to description and analysis, exist within the biological sciences. Functional biology is concerned with understanding how organisms function, whereas evolutionary biology is interested in why organisms are as they are. Many conflicts and misunderstandings in biology may be traced to the confusion between these two types of investigation. For example, Munson (1971, 1972) and Ruse (1971, 1972) advanced and defended conflicting claims about the evolutionary significance of adaptive traits. Neither apparently recognized that one was talking largely of physiological functions whereas the other of adaptational evolutionary significance.

Brandon notes that although the functional/ evolutionary distinction is useful in theory, it frequently fails in practice:

> We can always ask both functional and evolutionary questions of a given biological phenomenon. Furthermore, answers to both are necessary for a complete understanding of the phenomenon. . . . However the complementary relationship between the two has not always been recognized. This plus the fact that the same expression can be used to ask both functional and evolutionary questions can lead to great confusion. For example, in 1932 the geneticist T. H. Morgan was willing to answer, quite dogmatically, "Hormones!" to those who sought an evolutionary answer to "why are there sex differences?" (1981b:93–94)

Pointing out the difficulty of thinking clearly on confusing subjects Brandon also notes that a potential source of confusion exists even in the functional–evolutionary distinction: "Functional explanations may be evolutionary ones as well" (1981b:94). Mayr (1982) has suggested that we term the distinction as one between *ultimate* (evolutionary) and *proximate* (functional) causation, an amendment that should be fairly pleasing to all.

In light of the preceding discussion of adaptation and fitness, the criticism of the Darwinian view of evolution (see reviews by Ruse 1973 and Caplan 1979)—arguing that the fit survive and the survivors are fit—can be seen as the result of misunderstanding and oversimplification. Indeed, we do claim that the fit survive and the survivors are fit,[3] but the statement is not intended as explanation. Rather, natural selection is the conceptual framework in which we carry out our analysis; we cannot expect Darwinism to do our work for us and provide us with the *specific* reasons why changes in relative fitness occurred in any evolutionary case-study. Natural selection is what Brandon (1981a:432) has called an "organizing principle" or "schematic law" that serves to "structure particular biological explanations of differential reproduction." The evolutionary scientist is concerned with several interrelated issues: why organisms exist with specific traits; how traits interacted with the past environment such that survival and differential reproduction occurred; and what is the ultimate causation for observed change.

The Unit and Level of Selection

Echoing the babble of the functional–evolutionary debate and, in part, related to the same problem, neo-Darwinian theorists have long been involved in heated

---

[3]I must mention the cute proposal by Williams (1973) that by *not* defining fitness, that is by making it a "primitive term," we can avoid circularity in our formal axiomatizations of evolutionary theory. This may be formally correct, but it is heuristically useless.

controversy[4] concerning the level at which evolution occurs—What is selected? Genes? Individuals? Populations? Species? I know of few areas in biology in which scholars spend as much time talking past each other as in this particular debate. Understanding many of the conflicting arguments requires one to constantly redefine terms because most authors seem to apply them in radically differing ways; making sense of the controversy requires a stealthy approach from the rear rather than a frontal assault.

Mayr (1963, 1976, 1982) has properly stressed the revolutionary nature of natural selection theory. By discarding the concept of specific *types* or *essences* for biological entities, Darwin was able to relate long-term change to the variation existing within a population of related individuals. Mayr notes that, unlike essentialists:

> Population thinkers stress the uniqueness of everything in the organic world. What is important for them is the individual, not the type. . . . There is no "typical" individual and mean values are abstractions. . . . [The] uniqueness of biological individuals means that we must approach groups of biological entities in a very different spirit from the way we deal with groups of identical entities. . . . The differences between biological individuals are real, while the mean values we may calculate in the comparisons of groups of individuals (species, for example) are manmade inferences (1982:46–47)

What does Mayr mean by "inferences"? He cannot be implying that the numbers or descriptions so inferred have no existence in reality; that they are unlike, for example, the measured fecundity of a pair of organisms. Instead, he must be claiming that these types of values are a second-stage or "reflective" property of the group. Questions of "individualism" versus "holism" are largely irrelevant here logically (Sober 1981), and because of the manner in which higher order groups are constructed. Given the historical, evolved nature of all taxa, we must recognize that groups do not embody abstract quantities that serve to define uniquely that group. Instead the characteristics defining the group are products of its common evolutionary descent. As Mayr puts it (1982:56–57), "the 'classes' of the biologist often are not equivalent to the 'classes' of the logician." We create our classes on the basis of information concerning "real" individual organisms; these organisms form the data base for any population study. *After* a class is delimited we can analyze the population and develop statistics reflecting its properties. Having defined the higher order phenomenon, we may analyze it and—this is critical—by redefining the higher order group (say, on the basis of new information concerning its phylogeny drawn from a study of the traits of related "real" organisms), we are forced to reconstruct any statistics originally produced for the group. Higher order traits are properties of a group, but they are "the inferences" secondarily dependent upon the construc-

[4]For an introduction to the controversy, see discussions by Allee *et al.* 1949; Cody and Diamond 1975; Dawkins 1976; Emerson 1960; Haldane 1932; Lewontin 1958,1962,1970; Maynard Smith 1969, 1976; Slobodkin 1964; Sober 1980, 1981; Stanley 1975, 1979; Wade 1978; Williams 1966, 1971; Wilson 1980; Wimsatt 1980, 1981; Wright 1945; and Wynne-Edwards 1962.

tion of the group. Therefore, from a populational point of view, adaptations or any other trait of a group cannot be defined unambiguously as a property of the group because it is always possible to reconstruct the group to bring it into closer harmony with what we know about individual organisms. Group traits are not "real" properties of a group because the group itself is subservient to our observations of real individual organisms. Individual organisms lived, died, and differentially reproduced. Individuals were the focus of selection and it is only a matter of convenience to call higher order groups (e.g., species) *units* of evolutionary change.[5]

All of this does not deny that group phenomena may have an effect upon evolution. Numerous demographic, social, and ecological effects arise from the interaction of organisms with other organisms and with the environment. It would be silly to demand that we call such phenomena *individual* rather than *group* or *social* traits. Yet we must be careful in our use of language, especially as we begin to speak of the selection of group traits, so that we do not fall into the error of seeing the group as something transcending the individuals comprising it. A group is only individuals interacting with each other within the greater environment. Sober (1981:113) has pointed out that this interaction underlies, and limits, permissible models for individual and group selection.

> Individual selection does not require an atomistic view of the organism; it does not require one to ignore the fact that organismic fitness is context sensitive. . . . Similarly, group selection does not require a reification of the group; it does not force one to suppose that groups are something above and beyond the interactions of their member individuals and the environment.

Individuals may experience differing selective pressures because of the group they are in; groups may be differentially selected because of the group-related traits of the constituent individuals. But how does this reflect upon the probability for and importance of group selection?

One need not reify groups to posit group selection; however, if a group fitness component *totally unrelated to individual fitness* is used to elucidate evolutionary change, we may justifiably claim that emergent properties of groups are being proposed. Emergent properties, no matter how convenient for evolutionary analysis, must be demonstrated, not merely posited. Recalling our earlier discussion of populational thinking and "real" individuals, demonstration will be most difficult: How are we to separate the emergent properties of a group from the inferential manner in which groups are defined? Although the traits of a

---

[5]I do not really disagree with certain presentations of species as pseudoindividuals (Ghiselin 1969, 1974; Hull 1976), and it is useful to remember that individual organisms do not "evolve" any more than species "develop." In this sense we might call the species a unit of evolution. Still, I would prefer to keep thinking of species as species and save the word *individual* for organisms. Especially within the already chaotic world of group and individual selection, our abilities to make subtle distinction may end up leading to greater, rather than less, confusion.

group must change (if we are to recognize evolution), the traits have not changed without change occurring at the same time in the constituent individual organisms. At a more practical level, group selectionists have, on the whole, been unable to demonstrate convincingly group traits that might be examples of group adaptation (see an extended discussion in Williams 1966). Tactically, the best way to demonstrate group adaptation would be to stress traits that seem unable to arise by means of individual selection. Thus, a tradition has arisen holding that group selection must oppose the direction of individual selection. (Of course, this opposition need not be true [Wade 1978; Wilson 1980] because, even when group and individual selection work in the same direction, quantatatively different results may arise.) Insomuch as a tactic of antiindividual selection is followed, group selection relegates itself to being merely the resting place for the current "hot potatoes" in individual selection theory—and as each "anomaly" is resolved, group selection becomes that much the poorer.

Part of the confusion that exists in the literature of group selection can be traced to the simple fact that individual selection does yield groups. Individual and population are inextricably bound together within the populational view of evolutionary change by means of the natural selection of variant individuals through differential fitness. Adapted groups arise from adapted individuals. The adapted population is as much the result of common evolutionary descent as any other trait present in the population. Because evolutionary processes must be context sensitive (evolution occurs within a specific environment and other individuals are part of ego's environment), it is to be expected that traits conditioning, or arising from, the interaction of individuals will be subject to natural selection, and therefore that such traits will evolve. This observation neither proves nor disproves the validity of group selection, and if it were all that was meant by "group selection," little disagreement could result.

Group selection as generally understood is evolution by means of the differential extinction or proliferation of groups of related organisms.[6] Groups may represent anything from families (kin selection) to whole populations (interpopulational selection). Kin selection, because of the close genetic relationship existing within families, is largely inseparable from individual selection and generally is not considered to be a manifestation of group selection, a position with which I can agree. Most of the heat generated in the discussion of group selection comes out of attempts to understand the relationship between individual selection (accepted by all) and the selection of population units. All agree that natural selection over time results in a population having characteristics different from those found earlier. These changes in traits summarize the action of evolu-

---

[6]In keeping with this definition, I will not consider here evolutionary schemes based upon the selection of species (Stanley 1975; 1979; Van Valen 1975) or higher order groups such as ecosystems (Dunbar 1960; Wilson 1980).

tion. The individual selectionist claims that populations must be viewed as statistical results of selection at the individual level. The group selectionist claims that characteristics of the group may, under certain conditions, chart an entirely different evolutionary course.

The differential selection of groups is based upon the rather commonsensical idea that populations with maladaptive characteristics are likely to go extinct and to be replaced by related populations lacking these maladaptations; or, conversely, that groups with adaptations optimal for the survival of the group are more likely to prosper. Group adaptations, even those that appear to be inimicable to the "best interests" of individuals, would spread because of their enhanced group fitness. Mathematical models for group selection, however, are generally incompatable with what is known of individual selection because evolution selects "selfish" individuals. Nevertheless, changing basic assumptions inherent in the traditional models permits us to portray the evolution of group-related characteristics in a more-or-less believable manner (Wilson 1980). For our purposes, it is important to note that *all* modeling for group selection, both pro and con, has thus far proceeded solely within a genetic framework.[7] The functioning of mathematical models for the group selection of culturally transmitted (learned) traits would doubtless be both interesting and illuminating.

Group selectionism always has had strong adaptationist leanings. Wynne-Edwards (1962) was basically concerned with attempting to explain traits such as *altruism* that benefited the group more than the individual. David Wilson (1980:1) goes even further and attempts to develop a genetic model that could explore questions such as, "Are we justified . . . in saying that an individual exists for the function it performs in its community? Are there any aspects of an individual's behavior and morphology that can be understood only in terms of its role in maintaining a larger adaptive unit?" Pointing out the strong psychological basis for wanting positive answers to such questions, Wilson (1980:1–2) continues with the observation that

> The idea that an individual is a cell in a larger "superorganism" has a long history, but not the kind that can be used to support an evolutionary hypothesis. . . . We possess a very strong philosophical bias towards seeing purpose and order in nature above the level of the individual, but there is nothing in the first principles of natural selection to justify such a bias. . . . At the heart of the philosophical bias is an interesting scientific hypothesis that might well be reconciled with evolutionary theory. If the superorganism concept is not a first principle of natural selection, it may still be an emergent property.

The adaptationism, directionality, and emergent character of group selection has brought it into disrepute among many evolutionary theorists. Levins (1975:48–49) may be drawn upon to respond to the observations made by Wilson.

[7]The favorable reception given group-selected adaptationism by the social sciences compared to the generally hostile response to individual selectionist sociobiology would make a most interesting story. More is at stake here than either "adaptationist programs" or "genetic reductionism."

Despite assertions that communities evolve to maximize stability or efficiency or information or complexity or anything else, there is no necessary relation between evolution
within the component species and evolution of macroscopic community properties. Yet
such claims are frequently made, and seem to be attractive to biologists. Perhaps the
reason for this is a frequent reference by biologists to a philosophical framework that
seeks harmony in nature. Or it may be the transfer to ecology of the equally invalid
Adam-Smithian assertion, in the economics of capitalism, that some hidden hand con-
verts the profit-maximizing activities of individual companies into some social good.

Because group selectionism generally implies the evolution of adaptive and
emergent traits, this further implies a view of the world in which such traits are
accepted as observable. Thus it is not surprising that much of the recent group
selectionist theorizing has derived from the work of the systems ecologists.
Downhower (1979:vii) has pointed out that current ecological approaches may be
distinguished on the basis of their unit of analysis. One group of workers takes
the ecosystems as the fundamental unit: "this approach attributes homeostatic
properties to the ecosystem and is an application (to higher order groupings) of
the concept of the individual as an integrated, interdependent unit." A second
group focuses upon interactions between the species inhabiting a region and
stresses competition, resource utilization, and community structure. A third
group, important in its effect upon social scientists, takes a population as the
analytic unit and tends to "interpret the behavior and activities of the organisms
as adaptations that ensure the survival of the population or species" (Downhower
1979:vii). For the evolutionary biologist who attempts to keep ultimate causation
in mind, these functional approaches have limited utility. Without denying that
they may be extremely useful for summarizing and arranging observations, he
considers them incapable of explaining change. It is perhaps not surprising that
much of the foundation for modern evolutionary biology was laid by paleontolo-
gists who were seeking as much to explain extinction as adaptation.

I believe that much of group selectionism represents an attempt to integrate
the role played by the environment into individual selectionism. Perhaps in part
because of the role of Mendelian thinking in the creation of neo-Darwinism, a
tendency has existed to treat individuals as timeless and spaceless entities differing only in the possession of specific alleles. In a strange irony, the historical
uniqueness of the individual is here lost—the individual becomes the repository
for identical types of particles (genes) and these nice identical particles become
the real "stuff" of evolutionary change. In a sense, genetic reductionism is an
attempt to retreat to the comfortable typological thinking of pre-Darwinian times.
Group selectionism stresses the role of the environment (composed in part as the
environment always is by the other members of a population) and provides a
useful antidote for genetic reductionism. Still, if used at all, group selectionism
should be used as a way for discussing, in an efficient manner, the reproductive
success of populations comprising individuals who are all evolving by means of
changes in relative fitness (again, not necessarily adaptation!).

The environmental aspect of group selection is made clear in Sober's (1981:107) definition of group selection: "Group selection acts on a set of groups if, and only if, there is a force impinging on those groups which makes it the case that for each group, there is some property of the group which determines one component of the fitness of every member of the group." Sober goes to some pains to claim that "the definition should not have the consequence that group selection exists whenever fitness values are context sensitive . . . [that is,] whenever fitness values of organisms depend on the character of the group they are in." However, his definition should function rather like context-sensitive selection in that any characteristic of the group—taken as it must be in the total environment—will ultimately be cause for fitness differences. In the real world, the group will have numerous forces acting upon it, many of which are not intrinsic or causually related to group properties, and evolution will incorporate the interaction of the totality of these forces.

To explain my point, I will mention an example of group selection discussed by both Sober (1981) and Lewontin (1970). Biological control by means of a viral disease was attempted to eradicate feral rabbits decimating the Australian countryside. At first the disease killed more than 99% of the rabbits it infected, but in time the fatality rate fell. It was discovered that, as expected, the rabbits were becoming resistant to the disease (classic individual selection). However, it was also found that the virus was actually becoming less virulent. Why would the virus evolve to become less damaging to its prey?

To understand this, we must consider the relationship between the virus, the vector for transmission of the disease, and the rabbit. The virus is spread from rabbit to rabbit by mosquitoes. When a rabbit dies, (1) the virus can no longer reproduce, and (2) any virus in that rabbit will not spread to another rabbit because mosquitoes will not feed upon a dead animal. The entire virus species is divided into many local populations (groups) and being a member of a given group has far-reaching effects upon the fitness of the viruses in that group independent of their particular individual characteristics. Viruses of any virulence sharing a rabbit with highly virulent strains will be extremely unlikely to spread because the rabbit dies quickly. Conversely, those rabbits inhabited by an avirulent strain will contribute disproportionate number of the infections because the rabbits will live longer. Viruses inhabiting a rabbit colonized by avirulent strains will have a positive group fitness. This group fitness is an independent variable because individual selection is likely to select for the ability of any given virus to reproduce faster than other viruses and to be more virulent. Sober (1981:109) concludes that "less virulent strains . . . are 'altruists.' By being less virulent they reduce their expectation of reproductive success within the population they are in, but thereby increase the group's chances of survival and reproduction."

But the matter should not rest here. Lewontin (1970:15) notes that "it is very likely that these observations are a model of the general evolution of avirulence in which the death of the host causes the death of the parasite. One could predict that the introduction of a flea [as a vector] . . . would prevent the evolution of avirulence.'' Because fleas will stay on the animal and leave when it dies, taking with them viruses of *any* virulence, no group advantage would accrue to avirulence. Indeed, we might even predict that group and individual selection would now be mutually reinforcing with the most virulent strains contributing the greatest numbers of future infections (at least until such time as a balance is reached with the vector's dispersal abilities). We might therefore note that group selection, at least in the traditional sense of opposing individual selection, would now be hard to detect. But of far greater importance, it is the vector, *not* the virus and *not* the group structure of the viral population, that is evolutionarily significant. The vector is part of the context that determines significant components of fitness of the group. The ''property of the group which determines one component of the fitness of each member of the group'' (Sober 1981:107) need not be, in fact, a property of the group at all. Here the property has been dependent upon the vector. One may wish to define the vector as part of the viral group ''essence'' but this makes as little sense as trying to force evolutionary events determined by population structure and interpopulational interactions into the tight girdle of individual selection by the claim that all such events ultimately enhance the individual's self-interest.

To the extent that group selectionism attempts to identify and describe phenomena in the larger environment (including other individuals) that serve to fuel evolutionary change, the approach is not without merit. But if group selection is actually to function like individual selection, it must be open to all of the subtle effects brought about by differential fitness. We cannot posit inevitably adaptive functions for any group-related trait we describe. We cannot attribute teleological motives to group selection any more than we can to individual selection. After all, when we are speaking of group selection we are only speaking of individual selection in the context of population structure and the interaction of populations with the larger environment. Finally, we cannot attribute any greater prescience to group selection than we do to individual selection. For example, some group function may have contributed to the spread of the elm bark beetle in the United States. But this has not kept the population adapted to its resources. Instead, by contributing to the wholesale destruction of its resource base, the beetle is fast on the way to extinction. Likewise, the proverbial march to the sea of the lemmings has a clear group stimulus. But it cannot be held to be a device for population control: Lemmings can swim most streams; when they get to the ocean they just keep paddling. So with evolution at any level—the ultimate effects of any mode of selection can neither be anticipated nor avoided.

Strangely enough, by turning its fundamental prejudice for group adaptation and harmony upside-down, group selection may yet make major contributions to evolutionary theory. Although group selectionists place great emphasis upon seemingly adaptative and functional communities and systems, this observed harmony is hard to reconcile with the fossil record. The traditional group selectionist has tended to believe that the harmony and functionalism of the ecosphere has resulted from a loss of less "perfect systems"—"if groups and communities were not so well adapted, so functional, they would soon go extinct." Yet, in fact, this very observation of the inevitability of extinction must be made for the vast multitude of species that have existed. The ecologist and the palentologist, the ethnographer and the archaeologist, see very different worlds. Perhaps a lack of sensitivity to the larger group and environmental factors has caused us to stress the extrinsic causes of extinction—catastrophe, disaster, competition, disease. It is possible that the uncoupling of fitness and adaptation is directly related to intrinsic relationships emerging from the interaction of members of a species or a community. (Of course such properties do not really "emerge," but have been there all along.) By attempting to explain the totality of factors that enhance or reduce individual survival and reproduction, we may be led to consider the importance of population structure and dynamics. At times it may well be easier to speak of "group" or "populational" selective factors, but it does not matter so much what we call it as how well we understand it.

Natural Selection and Cultural Change

Because cultural change involves the replacement of one set of cultural traits by another, it seems likely that a natural dynamic underlies the process: There is by definition (although by definition alone) differential survival and reproduction of traits over time within cultures.[8] A Darwinian analysis of cultural change is theoretically possible: One can analyze cultural change in terms of the differential fitness introduced by alternative cultural traits into individuals and then, as statistical summaries, into the population. However, in doing so, one must recognize that *culture* must receive a rather specific meaning restricting it to the information learned by individuals from other individuals "which affects or potentially affects phenotypic behavior" (Richerson and Boyd 1978:129). Although this definition would seem to be so broad as to exclude little of importance, this is not the case. A focus upon learned and performed behavior must reduce the significance of reflective, symbolic, and aesthetic aspects of human

---

[8]Of course, cultural traits, like physiological or morphological ones, do not exist independently of the individuals possessing them or acting them out. Nevertheless, it is and will continue to be heuristically useful to treat the traits as if they had an existence of their own—nominalism can simplify discussion.

cultures. Leeds and Dusek (1981:xxi) criticized sociobiology for ignoring these components of culture and the charge also may be leveled against *any* selectionist model for cultural change that is solely behavioristic:

> a definition of culture *merely* in terms of "learned behavior" is wholly inadequate: all cultural behavior is learned, but not all learned "behavior" (even among humans) is cultural, and certainly all cross species learned "behavior" is *not* cultural. It lacks a key aspect which only humans have in any developed form: symbols and symbolic structures and processes. This key aspect is scarcely examined not only in the entire sociobiology literature . . . but also in the works of the critics . . . which are often vague or ambiguous as to the nature of culture.

The sociobiologist or cultural selectionist might respond that symbolic processes are included in the analysis of cultural change if they have effects upon cultural fitness; that is, if they contribute to behaviors affecting phenotypic fitness. Yet, until such an analysis is actually done on human cultural symbolism, the claim will ring hollow. They might also note that symbolism underlying behaviors are not necessarily correlated with the performance of the behavior itself—that symbolic structures and processes are only so much evolutionary "noise" bearing little relationship (except insofar as they influence behavior) to significant evolutionary events. The symbolicist, of course, would then feel obligated to query why symbolism is such a ubiquitous aspect of human culture. Finally, the selectionist might claim that the definition of culture is a functional one predicated by the model used—we define culture as learned behavior because it is convenient for Darwinian analysis. This is of course relatively unassailable. Culture here becomes an undefined primitive term established by means of the independent variables (behaviors, fitness changes, etc.) used to measure cultural change. This is fine as long as the major components of culture as we can observe it today are not neglected. Unfortunately, symbols do seem to be highly significant in extant cultures. Thus the conflict between behaviorist and symbolic approaches bodes no early resolution. Still, it would seem incumbent upon the symbolicist to come up with a definition of culture that can realistically integrate change with the important variables described. This is no easy task. Defining culture in unambiguous terms is a Sisyphean labor and we stand a good chance of being crushed by any simple definition we propose.

So, instead of seeking to define culture, it is appropriate to discuss several aspects of culture as they bear upon the study of cultural change. Culture in its most general sense functions as a behavioral means by which animals escape genetic predeterminism. By *genetic predeterminism* I mean the necessity for an animal to evolve specific genetically determined responses for each and every situation in which it might find itself. Behavior, as part of the selected phenotype, has been of major importance in the evolution of *all* animals because of the flexibility in response systems that it permits (Bonner 1980; Pulliam and Dunford 1980; Wolsky and Wolsky 1976). The potential for variable behavioral response will be of great benefit to any organism. Rather than having to code in the genome for each interaction and situation it might encounter, the animal may learn from the experiences of other animals. Given a genetic capacity for culture within a species, this capacity will *act* like a pleitropic gene that expresses itself in different manners under varying environmental conditions (Richerson and Boyd 1978). This capacity for culture must be more than the ability to learn from previously encountered situations. Culture is more than learning, although learning is a prerequisite for culture. Culture must include a structured mechanism permitting the extrasomatic transmission of information from generation to generation.

Culture must be viewed as a system of inheritance separate and distinct from the genetic system in all ways save the genetic capacity for culture. The rules governing cultural inheritance differ vastly from those governing genetic evolution. Genes are transmitted equally by both parents to the offspring at the time of fertilization and no further change or modification of the genetic message is possible after this time. Culture is not bound by consanguinity, specific timing of inheritance, or need for unitary transmission. It may be transmitted between individuals having no close genetic relationship. Cultural inheritance may occur at any point in an individual's life. Culture is not transmitted *in toto* and individuals may possess only a portion of the total available cultural information. Because cultural and genetic systems are transmitted in radically differing ways, it is likely that fitness changes arising from culturally transmitted modifications of the phenotype may be totally unrelated to fitness values for aspects of the phenotype established by genetic inheritance (Feldman and Cavalli-Sforza 1975, 1976). Boyd and Richerson (1980:104) also point this out, noting that "phenotypes which ensure genetic reproduction will not necessarily ensure cultural reproduction and vice-versa." One possible outcome of the independence of cultural and genetic inheritance systems is an actual conflict between fitness changes favored by each system. "On balance, although the equilibrium capability for culture must be the best genes can achieve, many cultural influenced phenotypic traits can be at the cultural rather than the genetic optimum" (Richerson and Boyd 1978:129). Such a conflict growing out of differences between genetic and cultural inheritance provides compelling evidence that genetic-determinist models for cultural inheritance and change must be rejected because cultural traits may be preferentially propagated despite a lowering of genetic (as opposed to cultural) fitness values.

I hope I am making one point abundantly clear: The details of cultural behavior in any animal should not be seen as genetically determined. Culture is permitted by genetics and, as is noted by both sociobiologists and their antagonists, there are times when specific behaviors may enhance the survival of gene pools. But all of this relates to the study of cultural change in only the most peripheral way: what interests us is the manifold variations cultures assume. To approach culture *only* in genetic terms is to fly in the face of the fundamental observation concerning cultural behavior, namely, that in its various particular manifestations, culture is *not* genetically transmitted. Proof of genetic determinism for a hypothetical cultural act would *not* prove that culture is genetically transmitted; it would only serve to remove that particular act from the realm of the cultural and place it in that of the instinctive.

If culture is permitted but not determined by genetics, how are we to reconcile the genetic capacity for cultural behavior with the nongenetic mode of transmission of specific cultural acts. It is vital to recognize that cultural behaviors of even the simplest kinds have two discrete (although interrelated) aspects: a fundamental genetically determined basis and specific manifestations. For example, imprinting in young birds is an instinct (genetically determined), but the *object* on which the imprinting is focused is dictated largely by experience, circumstance, and learning. It is the object of the imprinting, not the existence of an imprinting instinct, that may be the proximate cause of the evolution of bird species over time. Much of the controversy and confusion about the applicability of Darwinian analysis to cultural events is based on a lack of attention to this fundamental observation. The types of cultural changes in man that we seek to understand are *all* within the genetic potential of *Homo sapiens*—thus we may safely ignore the genetically predetermined aspects of the change we are seeking to explain.

But if we accept the fact that the specifics of cultural behavior are not under genetic control, have we not therefore lost the rationale for an application of natural selection to the understanding of cultural change? Can we have sustained phenotypic variation without genetic differentiation, or does natural selection act ultimately only upon the genome? This is a highly complicated and subtle question and one that is not to be answered to everyone's satisfaction. Certain general observations may, however, direct us to a deeper understanding of the question.

The first and most obvious is that natural selection seldom acts *directly* upon the genome. Instead, selection is of phenotypes—the expression of the genome as modified by ontogeny and the environment. Hereditable behavior, as much as morphology, is part of the phenotype and would thus appear to be, by definition, subject to natural selection.

It is important to recognize at this point that our data base for the evolutionarily significant aspects of the genetic control of morphology and thus of morphological variation is far from what we might wish it to be. Sufficient provocative data are present that the applicability of a selectionist model is not to be seriously doubted; yet, we are far from proving that the selection of alternative alleles is the main motive force behind *all* evolutionary change. Instead, everywhere we turn, practical and inherent difficulties prevent us from really getting our teeth into the nature of genetic variation. To illustrate this point, I can do little to improve upon the discussion of this problem by Lewontin (1974:21, 22, 23, 19–20):

> The evolution of species consists of the gradual accumulation of very small changes in physiology, morphogenesis and behavior. . . . Evolutionarily significant genetic variation is then, almost by definition, variation that is manifest in subtle differences between individuals, often so subtle as to be completely overwhelmed by effects of other genes or of the environment. . . . We see here the fundamental contradiction inherent in the study of the genetics of evolution. On the one hand the Mendelian genetic system dictates the frequencies of genotypes as the appropriate description of a population. The enumeration of these genotypes requires that the effect of an allelic substitution be so large as to make possible the unambiguous assignment of individuals to genotypes. On the other hand, the substance of evolutionary change at the phenotypic level is precisely in those characters for which individual gene substitutions make only slight difference as compared with variation produced by the genetic background and the environment. What we can measure is by definition uninteresting and what we are interested in is by definition unmeasurable. . . . Population geneticists, in their enthusiasm to deal with the changes in genotype frequencies that underlie evolutionary changes, have often forgotten that what are ultimately to be explained are the myriad and subtle changes in size, shape, behavior and interactions with other species that constitute the real stuff of evolution. . . . A description and explanation of genetic change in populations is a description and explanation of evolutionary change only insofar as we can link those genetic changes to the manifest diversity of living organisms in space and time. To concentrate only on genetic change, without attempting to relate it to the kinds of physiological, morphogenetic and behavioral evolution that are manifest in the fossil record and in the diversity of extant organisms and communities, is to forget entirely what it is we are trying to explain in the first place.

When we come down to trying to establish that a given trait, generally accepted to be under genetic control, is in fact genetically determined we run into the practical problem of the impossibility of separating genotype from phenotype. As has been stressed throughout this discussion, natural selection acts upon the phenotype—actual traits in a real environmental setting. Yet, as Burian (1981:51) has put it,

> the gap between genotype and phenotype is immense. . . . To have an effect on the adult phenotype, a gene must interact with other genes, with the chromosomal structure of the cell, with various components of the nucleus, with the chemical make-up of the cytoplasm, with the physical conditions affecting the cell, with the timing of differentiation and of development. . . . Thus *all* development is epigenetic—i.e., involves an inextricable mixture of genetic and non-genetic factors. . . . The dependence on multiple interactions and multiple pathways reinforces the point that if one changes the genetic or the environmental background, a given gene or gene complex may yield startingly different effects. This substantiates the major point . . . *the within population correlation between genotypic and phenotypic characters does not measure the degree to which the genes or genotypes in question cause the trait or traits in question.*

The Darwinian model is applied to the study of genetic variation and evolution despite major difficulties inherent in the proof of underlying assumptions.

The model is not perfect, and it presents numerous practical problems even in population and evolutionary genetics. Yet it is the only approach that offers any hope of making sense of the variation existing in nature. The position of the geneticist studying evolutionary change is not as far from that of the social scientist studying cultural change as it might appear. The Darwinian model is utilized not because it is the only possible approach but because despite its problems it proves to be the most interesting and productive one.

Both theoretical and practical considerations dictate (1) that natural selection be approached as more than a merely genetic phenomenon and (2) that culturally induced behavior may be of major evolutionary importance. The strongest possible case for the significance of the selection of behavior would be one in which behavioral changes precede genetic change. Such an evolutionary tendency exists and has been long recognized under the name of the Baldwin effect, after J. M. Baldwin (1896), one of the first biologists to appreciate its significance. The Baldwin effect, as described by Simpson (1953:183), holds that "adaptation may start as a matter of social preference in what we now call a deme or as an acquired reaction in individuals and may later become genetically fixed by selection." Interest in this phenomenon has continued since it was first advanced and it has been studied both in nature and experimentally (e.g., Broadhurst et al. 1974; Manning 1975). Much of the work relevant to our purposes has been done on the ethology of birds and the effect of imprinting upon speciation provides useful examples.

Imprinting in its various forms may yield conditions under which speciation is initiated. The variants in feeding behavior and mate selection conditioned by the early experience of the birds may act to reduce the probability of mating among the various behavioral groups of a species. One type of imprinting recognized by Immelmann (1975) is habitat imprinting. This form of early learning by birds tends to divide the resources of an environmental cline into subhabitats within which subpopulations of the species are specialized. Preferences for one or another subhabitat are culturally, not genetically, transmitted. Isolation within the various subhabitats will, of course, be an expression of the way the subhabitats are related but will offer the potential for the division of a population into a mosaic of subpopulations with culturally established habitat-linked differences. Another type of imprinting, sexual imprinting, will reinforce the segregation of the population by promoting preferential matings within habitat-linked subpopulations of the larger population. These cultural preferences, however, will be of no significance in the genetic evolution of species unless there is genetic change in the subgroups. But if genetic change occurs, we may speak of an adaptive radiation of species based upon differences in the behavior of individuals initiated and maintained by cultural learning. Emlen (1973:376) provides the example of two gulls that coexist in England. The herring gull, *Larus agentatus,* is distributed westward from England, whereas the lesser black-backed gull, *L. fuscus,* is distributed to the east. In England, the species breed sympatrically and, although hybrids are completely viable, they are almost never found. However, if we trace the morphology of these species throughout the range of their worldwide distribution, we find that adjacent populations intergrade to produce a complete range of transitional forms. The "species" are but the morphological extremes of one taxon that happen to meet in England, where behavioral mechanisms prevent their interbreeding.

Reproductive isolation brought about by behavioral mechanisms, even that associated with rather striking morphological and thus genetic divergence, may become ineffective under certain conditions. Immelmann (1975:51) has reviewed one such case:

> In the arctic Snow and Ross Geese, *Anser caerulescens* and *A. rossii,* competition for nesting-sites in large mixed colonies has been observed to result in mixed clutches. Young hatched in these nests obviously become imprinted to the wrong species, and this has led to frequent interbreeding between the two species; not less than 1400 hybrids are estimated to have been born annually in recent years. They are known to be fully fertile with each other and the parent species.

Thus, not only may learned behavior be implicated in speciation, but it may play a role in "despeciation".

Is this really the selection of behavior? I would argue that it is. Variation in habitat selection followed by habitat imprinting may permit colonization of new habitats by members of species (Immelmann 1975:246):

> If . . . an adult pair has settled in a marginal area of the species' distribution or if it has invaded a new type of habitat, the offspring will become imprinted on some of its characteristics and, for reproduction, will try to return to a similar kind of environment. If the habitat permits successful reproduction this may give rise to a new subpopulation that ecologically is more or less separated from the original stock.

It is ecological and behavioral separation of these subpopulations that is the proximate cause of genetic diversification in response to natural selective forces, but I think we are justified in calling behavior the ultimate cause. Further, it was the *fitness* that the behavior change introduced into the subpopulation that was to allow it to diverge genetically from the parental population—if the subpopulation did not prosper during its initial period of colonization, we would not have separate subpopulations (incipient species) in which genetic change could accumulate. Another way to explore the role of behavior in these types of speciation episodes is to consider what we would observe if we were able to view the diverged species at a very early stage in its divergence (before any genetic change had occurred). Would natural selection not be in operation simply because even incipient speciation had not (yet) occurred? If behavioral changes may precede genetic change (and its expression, speciation), it would seem that we could assert with a fair degree of certainty that the selection of various forms of behavior can ultimately cause and help maintain speciation. And if the genetic and morphological variation found within populations of any species is held to be the result of the natural selection of phenotypes, are we to exclude variation in cultural behavior from the same types of selective processes?

In summary, two points can be made about genetics and the evolution of behavior. (1) The genetic control of even morphological characteristics is not as simple and straightforward as is sometimes assumed. The applicability of Darwinian selectionism to morphological evolution does not rest on a demonstrable correlation between gene and form. Although this negative case is weak, it is important in reinforcing our understanding of the importance of the *phenotype* as the *level at which selection generally occurs*. (2) Behavior may serve as an isolating mechanism in genetic evolution. Thus, *genetic change is not the prerequisite of species evolution but is its dynamic basis and ultimate consequence.* The type of behavioral variation we are interested in studying in humans should not be expected to have genetic consequences: we are dealing with behavioral variation within a species.

Variation and Evolution

It is well-nigh impossible to overestimate the importance of the observation that the study of human cultural change is the investigation of variation, not speciation. Most of the errors justifiably attributed to sociobiology stem from its confusion of cultural variation with genetic change. Sociobiology logically does not have to involve genetic reductionism (the belief that genes code for specific behavioral traits), only the assumption that culture and genes necessarily act to the same genetic end, and therefore that cultural behavior may be meaningfully analyzed by changes in genetic parameters. It is indeed legitimate to ask why a cultural process should increase genetic fitness. Where is the direct feedback or common language linking the two systems? Sociobiologists have yet to confront this problem. Two fundamental errors have pretty well reduced most of the sociobiology literature to exercizes in futility. First, the sociobiological tradition has worked in a reductionist atmosphere thick from the confusion of cultural with genetic inheritance. As we have noted, not only are these systems independent, they may actually work in opposition. Second, most sociobiologists have assumed that genetic change (the ultimate result of species-level selection) may be taken as synonymous with variation. This latter confusion, although extremely unfortunate, is somewhat understandable given the intricate relationship linking selection and variation.

Variation among individual organisms is the raw material upon which natural selection operates. Yet variation, in the form of variant populations, is also the result of natural selection. Darwinism recognizes that evolution proceeds by means of the accumulation of variation over time. Evolution proceeds by the differential success of individuals possessing fit traits and the result is populations composed of successfully variant organisms. This insight allows the Darwinist to look at all evolutionary change—from that occurring within a generation due to differential reproductive success, to that occurring over vast spans of time and yielding the higher taxonomic orders—as the results of a single process. Evolutionarily significant variation is populational variation (itself the result of natural selection) that accumulates to yield evolutionary change. It is this simple microevolutionary process that permits the origin of higher taxonomic orders. A Darwinist cannot hold that natural selection is responsible for the origin of species and not the variation within species any more than that natural selection underlies speciation but not those processes yielding higher taxonomic groups. Species or higher order groups are levels of evolution that may be recognized post hoc, and although they may have unique characteristics permitting their classification at a certain taxonomic level, they have not arisen by any process other than that originally producing variant populations within the ancestral species. Variation, including evolutionarily significant cultural variation, can only be made explicable by means of the selectionist model. As we have noted previously, the acceptance of variation as both the cause and the result of natural selection underlies the Darwinian predilection for individual, populational, and nontypological thinking. As Simpson (1961:50) has put it: "Variation is not incidental or an 'accident' to be ignored at any level in taxonomy; it belongs to the very nature of taxa and is part of the mechanism of their origin and continued existence."

Darwin was unaware of the source of variation, as he admitted in the *Origin* (1859:12): "The result of the various, quite unknown, or dimly seen laws of variation is infinitely complex and diversified." Neither was he sure how variation was transmitted from generation to generation: "The laws governing inheritance are quite unknown" (1859:13). Nevertheless, Darwin correctly emphasized the importance of hereditable variation by noting that "any variation which is not inherited is unimportant to us" (1859:12). Yet, despite his ignorance of what we would call essential features of the neo-Darwinian model, he was still able to formulate and defend a successful evolutionary model. This bears upon the study of cultural change in that it gives us some confidence that the specific source of variation and the precise nature of cultural heredity (nice as it would be to have them) need not be known for application of the Darwinian model.

Darwinian evolution, predicated upon variation, differential fitness, and hereditability, requires only that evolutionarily significant variation be transmitted from generation to generation. As Mayr points out, Darwin was able to formulate his theory despite his ignorance of the processes that create variation and permit its retention through generations.

> Even at the present time variation and its causation are not fully understood. In the middle of the nineteenth century, the subject was enveloped in great confusion. How difficult this subject is becomes apparent when one realizes how bewildered even Darwin was, who had been preoccupied with variation all his life and who had thought deeply about it. . . . The most fascinating aspect of Darwin's confusions and misconceptions concerning variation is that they did not prevent him from promoting a perfectly valid, indeed a brilliant, theory of evolution. Only two aspects of variation were important to Darwin: (1) that it was at all times abundantly available, and (2) that it was reasonably hard. Instead of wasting his time and energy on problems that were insoluble at this time, Darwin treated variation in much of his work as a "black box." It was forever present and could be used in the theory of natural selection. But the investigation of the contents of the box, that is, the causes of the variation occupied Darwin only occasionally and with little success. . . . Fortunately for the solution of the major problems with which Darwin was concerned . . . a study of the contents of the box was unnecessary. (Mayr 1982:682)

Mayr (1982:682) notes that "In retrospect it is obvious that much of [variation's genesis and transmission] could not be clarified until after the rise of genetics." Boyd and Richerson (1980:101–102), however, raise another possible explanation for Darwin's seemingly overgeneralized concept of variation and inheritance.

> Darwin never rejected the possibility of inherited effects of use and disuse and in *The Descent of Man* frequently discussed the idea of "inherited habit" as being important. . . . It seems likely to us that Darwin was reluctant to abandon the Lamarckian notion of inheritance of acquired traits because of the obvious need for such an effect to account for what we now distinguish as cultural influences on the behavior of humans. Not wanting to consider humans radically distinct in this regard . . . and lacking any useful understanding of mechanisms of inheritance, Darwin simply avoided any distinction between what we now call genes and culture. . . .[However], our better understanding of inheritance mechanism and the clear distinction between genes and culture does not invalidate Darwin's broad claims about the efficacy of natural selection. Quite the contrary, selection may be the *only* fundamental mechanism for generating the ordered structures of biotic and cultural phenomenona.

Culture and Evolution

The nongenetic transmission of cultural traits does not prevent the selection of otherwise hereditable forms of behavior; the fact that human cultural behaviors are not the result of genetic differences does not imply that the traits could not have been the result of selection. Variation in human culture has some meaning—it is not merely the result of totally random and adaptation-neutral processes. All scholars who have worked on the problem of human cultural change have stressed the seemingly adaptive nature of culture. The problem has been in identifying the important variables controlling the expression of human culture and in finding some way of testing their potency. I would hold that the manifold adaptations of human culture are best seen as the result of change dictated by natural selection. However, selection, as it has been discussed here, has played a remarkably small role in discussions of cultural change. Richerson and Boyd (1978:130) point out that "although there is an enormous literature in anthropology and sociology which examines cultural evolution, very little of it is properly within the Darwinian tradition. Oddly enough, this comment applies to Social Darwinism and related ideas as well as explanations conceived in opposition to the Social Darwinists." In a similar passage, Dunnell (1980:37) concludes that "if evolution is taken to mean what it does in the sciences, it has yet to be systematically applied in either sociocultural anthropology or archaeology. The approach represented by cultural evolution is a social philosophy directly derived from the tradition of Herbert Spencer and the early anthropologists and is unrelated to Darwinian principles." And Blute (1979:46), in a paper aptly entitled "Sociocultural evolutionism: an untried theory," has demanded that "we have to be critical of sociobiology's relative neglect of the epigenetic sciences of the individual and, indeed, to reject firmly its implicit contention that there is no order to be found in history . . . except that imposed on it by evolving gene pools." We are finally beginning to realize that Social Darwinism was not Darwinian, cultural evolution has not been evolutionary, and sociobiology has confused society with biology: all in all, a rather amusing state of affairs. Yet, how are we to explain the long-standing confusion of *evolution* as used within the social sciences with the same term as used in biology? Part of the answer may be found in the history of the development of the concepts in their respective disciplines.

Spencer and Cultural Evolution

If theories of biological and cultural evolution share a common ancestor, this ancestor is probably Malthus. Malthus, a dismal cleric with a depressing message, held that life "was a struggle in which the best survived, and wealth and position were the rewards of virtue in this struggle [and] those who could not become petty capitalists and practice self-restraint were doomed to be periodically cut off by famine, plague and war" (Bernal 1967:644, 1060).

Without entering into the debate over specifically how much influence Malthus had on Darwin,[9] it appears clear that the effect of Malthusian thinking was radically different in the natural and social sciences. Both Darwin and Spencer were impressed by Malthus's argument and each sought to resolve it,

[9]See competing treatments by Freeman 1974; Ghiselin 1969; Gruber 1980; M. Harris 1968; Herbert 1971; Himmelfarb 1962; Kohn 1980; Schweber 1977; and Young 1969,1971.

albeit in radically differing ways. Darwin denied the Malthusian dilemma by turning the argument upside-down. In place of the doom meted out to the loser, Darwin stressed the change selection creates; progress was possible because of the struggle, not despite it. Furthermore, as Darwin continually stressed, the struggle was not simply (or even usually) played out in terms of survival, but rather by means of reproductive success. Ghiselin (1969:60) points out that "Darwin only used Malthusianism as a heuristic aid—as suggesting one type of mechanism which might be responsible . . . Malthus provided only a conceptual system or model, not the argument for an empirical proposition."

Darwin also sought to deny a second proposition of Malthus. Malthus, by his own admission, had sought to refute the claim of a "Perfectability of man and society" (Malthus 1803, quoted in Harris 1968:115). As a strict typologist, Malthus could not accept that humans might, en masse, transcend the limitations placed upon them by circumstance. Thus the poor must, by and large, stay poor and suffer the consequences; only the exceptional individual might rise above his appointed station in life. Yet implicit in this is a recognition of individual variation. Ghiselin (1969:59–61) expresses the debt Darwin likely owed Malthus:

> Prior to his reading of Malthus, [Darwin] was still thinking of species and varieties as mere abstract groupings of things characterized by particular attributes. He had yet to conceive of species as units of interaction composed of biological individuals. . . . Upon reading Malthus, his attention was drawn to the long-term effects of difference between individuals upon the composition of the population. . . . Treating social entities not as manifestations of abstract forms, but as the consequence of interactions among individuals brought about the scientific investigation of society. Human associations, like biological species, are populations, and conceiving of them as such was fundamental to the kind of thinking Darwin owed to his reading of Malthus.

Malthus, or the competition among individuals he highlighted, relates to two of the central propositions of Darwinian thinking: selection and individuation.

Spencer's response to Malthus must be seen as totally unrelated to Darwin's. Where Darwin used Malthus for his own ends, Spencer applied an already existing concept—that of orthogenetic evolution—to a resolution of the Malthusian dilemma. Both the model and the mechanism tell us much about the interpretation given *evolution* within the social sciences, and perhaps justify Harris's belief that it is Spencer, not Darwin, who should be credited with this conceptualization of evolution.

> The fact that needs to be established here is that Darwin's principles were an application of social-science concepts to biology. . . . The phrase "Social Darwinism" not only obscures our understanding of the functional matrix that inspired Darwin, but it also distorts the actual order of precedence between Spencer's and Darwin's specific contribution to evolutionary theory. . . . [Spencer] consciously attempted to prove that human nature, like everything else in the universe, was an evolutionary product. . . . he insisted not only that human nature is modifiable, but that it has undergone and will continue to undergo drastic changes in conformity with a universal law of development. . . . By

> trusting social life to the law of nature, Spencer held that human suffering would
> eventually be eliminated. In 1852, Spencer directly confronted the disparity between his
> own and Malthus's views on the perfectability of man. Out of this confrontation
> . . . [arose] the idea of progressive evolution resulting from the struggle for survival.
> (M. Harris 1968:122–123, 126–127)

But does Spencer's concept of evolution have anything at all to do with Darwinian evolution? In a word, no. Spencer's view of cultural evolution was progressive and transformational. It was the cultural manifestation of a general theory of change Spencer believed had a universal applicability. "By contrast, Darwin saw his hypothesis bearing only upon the biological evolution of species, on a transgenerational level, and only then in terms of differential reproductive success" (Nichols 1974:258–259). Spencer's evolutionism was based upon the adaptive integration of objects with the environment by means of the inheritance of directed, adaptive characteristics. It was simply not concerned with the selection of variant forms.

Under transformational evolutionary schemes, phylogeny recapitulates ontogeny (this is like trying to cut with the toothless edge of an old saw): evolutionary change is like the development of an organism; ontogeny is a metaphor for progressive evolution. Transformational evolution has long dominated Western thinking about organic change and to this day remains the prevailing usage outside of biology. We speak of the evolution of personality or the universe, the evolution of style in fashion or art, the evolution of ethics, technology, and warfare. Yet, in none of these cases do we usually speak of anything resembling evolution by means of natural selection. Evolution without natural selection may be possible (defined simply as "change" it is inevitable), but such evolution is certainly not Darwinian. It is bereft of the unifying mechanisms that make change comprehensible. Under the selectionist model, change only occurs because traits possessed by a reproducing object influence the ability of the object to survive and reproduce in a specific environment. Transformational schemes, in contrast, are totally amechanistic. At their best they may serve as a method for arranging observations (e.g., the cultural evolution of White 1958b), at their worst they presume entelechy as uniting and "explaining" transitions between states.

To understand the intellectual milieu in which Spencer's concept of evolution was conceived and developed it is necessary to consider first an important biologico-philosophical movement. *Naturphilosophie* was a romantic reaction to the Newtonian physicalization of the sciences that had occurred in the eighteenth century. In place of the frequently naive mechanistic interpretations of organic phenomena advanced by the physicalists, *Naturphilosophie* sought to substitute a transformational and developmental view of the functioning of the universe: nature was seen to be continually in motion with progress and change resulting from the motion. Change here was not to be seen as a strictly mechanistic process, however, and was seen instead as based upon the unfolding (literally, the *evolution*) of preexisting potentialities and/or the origin of new types within a greater, directed process. *Naturphilosophie* sought not merely to explain the organic world. Instead, it stressed the inherent unity of all laws: the evolution of the universe or the development of an organism could be understood by recourse to the same unifying principle. Indeed, the very image of the development of an individual organism came to be seen as the paradigm for the processes guiding all change. Simplicity leads to complexity, and the phylogenetic history of the organic world is the history of a primitive organism progressing to its highest manifestation—mankind. Such beliefs constitute a survival of the *scala naturae* concept of Aristotle (see Lovejoy 1936; Ritterbush 1964).

Evolution is not only acknowledged within *Naturphilosophie*, but becomes its central proposition. However, this evolutionism involves transformational and directed processes. It results from the

> progressive development of organic being under the influence of the ideal form. . . . a kind of unfolding of that which has always existed in the timeless world of ultimate reality. . . . To the *Naturphilosophie* of the last century, and to its present-day heirs as well, progressive change is not . . . a strictly natural process. Indeed, to call Goethe or Oken a forerunner of Darwin is like equating faith healers with practitioners of psychosomatic medicine. (Ghiselin 1969:81)

Under the influence of *Naturphilosophie*, Spencer, like many others, found evolutionism in its most general form an acceptable and perhaps self-evident proposition. There was a grandeur in its view of life—an ability to subsume *all* phenomena within a general model. Spencer, in the tradition of *Naturphilosophie*, sought to explain not only the evolution of society but that of the universe as well. Much of the *Naturphilosophie* literature was specifically concerned with the biological sciences; Spencer read widely in the natural sciences and published an influential volume, *Principles of Biology*.

The functionalism of Cuvier countered the evolutionism of *Naturphilosophie*. Cuvier, a French anatomist and paleontologist, had led the opposition to evolutionary thinking during the nineteenth century. Cuvier was a committed essentialist who placed great stress upon the variation found in nature. He stressed that few data supported continued belief in the *scala naturae*, and that it was difficult to find any steady increase in complexity or perfection. Instead, what we do see is a wide range of types of organisms, each type adapted to function within a specific environment. Cuvier, by means of a series of intermediaries, especially Von Baer, was to have an important influence upon Spencer.

Von Baer, like Cuvier an anatomist (and unlike him, but like Spencer, an anthropologist), had many ties to *Naturphilosophie*. Yet, when Von Baer sought to explain organic change, his specific approach was drawn in large part from Cuvier who had insisted that "organs should be studied functionally as shapes designed for performance, not as ideal series distributed to meet the requirements of philosophical visions" (Gould 1977b:59). In place of the naturphilosoph's view of nature as proceeding from the primitive to the advanced, Von Baer proposed that change occurred from the simple to the complex, from the general to the specific. Given the progressive interpretation that evolution was accorded in his day, this essentialistic view had little choice but to exclude evolution. Yet, despite the rejection of large-scale evolution, Von Baer could still be a committed teleologist. "The organic world was not only *zweckmässig* (. . . well-adapted) but also *zielstrebig* (goal-directed). Owing to the existence of goal-directedness, so [Von Baer] claims, adaptation precedes the formation of new structures" (Mayr 1982:516).

Spencer was working within the main intellectual current of his day. His contribution was to take the general evolutionism of *Naturphilosophie* and combine it with the functional developmentalism of Von Baer (Spencer's *Autobiography*, Bowler 1975, Young 1980). Evolution was progressive and structure was transformed, in an adaptive manner, by means of an evolutionary "law" that applied to everything from cosmology to the development of the human mind. In his earliest works, Spencer did not stress the progressive nature of the change that results from the process, but "within a few years, it had become . . . the foundation of his whole philosophy of development" (Bowler 1975:107). In keeping with his emphasis upon functionalism, Spencer placed considerable stress upon the interaction of the object and the environment. Progressive change, therefore, was not to be judged by external philosophical criteria; instead, adaptation was the basis for progress. Thus, Spencer could freely criticize writers such as Lamarck who had postulated idealistic progressive evolutionary systems. Because of its clear linkage between evolution and the environment (Carniero 1973b), Spencer's evolutionism does bear a superficial resemblance to Darwinian evolution. But this resemblance is likely to lead the unwary into confusion. We must remember that, for Spencer, adaptation was both the *result* and the *means* of evolutionary change.

We must recognize that Spencer was totally and unabashedly Lamarckian in his view of the means directing evolutionary change. Lamarckism, as we have seen, is not characterized by the inheritance of acquired traits (a process which need not yield appropriate adaptations), but is rather a "one-step" evolutionary process in which adaptation is the response of an organism to an environmental challenge. It was this strictly adaptationist mechanism that Spencer called upon to explain evolutionary change.

> since in all phases of Life up to the highest, every advance is the effecting of some better adjustment of inner to outer actions; and since the accompanying new complexity of structure is *simply a means of making possible this better adjustment;* it follows that *the achievement of function is, throughout, that for which structures arise.* (Spencer, 1899, quoted in Carniero 1973b:32; emphasis mine)

The teleology of adaptationistic evolution ultimately reduces it to orthogenesis. Spencer's definition of evolution likewise points out this bias: "evolution is an integration of matter and concommitant dissipation of motion; during which the matter passes from an indefinate, incoherent homogeneity to a definite, coherent heterogeneity; and during which the retained motion undergoes a parallel transformation" (1879, quoted in Mayr 1982:386). Evolution is the directional evolution of systems—selection is not required to make this type of evolutionary system function because any selection is merely the consequence of differing rates of systems' evolution.

Thus, although Spencer originated and spoke often of the "survival of the fittest," this type of selection was an *effect* in his model for evolution, not a *cause*. Because organisms were responding adaptively to environmental stimuli, those objects with less than optimal responses were simply removed, leaving the field to the best. Such a negative selection has never been denied by even the most rabid anti-Darwinists. Nobody believes a three-legged horse will get far. The critical issue of Spencerian evolution (and the one that remains with us today) is whether change may occur by means of adaptationist, Lamarckian processes.

Little doubt exists that Darwin owes no debt to Spencer and that Spencer in no way anticipated Darwin's concept of evolution by means of natural selection (Freeman 1974; Hofstader 1955; Medawar 1969; Nichols 1974). Indeed, as Mayr properly points out (1982:386), "It would be quite justifiable to ignore Spencer totally in a history of biological ideas because his positive contributions were nil. However, since Spencer's ideas were much closer than Darwin's to various popular misconceptions, they had a decisive impact on anthropology, psychology and the social sciences." The link between evolution as conceived in the social sciences and Spencer's concept of transformation and adaptational and progressive change has been established beyond any reasonable doubt[10] and it is not necessary to repeat all of the arguments here. Instead, I will focus upon why Spencerian evolution could have been accepted for so long as "evolutionary."

Several issues are significant in assessing the impact of Spencerian evolution upon the social sciences. First, we must recognize that Spencerian evolution grew out of the same historical and scientific milieu that produced Darwinian evolution. Thus, Spencer, at least to his contemporaries, was at least as respectable a theorist as Darwin. Second, Spencer used many of the same constructions as Darwin, such as "survival of the fittest" and "the interaction of individuals and environment." This has given the gloss of a valid evolutionism to Spencer's arguments. Third, when Spencer was rejected by the scientific community, the

---

[10]See discussion from several perspectives by Blute 1979; Carneiro 1967, 1972, 1973a, 1973b, 1974; Dole 1973; Dunnell 1980; Nichols 1974; Sahlins 1960; South 1955; Stocking 1968, 1974; and White 1959a, 1959b.

rejection was based in large part upon the racist and political implication of his theory (Stocking 1968), not the evolutionary mechanism he had proposed. Fourth, many of the intellectual movements in the history of anthropology have followed quite closely similar movements in biology. Anthropology has not been a discipline totally separable from biology. Fifth, and most important, most authors believing in cultural evolutionary theories have, in fact, subscribed to Spencer's model: cultural change has been seen as the result of an adaptive and adapting process.

The debt owed Spencer by the early anthropologists such as Tylor and Morgan seems clear insomuch as these writers accepted progressional stages in the development of human cultures (see, Blute 1979; M. Harris 1968; Opler 1962, 1965; Service 1981; South 1955). Based upon the concept of evolution advanced by Spencer, such a progress is not only understandable but inevitable. Likewise, the cultural evolutionism of Childe (1941, 1951), White (1949, 1959a, 1959b), Steward (1955) and Shalins and Service (1960, Service 1975) traces back to Spencer via the Morgan–Tylor link (Carneiro 1972, 1973a; Dole 1973). The Spencerian nature of their evolutionism was known to the cultural evolutionists. As Dunnel (1980:41) concludes: "the popular confusion between Spencerian and Darwinian views often compounded by critics of cultural evolution (e.g., M. Harris 1968:636) cannot be laid at the feet of the cultural evolutionists themselves. They know they were not Darwinian evolutionsts." The progressive nature of cultural evolution is sufficiently well entrenched that Carneiro (1972:249) could speak, without flinching, of the "devolution of evolution" from the original "precise, rigorous and systematic concept of Herbert Spencer" because of our modern, Darwinian denial of progress as the true measure for any evolutionary process. Even the cultural ecologists, with the stress they lay upon the adaptive, optimizing response of culture to changing conditions, are better seen as heirs of Spencer rather than of Darwin. As was noted earlier, cultural ecology has erred by taking a descriptive model from Darwinism and applying it to the explanation of cultural change. However, within the functionalist and adaptive evolutionism of Spencer, such a tactic forms a valid basis for explanation.

Given these observations, the confusion of authors such as M. Harris (1968, 1974) and Carneiro (1967, 1972, 1973a, 1973b, 1974) becomes comprehensible. Harris and Carneiro, believing that evolution in the social sciences is akin to Darwinian evolution, cannot help but also believe that credit for its invention rightly belongs to Spencer; their analysis is correct except in the assumption that cultural evolution has been, and generally is now, Darwinian. The functional and transformational evolutionism of Spencer has made it unnecessary for social scientists to call upon selectionism to explain the form and function of culture. Progressive change towards increasing complexity became the definition of evolution not only for Spencer, but also for most of the social scientists who were to follow him in time as well as philosophy. The interaction between culture and environment given prominence by Spencer permitted peaceful coexistence and commerce between cultural evolutionists and cultural ecologists. Culture, to take the name given it by an influential collection (Montagu 1968), became "man's adaptive dimension."

The confusion of Darwinian and Spencerian evolution is compounded by the fact that Spencerian evolution was not only the development of a nineteenth century scientific tradition, but similar evolutionary schemes were to remain strong in biology at least until the middle of the twentieth century. We tend to forget that the "triumph" of evolutionism predated Darwin and that, conversely, the "triumph of selectionism" postdated him by nearly a century. Our generation has been the first to be raised with Darwinism as a ruling paradigm in all of the natural sciences.

Mutationism and Lamarckism

Following Darwin's initial acceptance, a countermovement was established that caused Darwinian selectionism to lose much of its support within the scientific community. From about 1890s until the 1940s, two major non-Darwinian evolutionary schools—mutationism and neo-Lamarckism—appeared to have inherited the evolutionary mantle from the Darwinists. We spoke earlier about the difficulties encountered by Darwin in explaining the source and inheritance of variation. It was this very issue that permitted most scientists to deny natural selection while at the same time accepting evolution. As Provine (1971:9) explains it: "the critics were quick to seize Darwin's profession of ignorance on the production of variation as a loophole into which other possibilities could be inserted. Many claimed that the production of variations was directed and that the variations rather than natural selection, determined the direction of evolution."

Seen from the vantage point of neo-Darwinian selectionism, mutationism and Lamarckism may be viewed as heretical interpretations of, respectively, Darwinian fitness and engineering fitness. No evolutionist has ever doubted the existence of negative selection; instead, non-Darwinian theorists have argued that the selection of individual variation is less important than Darwinian theorists would hold—that the variation in nature is so great, and so much of it appears to be unrelated to adaptation, that most evolution must proceed without needing natural selection—or, that the precision and perfection of adaptation could not have arisen by means of a "blind" process like natural selection. In the first case, the mutationist holds that evolution is nothing but the history of a taxon. Selection may remove the maladapted, but the major characteristics of organisms are simply those that appeared over time during their evolution. (For extended treatments see Gould 1977a; Mayr 1982; Simpson 1953, 1956; Wolsky and Wolsky 1976). In the second case, the fit of organism and environment seems too great to be explained by means of natural selection. In contrast to the benign meaningless posited by the mutationist, the Lamarckian sees an order and purpose in nature; the demands of engineering fitness are so great that some type of direct adaptive response to the environment is required. Lamarckism and adaptationism are, by necessity, closely allied (Caplan 1981). The mutationist position, fundamentally, is that evolution has been guided largely by the particular type of mutation that has occurred in the history of any particular lineage.[11] For the mutationist, natural selection merely places limits on the evolution of organisms. It serves to remove the clearly unadapted, but it is not the process by which new species, or higher orders of organisms, arise. Although the Darwinian sees the gradual and continual selection of small traits as sufficient to account for evolution, the mutationist advances various types of mutations to account for the origin of new groups of organisms. Most of these could be described as basically *discontinuous;* thus new groups of organisms arise more or less in one step rather than through the numerous intergrading stages proposed by the Darwinian.

Extreme mutationism can do nothing but describe; it is evolution without a cause. It provides us no way of interpreting the existing form and function of organisms, no theory for relating survival or extinction to the traits an organism might possess. In its less extreme forms, however, mutationism is not so easy to discard. It is clear that not all the genetic or even phenotypic traits of organisms have been demonstrated to be adaptive. Thus certain traits, for example, those that characterize the higher taxonomic groupings of organisms may be seen as results of their ontogeny rather than of natural selective pressure. Differences in reproductive structure between various orders of plants or the different arrangements of internal organs in animals seem to be part of the nature of the organism under consideration. Form is not directly related to function; there is more than one way to produce the same result. It appears that these various ways of achieving the same end arose independently and, because they were not clearly nonadaptive, survived as examples of a more or less "drifing" evolution, perhaps inevitably carried along with those traits that *did* induce fitness.

That a given trait or series of traits appears to lack adaptive significance, however, does not preclude eventual discovery of such significance. Likewise, a trait that seems neutral as far as selection is concerned may yet come to be understood as a result of natural selection. All we can really say is that we do not understand how or why the trait arose. If we accept a priori that at least some traits are now selectively neutral, we are still faced with the problem of identify-

---

[11]The story has yet to be told relating the rise of the Boasian school of historical particularism to the development of evolutionary mutationism. Neither was antievolutionary; both arose at about the same time and in response to similar doubts concerning Darwinism; and many of their prejudices parallel to a degree that independent invention would seem highly unlikely.

ing them: Which traits in the constellation of characteristics that serve to separate organisms in our classification schemes are selectively neutral, and which are undergoing selection? And, if a trait is at present selectively neutral, how can we be sure that it has always been selectively neutral? Perhaps it arose because of selective pressures that later disappeared; if so, can we really call it neutral? If the trait were to disappear, would the organism be at a selective disadvantage? It is hard to call any trait truly neutral in terms of selection.

Thus the maintenance of even a modified mutationist position has many problems, all arising from the assumption that the form of a trait is not related to natural selection. Yet at the same time we are confronted by evidence that seemingly cannot be explained by simple natural selection; variations on a given theme exist to which we are unable to relate the survival or demise of individuals. This is especially clear in the case of cultural traits.

Cultural behavior presents such a bewildering variety of forms, and shows such plasticity and inventiveness, that it is hard to believe that the specific forms of many cultural patterns have any survival value. If we try to discover, however, whether these cultural traits were indeed irrelevant to the survival of individual human beings when they first arose, we are confronted with the same problems as in the study of morphological and developmental traits of organisms. We can never be sure that any given cultural trait is without adaptive significance; we cannot separate the adaptive from the nonadaptive; we cannot be sure that a trait, perhaps nonadaptive today, has always been so, and we cannot identify the relationships among traits. Like the traits used to define higher groups of plants and animals, cultural traits often constellate and can be used to create taxonomies of cultural groups or processes. If we restrict ourselves to the consideration of human cultural traits, we find that we can say virtually nothing about the survival value of cultural traits in general, for we are unable to examine any human society that lacks culture. The possession of a human culture serves, in part, to define us as a species.

For the moment, however, we are forced into a basically mutationist position as far as the origin of cultural changes is concerned: such changes arose *randomly*. This is not incompatible with Darwinism and is an inevitable outgrowth of our lack of understanding of the function and development of most forms of cultural behavior. The capacity for culture was surely not irrelevant to human survival and evolution—the strict mutationist position is untenable. It also essentially precludes causation: there is history and nothing else.

It is fundamental to the Darwinian view of evolution that the appearance of a new adaptation is unrelated to the environmental conditions at hand and is not to be seen as a response to them: the *appearance* of a trait is unrelated to the *need* for it. Mutation—the appearance of a new trait—is random, an unpredictable event not tied to demands for survival. Indeed, most mutations render the organism less fit. Very few traits give organisms possessing them a clear reproductive advantage. Further, a trait may increase the average fitness of a lineage without being of clear and unequivocal benefit at all times and for all individuals.

The non-Darwinian theory that argues from the precision of adaptation to depreciate natural selection is Lamarckism. Lamarckians reject the notion that evolution occurs when environmental conditions favor or discourage the traits already present in organisms. They believe instead that adaptations are caused by direct efforts made by individual organisms or by direct influence of the environment on the form of the organism.

Lamarckian evolution is thus either "evolution by intent" or a strict environmental causation: in trying to reach the higher branches of trees, the proverbial giraffe brings about an increase in the length of its neck; the plant that grows in a desert is forced by dry environmental conditions to develop a fleshy stem for the storage of water. The cause of variation is radically different from that posited by the Darwinian—variation becomes the adaptive response to the environment (*one-step evolution*).

It is essentially impossible to advance Lamarckian hypotheses within the biological sciences, for there is no known mechanism by which the efforts of an organism to achieve a particular end may become integrated with that organism's genetic material. As is frequently stated, there is no transmission of acquired characteristics by heredity. Differences that arise solely in the phenotypes of organisms have no effect on their genotypes. This is not to say, as noted earlier, that the Darwinian rejects selection of the phenotype or variability in phenotypic expression. The same organism may exhibit differing phenotypes, different expressions of the same genotypic potential, in different environments. For example, various plants may bear different types of leaves, depending on whether they grow under water or on relatively dry land. If transplanted from one habitat to the other, their new growth will be appropriate to the new environment. The phenotype has changed, but the genotype has remained the same; the genotype is plastic enough to respond differentially to the differing environment. For cultural behavior, this plasticity may be of the essence, but we should not confuse selection for plasticity with change *caused* by the action of the environment. There is no evidence that the environment is capable of forcing any organism, including people, to adapt to the demands of survival.

Lamarckian evolution is fundamentally teleological: both the environment and the individual strive toward adaptation. The mutationist is willing to accept mutation as the mechanism that creates variation in individuals. The random nature of mutational variation causes no problems because the tolerance of the environment for almost any mutation is very high; the specific nature of the mutation is simply not very important. The Lamarckian view of variation, in contrast, is conditioned by the need to produce an adapted organism, one that is capable of responding appropriately to a change in the environment. Lamarckian evolution has no mechanism for understanding variation; it substitutes for it the need for adaptation. Lamarckian environmental determinism is predicated upon intentional teleology because the environment somehow causes the organism to adapt in the "best" or "most appropriate" manner.

Within the biological sciences, as I have said, the overriding objection to Lamarckism is that it is incompatible with our present concepts of mutation and genetic inheritance. When we consider culture, we lose the firm foundation that Mendelian genetics has put under the study of biological variation and selection. Culture is a constellation of traits whose specific performance is dependent upon learned, rather than innate, behavior. There is no purely materialist and mechanistic general model to account for the origin of cultural variations and, because the spread of cultural traits is by learning and imitation, there is no dictum to prohibit the inheritance of acquired traits. Indeed, cultural inheritance is defined as the inheritance of acquired characteristics. However, it must be demonstrated, not presumed, that *changes* in cultural patterns are inherently adapting: the *mode* of inheritance (learning, imitation) is totally independent of the *type* of response made by a culture to a given situation. Spencerian functional evolution requires far more than the inheritance of acquired traits. It requires that changes in the state of a trait be appropriate to the situation at hand; that progress consist of increased adaptation; that selection be of minor importance compared to directed or orthogenetic cultural processes.

The extreme formulations of Lamarckism and mutationism, however, become understandable when we understand that they are attempts at explaining complementary parts of the total evolutionary pattern. Evolutionary change involves modifications of character states within a lineage as well as changes in the total diversity of life. Lamarckism speaks to an attempt to understand adaptive modification—Why do organisms show differing and frequently adaptive variations that fit them to the environment? In the same manner mutationism attempts to speak to the problem of diversity—How can there be so many different types of organisms? Thus, as Lamarckism produces adaptation, so does mutationism produce diversity. But, from the Darwinian's viewpoint, both adaptation and diversity are aspects of a fundamental process that is capable of explaining both.

We are stuck, at present, with an essentially mutationist view for the *origin* of cultural innovations. We have no mechanistic method for describing variation in individual behavior or the potential significance of such variation. The significance of the individual in the analysis of cultural behavior has only recently begun to be recognized (for a brief introduction, see Vayda and McCay 1975), and this recognition has yet to be applied to significant questions. Until we begin to incorporate the benefits and costs to individuals into our descriptions of social processes, the philosophical rationale for cultural selectionism will remain hypothetical.

Several important issues arise from the Lamarckian nature of the transmission of cultural behavior. Although, as we have seen, cultural transmission is not inherently opposed to a Darwinian interpretation of change, we know very little about it. How variable is cultural behavior from the point of view of potential for change? How do these variations become expressed in the choices made within a culture? How effectively are they transferred from generation to generation? What is the relationship between specific behaviors and the general social environment? Questions such as these are difficult to incorporate into a selectionist model for cultural change, yet they must be applied to the real world of cultural change before selectionism becomes a viable intellectual tradition.

It is easy to propose an "adaptive" or even a "fitness" component for almost any conceivable situation. However, this goes against the spirit of Darwinism, which is explanation, not mere description based on a tautological and inaccurate definition of natural selection. What we aspire to is an *explanation* of the forces and events involved in any evolutionary sequence. This is not to be achieved by means of "selectionist reductionism;" instead, we should adopt a case-study approach to the understanding of the selective components of cultural variation and change. Over time, such studies may well contribute to a synthetic theory of the evolution of cultural behavior.

Cultural Selectionism

Thus far, the literature treating the evolution of culture as a nongenetic yet selection-mediated process is pitifully small and is largely composed of historical or theoretical analyses attempting to point out that a cultural selective perspective is, in fact, feasible—case studies are sorely lacking. Yet, the discipline, despite its problems, is still mature enough that it may be recognized and characterized both in what it accepts and how it differs from contemporaneous approaches to the study of cultural function and change.

Cultural selectionism[12] is a biological approach to the study of cultural events that accepts the primacy of variability, heritability, and differential fitness in explaining the cultural evolution of organisms (especially *Homo sa-*

---

[12]In proposing *cultural selectionism,* I have been guided by the necessity for a new term to indicate a new concept. As we have seen, *evolution* as used within the social sciences has little connection with Darwinian evolution and is related to specific pre-Darwinian philosophical notions encapsulated and given the status of a theory of evolutionary change in the work of Spencer. As Dunnell (1980:36) points out, *cultural evolution* should be used "to designate this particular expression of sociocultural evolutionism and not as a general term for the evolution of cultural phenomena." Although I dislike the creation of new terms as much as anyone, I believe that in this case the formulation of a new one is warranted for two reasons: (1) cultural selectionism is based upon an entirely different set of premises than competing systems for the analysis of cultural change, and (2) few will notice the difference unless it is given a name.

piens). Cultural selectionism accepts that human behavior is determined by the interaction of two inheritance systems—the genetic and the cultural—and generally places its emphasis upon the latter (although it may still be concerned with the acquisition of the capacity of culture—what has recently come to be called *protoculture*). Although human cultural behaviors may enhance biological survival, the specific processes determining cultural traits have acted largely within the cultural, and not the genetic, inheritance system. Hence, it differs from sociobiology in the emphasis it gives to learned behaviors transmitted within the cultural context. Cultural selectionism denies the importance of genes for the direct control of specific cultural acts or that a correlation must exist between genes and culture such that changes in fitness may best be modeled solely at the genetic level.[13] Emphasis is also placed upon behavioral acts and not the predelictions, reasons, or symbolic functions underlying these acts. Although symbolic aspects of culture may be important insomuch as they affect the behavior of individuals, the selection of cultural traits will always occur at the level of the act. Thus traits may not be selected independently of their actualization.

Cultural selectionism, in rejecting developmental and transformational models for cultural change, is distinct from cultural evolutionism in numerous ways. Instead of seeing cultural evolution as analogous to biological evolution *in its results* (phyletic evolution), cultural selectionism stresses the similarity between biological and cultural evolution in terms of the *process involved* (natural selection). No law governs cultural change other than the differential cultural fitnesses that specific behaviors may induce. *Fitness* here must be accepted with its broad meaning and is not to be confused either with simple numerical increase or greater adaptation. Cultural selectionism also repudiates essentialistic and typological classifications of culture and cultural change. It accepts no dictum of progress in its explanation of cultural change whether that progress be measured by means of "increasing control over the environment," "increasing capture of energy," "sociopolitical complexity," "increasing sophistication, consciousness, or class differentiation," or "resemblance to Western capitalist economies." Directionality and diversification may result from evolutionary change in cultural systems, but they arise from an interactive process, not one of internal, orthogenetic, or self-directed change.

Cultural selectionism is close to cultural ecology and might even be described as a cultural ecological rejoinder to sociobiology, but it bears the same

---

[13]Proponents of sociobiology with tendencies towards genetic reductionism include: Alexander (1971, 1974, 1975, 1977, 1979a, 1979b), Barash (1976), Darlington (1969), Emlen (1966a), Friedman (1979), Hamilton (1964, 1975), Hartung (1976), Irons (1979a, 1979b), Kurkland (1979), Laughlin and d'Aquili (1974), Tiger and Fox (1966, 1971), Trivers (1971, 1972, 1974), Van den Berghe (1975), Van den Berghe and Barash (1977), Wilson (1975, 1978).

For an introduction to the "sociobiology debate," see *The Philosophical Forum* 13 (Nos. 2–3) Winter–Spring, 1981–1982, and Caplan (1978), and the references cited in them.

relationship to it that evolutionary theory bears to ecological theory within the biological sciences (as traditionally conceived): the difference here is between an interest in proximate and functional explanation (cultural ecology) and an ultimate and evolutionary interest (cultural selectionism). Although recognizing that evolution produces regularities, cultural selectionists see their significance in the opportunity presented for finding explanations for *why* regularities exist at all. Homeostasis or adaptation are not self-evident properties of cultural systems any more than they are of biological systems; they must be explained in a materialistic and mechanistic manner and cannot be assumed to be intrinsic to the system per se. Hence, cultural selectionism places its emphasis upon the fitness induced by specific traits, not on the adaptive and equilibrium processes that *may* result from selection. Nevertheless, the contact between these two approaches has always been close and will likely remain so: cultural ecology, after all, is studying the type of variability that will prove of greatest interest to the cultural selectionist, whereas the cultural selectionist is studying the processes that may help the cultural ecologist to understand the phylogeny underlying cultural variation.

Given the numerous sources of data in several traditionally distinct fields that the cultural selectionist must draw upon, it is not surprising that its proponents comprise a rather motley crew. One of the first modern[14] cultural selectionist papers grew out of an interdisciplinary study with representatives from anthropology, biology, mathematics, social psychology, and sociology. Gerard *et al.* (1956) supported the general applicability of Darwinian selectionism to cultural change by drawing largely upon a self-conscious analogy to genetic speciation. Interestingly, this paper is one of the few published that clearly considers the possibility of analogizing cultural change to subspecific rather than species-level selection; however, like most later authors, they end up stifled by their emphasis upon a species level and and by their genetic analogy for cultural change.

The psychologist Campbell was the first recent author to make a major impact with his reinterpretation of Darwinian evolution in terms of cultural change. In a series of papers (1956a, 1956b, 1960, 1965, 1970, 1972, 1974a,

---

[14]Cultural selectionism may be traced, if one so desires, back to Darwin—especially to his *Descent of Man* and the *M and N notebooks* (Barrett 1980). However, this is a rather pointless task; the clear enunciation of genetic evolution was a necessary prerequisite for the development of a truly cultural selective perspective. As we have already seen, Darwin was not in the position to make a nice distinction between cultural and genetic inheritance. Furthermore, the whole issue of the applicability of Darwin's work to the study of cultural change has developed such an overburden of emotional and ideological rhetoric (because of the irrelevant connection of natural selection theory with admittedly racist and reactionary—frequently fascistic—doctrines) that even attempting to bring up the possibility that cultural change may be described in materialistic and mechanistic terms almost forces one into the position of writing sentences such as this.

1974b, 1975, 1976) he argues that the concept of natural selection is sufficiently rigorous to be applied, without modification, to the selection of cultural traits, and that only those theories based upon variation and differential selection may be justifiably termed *evolutionary*. Cultural evolution, therefore, must be based upon an analogy to the process underlying biological evolution. "The analogy to cultural cumulations will not be from organic evolution *per se,* but rather from a general model for adaptive fit or quasi-teleological processes for which organic evolution is but one instance" (Campbell 1965:26). Campbell describes selectionism in somewhat more universal terms than other cultural selectionists and has argued its applicability to disparate subjects ranging from learning theory to epistemology. He believes that a selectionist model is congruent with observed changes in culture, such as new or increased adaptation, complexity, size, or social integration, and that it may also permit us to appreciate *all* cultural patterns by linking variation to selection and thus cultural change. In this, he stresses the importance of "blind" or random cultural variation as the key to understanding self-transcendence in cultural evolution. "One of the services of terms like 'blind' and 'haphazard' in the model is that they emphasize that elaborate social systems . . . could have emerged . . . without any self-conscious planning or foresightful action. It provides a plausible model for social systems that are 'wiser' than the individuals who constitute the society, or than the rational social science of the ruling elite" (Campbell 1965:28). Unlike most other cultural selectionists, Campbell has gone to some pains in attempting to relate cultural activities to individual perceptions, habits, and activities. He also stresses the frequently antagonistic nature of cultural and genetic evolution, holding that this may explain (perhaps by means of group selection) the evolution of social behaviors that are disadvantageous for most of the individuals within the society. This uncoupling of genetic and cultural evolution prevents him from falling into the naturalistic fallacy (claiming that what exists should be), because what has arisen by cultural selection may be changed by the same means.

Campbell has also stressed the importance of avoiding adaptationism when applying a functional model of evolutionary change to cultural evolution: "the wisdom of evolution is retrospective. If the environment changes, the products of past selection may be stupid" (1965:34). I would add to this the uncomfortable fact that many cultural changes directly involve changing the environment. Therefore, that which has initially been selected may change the environment in a manner that soon makes the selected trait "stupid." Campbell's speculations and reflections go far beyond our current understanding of cultural systems and the specific events that have brought about change in them, but they will likely provide interesting hypotheses for some time to come.

Alland (1970, 1972, 1973, 1975; Alland and McCay 1973), a medical anthropologist, argued for a two-systems model of cultural change at an early date, but has mostly spoken from a strong cultural ecological viewpoint with its concomitant tendency to confuse genetic and cultural fitness (although this may grow out of his utilization of adaptation to disease as a frame of reference).

The biologist Dawkins, in his popular volume, *The Selfish Gene,* advanced a clever model for describing evolution in which the "true" unit of evolution is seen as the gene—evolution is merely a way in which genes maximize their own representation within a population. In the last chapter of this volume, he extended this same concept to culture, proposing the term *meme* to refer to a theoretical unit of cultural transmission that would be selected in a manner analogous to genes. Cultural change is the result of selection between memes and favors those "memes which exploit the cultural environment to their own advantage" (1976:213). Although Dawkin's model is cute, it is essentially impossible to operationalize—cultural traits, although definable, are not material, and, although hereditable are not atomistic.

Cloak, who was trained as an anthropologist, has taken an approach to cultural change that is similar to Dawkins's. He has developed a model for cultural change (1975, 1976, 1977) that places emphasis upon presumed "cultural instructions" that underlie cultural actions. Although this permits him a wider concept of fitness than many other authors, and although the basis for his argument is empirically correct, he errs, like Dawkins, in totally misunderstanding the causality of natural selection: the cart of change-in-representation-in-traits is here placed before the horse of selection-of-phenotypes. As Cloak states (1977:51): "instructions are primary, and their phenotypic products are secondary, mere instruments by which instructions succeed." At times, of course, such an "instrumental-reductionism" may be useful in making a point or as a heuristic device: "maladaptive behavior may become widespread because the real unit of natural selection is not the organisms but the unit that replicates itself—the genetic or cultural instruction" (1977:51). However, a slavish insistence upon this model is bound to be counterproductive: "a cultural instruction . . . is like an active parasite that controls some behavior of its host. It may be in complete mutual symbiosis with the human host . . . [or it may make] the host behave . . . [in a manner] that results in extraorganismic self-reproduction of the [parasite] but not in survival or reproduction of the host or his conspecific" (1975:172). I believe we are dealing here with a strange, very modern form of idealism that often expresses itself in the system-theorists' jargon of constructed variables, compound words, abbreviations, and acronyms (CV-CW-2A). The purpose of this, of course, is to remove much of the messiness that dealing with real organisms involves—genes or memes become the neat identical particles of the physicists. But the effect of such a program generally is confusion in place of the accuracy for which it strives (see especially, Cloak 1976). Another outgrowth of instructional-reductionism is the belief that instructions are in some manner coadapted such that internally integrated systems of cultural traits may be presumed to exist. Two influences are probably at work here—an overblown analogy to gene integration and an adaptationist prejudice for cultural function, probably drawn from cultural ecology. In any case, the implication of belief in such integrated systems of cultural instructions is that selection occurs within the system of instructions, whether this system be subindividual or as complex as modern urban society, and it is at the "appropriate level" of systemic integration that selection must be understood. Yet, as Cloak admits (1976), defining the boundaries of the various subsystems, no less explaining selection at the appropriate level, is at present impossible. On the whole, Cloak appears to be attempting too much, given the highly limited model he promotes.

Durham, an anthropologist trained as an evolutionary biologist, has done much to develop a general model for cultural selection (1976a, 1976b, 1977, 1978). In his "coevolutionary model" of human biology and culture, he distinguishes *natural* (genetic) from *cultural* (nongenetic) selection and attempts to relate both to an increased inclusive fitness criterion for evolutionary change. In contrast to the model present in this volume, and to those of most other cultural selectionists, he does not accept the independence of the selection of genetic and cultural traits and instead argues that selected cultural traits will generally be genetically advantageous. As he concludes (1978:444), cultural selection "functionally complements natural selection by retaining in time those cultural variants whose net effects best enhances the inclusive fitnesses of individuals." I view this as an unproven and probably misleading assumption. Curiously, despite his continual stress upon inclusive individual fitness, Durham places great emphasis upon group selection, although he tends to see such selection as also working in concert with individual selection. It is likely that Durham's interest in both group selection and the complementarity of cultural and genetic selection resides in the great emphasis he places upon adaptation as the sine qua non of evolutionary change. But as we have already seen, functional approaches to evolution and the criterion of engineering fitness may yield errors if used as the exclusive measures of evolutionary change.

The archaeologist Dunnell (1978a, 1978b, 1980; Dunnell and Wenke 1980) has argued strongly for the applicability of a scientific approach, and specifically for a Darwinian model in the explanation of cultural change. Like Blute (1979), Dunnell stresses the nonrelation of cultural evolutionism in the past to Darwin's concept of natural selection, and maintains that cultural evolution must be recognized as fundamentally distinct from biological evolution if it is to be successfully applied to understanding cultural change. Dunnell's most original contribution lies in the correlation he sees between the size of groups and the types of traits that will be favored by selection. He believes that cultural traits are uniquely suited to group selection and relates changing parameters in selection to the rise of complex societies.

the appearance of complex society is a consequence of a shift in the scale at which selection is most effective. This is seen as the result of an increase in the amount of

information that must be transmitted from one generation to the next in order to reproduce the full human phenotype. The critical threshold is the amount of information
that can be transmitted reliably by the combined genetic and cultural transmission
mechanisms of single individuals. In this context, the development of nonkin-based
altruism and functional specialization will no longer be selected against . . . shifting the
scale at which selection has its greatest impact from the scale of the individual to that of
society. If this argument is correct in general outline and the assumption that innovation
is a random phenomenon is justified, then the occurrence of complex society on a global
scale will, among other things, correlate indirectly with population size. (Dunnell
1980:66)

This is an interesting conjecture and one that deserves far greater elaboration and
incorporation into the form of a general model including testable hypotheses.

Perhaps some of the most exciting work being done in cultural selectionism
has grown out of biometric approaches for modeling genetic change in populations. Especially important has been the series of theoretical papers[15] and the
volume published by the population geneticists Cavalli-Sforza and Feldman
(Cavalli-Sforza 1971; Cavalli-Sforza and Feldman 1973a, 1973b, 1976, 1978,
1981; Cavalli-Sforza et al. 1982; Feldman and Cavalli-Sforza 1975, 1976).
These workers have developed several models of cultural inheritance systems
and have sought to elucidate the dynamic properties of these systems for both
"pure" cultural selection and the interaction of genetic and cultural systems.
Richerson and Boyd (1978; Boyd and Richerson 1980; Richerson 1976, 1977),
theoretical ecologists, have also presented mathematical models for cultural selection, but theirs are based upon the equilibrium rather than the dynamic properties of the systems under consideration. These workers have provided the clearest
and most readily understandable rigorous treatment for the selection of cultural
traits, as well as an extremely articulate rationale for adopting a cultural selective
perspective (see especially, Richerson and Boyd 1978). The evolutionary biologists, Werren and Pulliam (1981), have recently applied a mathematical model
for cultural evolution to variations in human lineage systems. Contrasting sociobiological genetic models for kin-selection with their cultural transmission
model, they found that the cultural model was at least as consistent with observed
patterns. Of greater importance, they were able to deduce several conflicting
predictions from the two models, which are subject to empirical test. Little doubt
can exist but that modeling of the sort done by these authors will be of major
importance in gaining a new perspective on cultural change. As May (1977:13)
has pointed out: "formidable mathematical difficulties stand in the way of a
more full understanding of the interplay between cultural and biological evolutionary processes. . . . [Nevertheless] I believe that the incorporation of cultural
inheritance into the quantitative theory of population genetics, as begun by
Feldman, Cavalli-Sforza and others, is likely to open exciting new areas."

---

[15]A clear introduction to this work is found in Pulliam and Dunford 1980: Chapter 6.

Although the work of theorists such as those mentioned above holds great hope for establishing a scientific and experimental foundation under the edifice of cultural selectionism, a viable and active program of cultural selectionist investigation faces numerous obstacles. Perhaps the greatest problem at the present time is that cultural selectionism itself interfaces poorly with existing divisions in scholarly investigation and the academic departments and scholarly meetings these divisions reflect. A side effect of this lack of fit is that it is difficult for young scholars with a cultural selectionist commitment to find an appropriate position—joint appointments in, for example, anthropology and evolutionary biology are rather uncommon. Reflecting its hybrid origins, the literature of cultural selectionism is scattered in numerous journals[16] and cultural selectionists tend more to reference authors coming from similar backgrounds rather than to have a general familiarity with the work of others taking the same approach. Of course, all of these problems are to be expected in a young discipline and should quickly fade into insignificance as cultural selectionism develops into a maturity that its work thus far can only anticipate.

[16]The greater than usual diffusion of this literature can be illustrated by a listing of some of the journals used by cultural selectionists: *Human Ecology, Journal of Theoretical Population Biology, Behavioral Science, Annual Review of Ecology and Systematics, Annual Review of Anthropology, American Journal of Human Genetics, American Ethnologist, The Philosophical Forum, Psychology Review, American Antiquity, Journal of Social Issues, Journal of Economic Behavior and Organization, Social Biology, Philosophy of Science, and Interdisciplinary Science Review.*

CHAPTER 3

The Naturalness of the Human–Plant Relationship

Study of the evolution of plants under cultivation has served as a meeting point for social and natural scientists for much of the past century and, derived from this synthetic approach, the development of the "centers of agricultural origin" theory was one of the great accomplishments of the first half of this century. As early as 1853, Schleiden had pointed out that man's crop plants and, especially, the weeds of agricultural fields could be used as a guide to the history of man's movement across the globe. The first worker to catalogue centers from which the species of cultivated plants had diffused was de Candolle (1886). Adopting an approach that is striking in its modernity, he argued that the history of cultivated plants could be understood only by combining the views of the botanist, the archaeologist, the historian, and the linguist. He identified a number of world regions whose wild species he considered likely progenitors of given domesticates. De Candolle concluded that agriculture was not chosen, but was forced upon the society, most often by a lack of sufficient food for a growing population. Like modern workers such as Boserup and Cohen, de Candolle regarded population pressure as the most important factor demanding that a society "choose" an agricultural way of life.

De Candolle recognized that the processes contributing to the development of cultivated varieties were not the same as those that first brought a plant into cultivation:

The difference in value, however great, which is found among plants already improved by culture, is less than that which exists between cultivated plants and others completely wild. Selection, that great factor which Darwin has had the merit of introducing into science, plays an important part when once agriculture is established; but in every epoch, and especially in the early stage, the choice of species is more important than the selection of varieties. (1886:3)

Vavilov (1926), taking up the work of de Candolle, stressed that the major cultivated plants of today cannot be assumed to have arisen directly from wild plants still extant, no matter how close the resemblance. He believed that many of the wild species might well be fugitives from cultivation and that it was misleading to assume that a living species is a good representative of the progenitor of even a closely related plant. He reasoned that the regions that today have the greatest number of varieties within a cultivated plant species are the likely areas of that species's original domestication.

Vavilov believed that as a crop spread from its area of domestication it encountered successive environmental and cultural filters. He hypothesized that these filters reduced the genetic diversity of a species as it moved from its original center of origin by selecting against certain characters that had been present in the plant in its homeland. There would also be a founder effect on removed gene pools: any selection from the originally domesticated species that traveled from the original site of domestication would carry with it only some portion of the total diversity that was present in the larger population. Although Vavilov recognized that diversity might appear as a secondary development in centers far from the original one, he gave the matter little attention.

Vavilov also developed the idea of *primary* and *secondary domesticates,* primary domesticates being the first species brought under cultivation and secondary ones being derived from the weeds inevitably present in any field crop. When the domesticated crop is introduced into a new environment to which it is ill-adapted, weeds have the potential for developing into new crops by being harvested in ever-increasing proportions over the original crop plant. This distinction recognizes two different modes of selection. Although Vavilov did not develop a general theory for the evolution of primary domesticates, he did note that the selective forces creating secondary domesticates are to be found both in the processes of agriculture and in environmental influences on the total cultivated crop, including weeds.

Vavilov held that some plants moved from their original homes *before* becoming part of an agricultural system. Thus it was possible that they could be incorporated into cultivation at different times and in different societies. Some of these plants, such as hemp, he characterized as weedy camp followers that "to a certain extent became cultivated almost independently of man's will" (1926:236).

Unfortunately, little thought was given in the past century to the problem of the primary domesticate itself. Thus much of the theorizing about agricultural origins has been burdened by a confusion that is based on a misinterpretation of Vavilov. Later workers consistently assumed that the selective forces that formed the secondary domesticate could be held to account for the evolution and the origin of the primary domesticate. Of course this is both illogical and absurd— the techniques of agriculture cannot account for the plants that were used in early agricultural systems. The plants first cultivated by man had to arise by means of some other process.

The success of Vavilov and his students in establishing centers of origin for major cultivated plants caused this approach to dominate the thinking of many scholars. Perhaps the best-known proponent of the centers-of-origin theory is Sauer. He suggested that agriculture would have begun in a region with diversified terrain—only this type of environment could provide the many plants necessary for agriculture. Sauer believed that clearing of grasslands would have presented insurmountable problems to early man and therefore that agriculture would have had to begin in woodland environments. (He rejected floodplains largely on the basis of the Vavilovian postulate that centers of cultivar diversity are mountainous regions.) Sauer also believed that preagricultural peoples would have had to be sedentary, because agriculture requires this, and relatively well fed, to allow for the unprofitable experimentation that would have had to precede the development of agricultural systems: "People living in the shadow of famine do not have the means or time to undertake the slow and leisurely experimental steps out of which a better and different food supply is to develop in a somewhat distant future" (1969:21). Finally, "the inventors of agriculture had previously acquired special skills . . . that predisposed them to agricultural experiments" (1969:22).

Having listed the characteristics of the area in which agriculture could begin, Sauer then identified the area that best fit his criteria—Southeast Asia. He argued that farming grew out of a culture "tied to fishing in this area" and considered it "the world's major center of planting techniques and of amelioration of plants by vegetative reproduction" (1969:25). Sauer believed that all farming techniques and many plants were originally dispersed from this one great hearth and that grain farming was a secondary development.

Isaac (1970) has taken a rather different view of the ultimate hearth of agriculture. For him, agriculture began in the Near East: the plow was not developed from the "lowland tropical hoe," vegeculture is an adaptation to the climatic demands of the tropics, and the idea of agriculture spread out in waves from the Near East. Isaac maintains that cultural similarities found in the Upper Paleolithic–Mesolithic in the New and Old Worlds support his view that the "Old World migrants were the 'incipient' domesticators of the New World" (1970:74).

The centers-of-origin approach has been under attack for several generations; almost every assumption and method of this school has been effectively refuted (see Schieman 1939; Harlan 1970, 1971, 1976; Zohary 1970). Although Vavilov is generally recognized as its inspiration, his work was based on different assumptions. He had a rather different view of the time span over which domestication occurred than is commonly accepted today; he sought the origin of cultivated plants in "the remotest past, for which the usual archaeological periods of five to ten thousand years are but a short term" (Vavilov 1926:224). Attempts to identify his centers with archaeological sites show a poor understanding of his work. Moreover, Vavilov did not assume that a particular hearth could be identified as the source of the *idea* of agriculture—he was a botanist and was not concerned with such concepts. It should also be remembered that Vavilov's research was motivated, at least in part, by the need to find locations of high varietal diversity that could be exploited by modern plant breeders. It is thus understandable that he concentrated on the study of those field grains that are of such importance today. This "centers-of-diversity" aspect of Vavilov's work was being stressed even as the centers-of-origin theory was beginning to fall from favor.

Vavilov had no way of knowing whether the region that showed the greatest divesity of cultivars was a center of origin or an area of diversification, and the basic problem with his approach is that most such regions of diversity probably are the latter. They are frequently regions in which environmental barriers prevented the transfer and introgression of crop plants, and new varieties developed especially rapidly in isolàtion. What little archaeological evidence we have at present does not indicate that these were necessarily important areas of plant cultivation during the early stages of settled agriculture.

Although the centers-of-origins approach can be rejected as overly simplistic, it must be given credit for focusing our attention on the cultivated plant as a source of evidence about the process of domestication. Unfortunately, as this approach developed, the basic assumptions made by social and natural scientists about the evolution of the cultivated plant became indistinguishable. Both viewed agriculture as a particular stage in the development of a culture and agreed that the stage could be localized in time and space. The question of *how* agriculture arose was displaced by the question of *where* it arose. Emphasis was placed on environmental conditions, and dichotomies such as vegeculture versus seed culture, hoe vs. plow, and tropical vs. temperate became basic explanatory symbols in much the same way as matriarchy and religion were used by the cultural theorists to explain fundamental differences between human societies. Unfortunately, many of the earlier insights were lost in the unification of social and natural views of agricultural origins. Without a doubt, the most important loss was of Darwin's view of domestication as a natural, evolutionary process.

Biology and Intent

As we have already noted, Darwin used the existence of cultivated plants as a major proof for his interpretation of the natural selection model for evolutionary change. There is a certain irony in the fact that whereas we have accepted his basic model, we have rejected one of his major examples. As I have pointed out, Darwin distinguished between *methodological selection,* the systematic modification of a plant variety or animal breed according to a predetermined standard, and *unconscious selection,* the preservation of valued individuals and the destruction of individuals of less immediate value with no intention of altering the breed. This distinction he summarized in one word—*intention.* His interest in unconscious, unintentional selection has been overlooked by most scholars working on the problem of agricultural origins and the domestication of plants. Instead, as we have seen, the paradigm of consciousness has been applied to the elucidation of this problem. I argue first of all that intent, being unverifiable, has no standing in the scientific explanation of cultural change. I show that intentionality is not a prerequisite for the domestication of plants and, indeed, may be counterproductive. In this I am not denying that people act and that their actions may be motivated by immediate goals. People, likely any other animal, will not choose an obviously inconvenient, difficult, or inefficient subsistence strategy. This is not to claim, however, that the strategy chosen need represent the ''best'' or ''most efficient'' one—it may only be judged given the perceptible options in any place and at any time. People could not intentionally *domesticate* a crop. However, they could, and surely did, *favor* those individual plants that were most pleasing or useful to them. This is not a picky distinction; it represents the essence of my argument for it encapsulates the distinction between *evolution* (the result) and *selection* (the means).

Culture, in many ways the collective manifestation of consciousness, exists and produces effects, and yet it does so only by virtue of the actions of individuals, none of whom possesses the culture in its totality. In essence, culture is a metaphysic—albeit a fascinating and powerful one. Indeed, I would hold that the paradigm of consciousness I am discussing here is one of its particular manifestations; it is part of the mythology of our culture to believe that we are in control of our destiny. This myth, properly limited, no reasonable person can deny: people invented agriculture, just as they invented chocolate mousse, atom bombs, and pornography. Such statements are totally true and absolutely trivial. Inventions are merely the events, behaviors, and material objects associated with human activity. This tells us nothing we did not already know.

The invention of agriculture, however, faces major problems because of the biological systems of its constituent plants. Cultural teleology is not applicable to the biological problems we confront in trying to explain the origin of agriculture. Even if the notion of cultural consciousness were to be accepted, it would be incapable of accounting for the specific evolutionary events that had to accompany the evolution of cultivated plants. The long-term effects of plant breeding or of other forms of environmental manipulation are essentially impossible to predict. Actions taken within dynamic evolutionary systems cannot be reliably predicted. Any change in the environment presents new opportunities for evolutionary change; thus, even if we were able to analyze a situation in its totality, we would be unable to predict in detail the effects of any action because, once the action had been implemented, the situation would no longer be the one we had studied.

For example, a certain strain of male-sterile maize was discovered in the 1960s that permitted more economical production of hybrid corn seed. However, unknown to the plant breeders, this maize was also susceptible to a previously unimportant fungus, and it passed this trait on to its offspring. The economic advantages of using this strain were great, and in a short time almost all the hybrid corn being produced in this country was based on lines with cytoplasm T. From the point of view of the fungus, there was an almost unbroken field of vulnerable corn plants stretching from the Gulf Coast into Canada, and the fungus took advantage of it. The result was a disastrous maize-crop blight that moved north as the season progressed. There was no way short of precognition for the plant breeders to have anticipated this. In a fundamental sense, teleology requires precognition—the knowledge that actions taken to a desired end will be *effective* in attaining that end.

The logical problems attendant on intentionalistic argument do not, however, invalidate it. This invalidation arises from consideration of the requirements of an agricultural system. Intention cannot be accepted as an explanation of domestication because of what we know about the biology of domesticated plants, wild plants, and weeds.

If we assume that people intentionally invented agricultural systems and fitted cultivated plants into that system, we must face a fundamental problem: the response of the plants would have been both unpredictable and slow. This follows because the plants with which any agricultural system would have had to start are wild and adapted to the conditions of survival in the wild: many of their traits are ill-suited to survival in an agroecology. For example, it is likely that only a small percentage of the wild seeds sown in an "invented" agriculture would come up the first year, the rest remaining dormant. In the wild, delayed germination is an adaptation; the plant will not become extinct merely because of the loss of one year's seed crop. We may presume that some plants of the species in question, at sometime in the past, had simultaneous seed germination, but one bad year would have been sufficient to eliminate whole lineages of such plants from the general genetic pool of the population. Disasters, by definition, do not occur every year; they are the extremes of normal environmental variation. Adaptation is to both normal and disaster conditions.

Even if our hypothetical inventor of agriculture had been willing to accept low germination in the wild plant he had begun to sow, the seeds harvested would not have had any higher germination than the original wild seed. The plant cannot know that it is being protected from the extremes of the environment. It still has its old adaptations; and there is no way for it to change merely because the conditions have changed. Given enough time, however, the trait might have changed through natural selection. Mutations are inevitable in plants, and one mutation is the loss of seed dormancy. Seeds with this "desirable" trait could not have been identified before planting, but, naturally and inevitably, this lineage would have come to dominate the planting by the differential survival and reproductive success of various individuals. Over long periods of time, plants showing simultaneous germination could have comprised an ever-increasing proportion of the harvest; in the lineage that had lost seed dormancy, all the seeds, rather than just some, would have come up every year. Eventually the plants exhibiting simultaneous germination would have dominated the population.

It is difficult to say whether the loss of delayed germination would have been as successful an adaptation for people as might appear at first glance; a great deal depends upon the agricultural context in which the sowing occurred. For example, if the environment were subject to occasional drought and agriculturalists had not already adopted irrigation, one year of drought would have been sufficient to cause the extinction of the developed nondormant grain. With the extinction of this grain, people would have had to begin again the process of selection from the wild grain. Just how long effective selection for loss of dormancy in seeds would have taken is indeterminable. The point is that the mere act of sowing need not be immediately or continually rewarded with success and a pattern of success would be necessary for the invention of agriculture.

Domestication must proceed within the confines of the genetic system of the plant and by selection occurring within the agroecology. Even today, many agricultural plants are not "completely" domesticated. For example, even after millennia of interaction with man, not all varieties of wheat are characterized by a totally indehiscent rachis, the ne plus ultra of domestication in the small grains. Similarly, in oats there is a tendency for the ripe grain to shatter, reducing the total yield that can be harvested (see Schwanitz 1966:34). Most soybeans have somewhat dehiscent pods that make prompt harvest essential (Smartt 1978). Several legumes originally domesticated in the Middle East and commonly grown there today show a range in the effectiveness of their seed dispersal mechanisms (Ladizinsky 1979). The lentil (*Lens culinaris*) shows "classic" indehiscence of the ripe pod, apparently the result of a single recessive gene. The chick pea (*Cicer arietinum*), however, has a pod indehiscent even in the wild state. Dispersal of the wild species is dependent upon an eventual shedding of the pods from the plant, followed by dehiscence and scattering of the seeds. The cultivated species does not shed its pods as readily nor are they as likely split open. Apparently the change in dispersal morphology is the result of several genes acting in concert. Ladizinsky notes that "it can be concluded that chick pea domestication was not a clear cut event as in the lentil, but rather a gradual selection for types with less tendency for seed dispersal'' (1979:286). Finally, fenugreek (*Trigonella foenum-graecum*), widely used as a seasoning, vegetable, and forage crop, has totally indehiscent and nonabscissing pods, even in the wild state. Dispersal in the wild is dependent upon decomposition of the pods in the season following maturation. Here the cultivated type may be distinguished from the wild only by its erect growth habit, the greater number of pods per plant, and pod length and size. "Natural" dispersal mechanisms characterize most of our cultivated ornamental plants and many of our vegetable forage and fruit crops. Clearly, domestication cannot have and has not had inevitable effects on the plants involved in the relationship. Furthermore, many weeds show the desirable (to people) trait of indehiscence, but in this case we have long accepted that the indehiscence is merely a response of the plant to the conditions of tillage. It is absurd to claim that man invented weeds or that weeds are indispensable to the cultural adaptation of agriculture. For these indehiscent weeds, people have become the dispersal agent, and they are threshed and sown along with the crop. We are thus forced to consider the possibility that morphological change in crop plants also is a response to selective forces and that the desires and intentions of people are fundamentally irrelevant.

Returning to the problem of the gap between the invention or adoption of agriculture as a means of adaptation and the time required for morphological change, we might consider the domestication of a tree fruit such as the apple or pear. Here, the wild fruit is already in a relationship with nonhuman animals that disperse it, and the morphology is adapted to these modes of dispersion. Wild pears are small and frequently taken by birds. The development of the pear from something the size of a blueberry to the gigantic fruit of today surely took many, many generations, and it is unlikely that this development was meaningful in its initial selection and introduction into cultivation. Rather, the change in size probably preceded cultivation behavior on the part of people.

Any theory that calls for the invention of agricultural behavior requires either immediate or long-term rewards. Long-term rewards can be excluded by excluding precognition. Short-term rewards require the invention of more than one agricultural behavior at a time. For example, in certain environments, sowing could not continue to be rewarded without irrigation. Likewise, the cultivation of any plant requires that the mere act of cultivation place the cultivated plant at a selective advantage (to people) over the wild plant—otherwise people would be content with the wild one. This advantage could occur in two ways: (1) The environment under cultivation might in some way "ennoble" the plant. Although weeding and fertilizing could do this, they would, like irrigation, have to arise concomitant with planting. Again, as in the case of irrigation, we are faced with the improbability of a system's arising as a human invention or adaptation. (2) The plant might undergo genetic changes that increase its yield. Increase in yield is not, however, amenable to the simple type of mass selection that must be assumed for early agricultural systems.

Many of the mechanisms advanced to account for the development of the cultivated plant presume the existence of agricultural systems. One of these is the saving of the best seeds (Zohary 1969), the type of artificial selection that could account for the indehiscent cob in maize.

Maize is a wind-pollinated plant. Thus, if humans had decided to save the best seed from their early cultivated maize harvests in an area in which the wild progenitor species was growing, introgression would have been considerable and continuous between the two populations. If cultivated maize had been cross-pollinated by wild maize, the offspring would have shown traits of both. This would have been significant if the trait showing mixing had been the trait for which people were cultivating the plant. Wild maize must have had a method for dispersal. Thus the seed of the cultivated maize might well have had the wild trait for dispersal. The magnitude of this introgression would depend, of course, on the relative abundance of wild maize in the area of early cultivation. If it were sufficiently abundant, it seems likely that many of the offspring of even the best plants would not have been of the cultivated type, at least in terms of indehiscence. Thus one wonders why people intent on inventing agriculture would have continued saving the best seeds if those seeds were not reliably producing what people expected.

There are four possible ways of attempting to answer this question. (1) If the indehiscent trait in maize was cytoplasmic, it would have bred true in the female line. This is a possibility that cannot be dismissed until we know more about the evolution of maize from its wild progenitors, but from what we know about the genetics of maize and teosinte (a close relative, ancillary to or derived from maize) it does not seem likely. (2) The wild maize might have become extinct. When cross-pollinated by the cultivated maize, it would have picked up the indehiscence characteristic of cultivated maize and then, lacking a natural dispersal mechanism, eventually perished. (Indeed, it is just this type of introgression that Flannery [1973] cites for the extinction of a theoretical wild maize.) As long as there was wild maize, the problem would have persisted for the cultivator of indehiscent maizes, but this need not have stopped people from saving seed for planting. Even today, introgression between teosinte and maize is common in Central America, but it is difficult to call humans under these conditions an intentional agent in the invention of the maize–teosinte complex. (3) Humans might have removed the cultivated crop from the area in which wild maize was plentiful. Maize, however, like numerous other crop plants, suffers greatly from inbreeding depression. Isolating a small sample of the crop would have decreased both vigor and yield. Under conditions of relative isolation, the saving of the best seeds might also have been counterproductive: like introgression, inbreeding depression is dependent on statistical considerations, but it nevertheless places limits on what would be an "effective" saving of the best seeds. Inbreeding depression and introgression point out another important fact in the evolution of out-crossing crops such as maize: the "superior" plant that would be the best choice for intentional propagation would be characterized by its particular *combination* of genetic traits, and these would not be passed, as a unit, to the offspring. (4) The microenvironment of the plant might have been particularly favorable. Clearly, however, saving the best seed in a field from a plant that happened to be growing in a favorable location would have no effect whatsoever on its offspring under normal conditions.

Any intentionalistic or adaptive theory for the origin of agriculture must somehow deal with questions such as why people have *not* domesticated certain crops, why other domesticated crops seem to have entered agricultural systems at a distance from their wild progenitor species, and why weeds exist. We must also try to understand why certain cultures have never depended on agricultural production for their subsistence.

Adaptive and intentionalistic approaches to the origin of agriculture have made taxonomic decisions that imply that the forces responsible for the pristine development of agriculture will be found only in situations in which that development occurred. Cultural ecologists hold that agriculture developed as an adaptive response to stresses arising in specific historical situations; cultural evolutionists believe that agriculture arose as the historical situation itself unfolded and people, or their culture, came to recognize the benefits agriculture could bring. A discontinuity is posited between agricultural and nonagricultural behaviors. Attention is focused on the history of agriculture, and it becomes irrelevant to ask why agriculture did not appear under certain conditions. If agriculture did not appear, it is because the stresses requiring adaptation or the historical developments causing invention were not present.

One effect of this situation is a certain circularity in our discussion of agriculture. How do we know where to look for the conditions that created agriculture? We look to the places where it has already arisen. This would not be so bad except that all the cases discussed are known only from archaeological remains. Intentionality itself clearly leaves no evidence, and the ecological conditions creating situations that require adaptation leave at best an ambiguous record. The investigation of our model is complicated by the belief that agriculture arose in a historical setting and that the forces causing transitions have disappeared with the rise of agricultural behavior.

Far more significant is the problem that it is necessary to multiply causes to account for the origin and subsequent development of each different "type" of subsistence behavior. Why does the intent-to-agriculture occur in some cultures and not in others? Why does adaptation take one form in one environment and different forms in other, comparable environments? Agriculture occurs today in areas in which it surely did not arise. Thus we cannot claim that the environment, in some sense, prohibited agriculture as an adaptation. If agriculture was an adaptation, are we to assume that other, nonagricultural forms of behavior were likewise adaptations? If so, how do we judge the relative adaptiveness of differing cultural adaptations? Clearly, agriculture was preceded by a nonagricultural adaptation. Why did the culture change? It changed because the ecological situation, the stresses placed upon the society, changed. How do we know the stresses changed? Because the culture changed its subsistence pattern. Why did the culture change in one direction and not another? Because it was adapting to particular stresses in the environment. How can we identify these stresses? We look to what the adaptation was, and then we may identify the stresses for which the adaptation was appropriate. How can we say that the response was appropriate? Because culture is a human means of adaptation, cultures that do not adapt cease to exist: they become extinct. Only the adaptive cultures remain.[1]

Thus every cultural adaptation can be seen only in terms of the specific ecological conditions that caused it. Our attention is directed away from processes shared by cultures. The ex post facto judgment of adaptation can interfere with the recognition of general forces influencing cultural change. These general forces, unlike purely historical interactions, are still working and can be studied in the field even today. Of course, adaptations do exist, and they are a legitimate and important field of inquiry. But we cannot let the judgment of adaptiveness allow us to ignore other nonadaptive aspects of culture, nor can we stop with adaptiveness in our analyses.

The case of nonagricultural societies is in many ways analogous to the problems presented by weeds. Weeds cannot be considered products of man's intention, nor as adaptive within agricultural systems. If we are to explain their existence, we are forced to do so separately from our explanation of cultigens. Man either invented agriculture or culture adapted by utilizing it; weeds then took advantage of the new ecological niche people had created. It thus becomes necessary to hold that weeds are in some manner fundamentally different from cultivated plants: not only are weeds judged different by people, but they evolved by means of a different process.

If we view agriculture as an invention or adaptation, then we must accept that this invention or adaptation was directed only toward useful plants. Within this view, we must deal with weeds as opportunistic plants that might be described as parasites on agricultural systems. Yet if agriculture arose by means of a process that was acting with equal intensity on both crop plants and weeds (that is, if the process behind the domestication of plants and the origin of agriculture was not capable of discriminating between these two types of plants), it would be

---

[1]At this point a bit of social Darwinism could easily creep into the argument; the culture that spreads most is the most "adapted."

very difficult to call that process either adaptive or intentional. If domesticated plants arose by means of an interactive process between humans and plants, it is absurd to speak of either intent or adaptation. We might equally well describe the evolution of domesticated plants by saying that the plants chose humans to protect and disseminate them, or that the plants adapted by using humans to increase their own fitness. Would we then have to discover the ecological imbalances in the demographic structure of ancient plant populations that forced the plants to adapt in this way?

The problem of the nondomestication of certain plants presents a challenge to adaptive and intentionalistic argument that is similar to the one presented by weeds. If certain plants were not domesticated, there was either no need or no intention to domesticate them. This is not very enlightening. It gives us no understanding of *why* the plants remained undomesticated, whereas in other situations domestication did occur. As in the case of weeds, our attention has been diverted from people's interaction with all the plants in their environment to focus on the cultivated plant alone. Again it might be instructive to reverse our point of view and ask why certain plants did not choose humans to start agricultural systems, or why certain plants did not adapt to their environment by using agriculture to increase their fitness. Although such a reversal seems silly, it nevertheless does point out that people are not the only organisms benefiting from agricultural systems, and it can allow us to begin to understand that nondomestication and domestication do not have to be mutually exclusive options. The same dynamic may underlie both.

Acorns have been used as both a staple source of food and as a major alternative source of food in many cultures throughout the world. Although the acorn-based cultures of the west coast of North America are the best known, acorns have also been extensively utilized for human food in both Europe and the Near East. Indeed, it is likely that the acorn has been utilized wherever it is present. It is a large, conspicuous food source that needs only minor processing (such as leaching) before cooking. In certain species of oak, the acorns are palatable without any processing. In many ways the oak seems ideally suited to form one of the basic crops of an agricultural civilization. The fact that it is a tree is not necessarily significant; olives, various tree fruits, such as apples, pears, cherries, and persimmons, numerous palms, breadfruits, hickories, pecans, walnuts, and the ramon nut have all contributed to agricultural systems without being herbaceous plants.

The acorn, however, has never been recognized as having evolved into a cultivated form. If an intentionalist process—adaptive or inventive—underlies the origin of the domesticated plant, how are we to account for this? I find it improbable that the preconditions required for an adaptive origin of agriculture have never arisen in the context of an acorn-based subsistence. We might imagine that the oak was capable of providing a sufficiently large harvest that the conditions of stress favoring the development of domestication would not have arisen. However, extensive stands of wild wheat in the Near East are quite as productive as many of the older varieties of wheat (Harlan and Zohary 1966; Harlan 1967), and this did not prevent the domestication of wheat.

From another viewpoint, however, we may conjecture that definite biological reasons may underlie the nondomestication of the acorn. Oaks had already established highly evolved dispersal relationships with other animals by the time people arrived in the areas where oaks grow. Squirrels not only harvest acorns, but also "plant" them. We can call the planting behavior of squirrels "hoarding" and note it as merely incidental, but as far as the propagation of the oak is concerned the distinction is meaningless. The squirrel has an agricultural relationship with oaks, and fitness has been maximized in the oak phenotype by means of adaptations to this behavior. It would be extremely difficult for humans to upset the squirrel–acorn relationship.

It might be claimed that the appropriate mutation has not occurred that would permit humans to establish an agricultural relationship with the oak. Although this is possible, I am hard pressed to imagine any sort of morphological change that would exclude the squirrel from feeding upon acorns. Human activities would have to act as a strong selective force to permit the divergence of the original oak gene pool into two groups—the original squirrel-adapted taxon and the human-adapted taxon.

Accepting the primacy of appropriate mutation in the plant would create further problems for any intentionalistic model for the origin of domesticated plants. There is a conflict between claiming that humans chose to utilize a form of subsistence based on cultivated plants, and the position that domestication must be based on the appearance of appropriate mutation in plants: humans interact with exclusively wild plants only until a beneficial mutation appears. Yet these mutations must frequently be fairly extensive (for example, in many grains, simultaneous germination plus indehiscence) before they offer advantage. Also we must devise a means of "storing" mutated plants until such time as appropriate agricultural techniques arise. The evolution of cultivated plants becomes, at least during the early stages, a situation in which humans must wait until advantageous mutations appear in plants. Oddly enough, by trying to maintain intentionality we *reduce* the importance of the creative role (in an evolutionary sense) that humans have played in the evolution of cultivated plants.

The same type of analysis briefly sketched above for oaks can and should be extended to many other plants, both domesticated and nondomesticated. This kind of approach must look to all the relationships between plants and animals in the particular environment, giving attention to the actions of people and how these interact with the biology of the plants on which humans feed. We must begin to understand how people have acted as an evolutionary selective force. For example, an analysis of the nondomestication of wild rice will have to consider the dispersal activities of songbirds that feed on the grain before it ripens fully and before it falls from inflorescence, and of waterbirds that largely feed on the ripened grain that has fallen to the bottom of the pond. Such an analysis will have to consider the seed biology of wild rice: if wild rice dries out totally after it is ripe, it will not germinate. Then the analysis will have to consider the various ways in which people harvested wild rice: were these methods conducive to the establishment of domesticated relationships? If so, can we identify morphological traits in the plant that are results of these practices? Again, useful information on the domestication of crops might be found by looking to the nondomestication of timber trees. Only in the very recent past have humans begun to domesticate timber crops. It is probably significant that the part of the plant used by humans is the vegetative growth, which usually has no potential for reproduction. Therefore humans have been only a consumer or predator of these plants. There is little chance that humans' actions would increase the fitness of certain morphotypes in the plant population.

Further biological information indicating that crops might have evolved through natural, unintentional processes can be found in the genetics of plants. We tend to assume that any distinctive morphotype of a cultivated plant is, like the cultivated plant itself, the product of intentional selection and recombination of traits from an ancestral gene pool by man. In many cases this obviously has been true, especially in the recent past, but the situation is not so simple as might be imagined at first. For example, a South American race of maize, *kculli,* is characterized, among other traits, by a remarkably distinct set of color characteristics in the ear. The innermost part of each kernel is white and floury; outside this is a purple and/or brownish layer, and the outermost layer is a deep, cherry red. Although the details of the genetics involved are somewhat complex (Mangelsdorf 1974:114–115), they clearly indicate that this combination of characteristics could not be selected *as a group* by combining the various traits from a more generalized population of maize. The phenotypic effects seen in this race are not caused directly by genes, but instead arise from the interaction of several genes. Because none of the genes by itself will produce the phenotypic effect, the complex of traits characteristic of this race cannot have arisen by means of simple selection. The combination of genes had to be made before the phenotype could express itself; only then could simple selection have served to rogue the population of off-types when they occur.

Hybridization has long been recognized as a major factor in the evolution of cultivated plants. A historical example is the origin of the Corn Belt dent varieties of maize, which arose through hybridization of the northern flints and the southern dents when they were carried to and grown in proximity to each other during the settlement of the Middle West of the United States. Much the same process has occurred in the development of the many varieties of maize that have been grown in the New World, even during pre-Columbian times (Mangelsdorf

1974:121–122). Hybridization has also been proved for the origin of our bread wheats in ancient times. Even today, the effects of hybridization cannot be predicted. Plant breeders can only bring gene pools together and select desirable offspring. Thus the role of intentionality in the production of hybrid crops has, justifiably, been ignored. Yet although we can accept the natural hybrid development of cultivated crops, we still tend to believe that the progenitors of those hybrids had to enter into their relationships with people as a result of human intention.

Another puzzle created by domesticated plants for any intentionalistic concept of plant domestication arises from the observation that certain cultivated plants seem to have been domesticated outside the region to which they were indigenous. Of course it is frequently difficult to be sure that the location of nativity for a particular crop has been correctly determined; determination of the area to which the progenitor of the crop was native is also extremely difficult. Nevertheless, at least a few cases exist in which it appears that domestication proceeded in an area removed from the ancestral home of the species. For example, the sunflower (*Helianthus annuus*) is almost certainly a native of the western region of the United States and provided an important source of wild food to the early inhabitants of that region. "It has been postulated that, in time, the sunflower became a camp-following weed and was introduced from the western to the central United States. Somewhere in the latter area the sunflower appears to have been domesticated and, as a domesticated plant, was carried both eastward and to the southwest" (Heiser 1973:37). The tomato is apparently descended from a wild species of *Lycopersicon* indigenous to the Andes, yet within the Andes there is no evidence that it was known as a domesticated fruit. Instead, all the evidence, linguistic, biochemical, and historical, points to the domestication of the tomato in Central America and Mexico (see Rick 1950, 1958, 1978). A similar history of migration of an ancestral form followed by domestication is given by Heiser (1969:39) for the chile pepper: "In all probability [the ancestral form] migrated from South America, either by natural means or as a weed along with man, and reached Meso-America at an early date where in one or several places it became the special object of man's attentions and was domesticated." Hymowitz (1972) gives an example of the introduction of a wild bean species into India during historical times and its subsequent domestication there. He uses the term *trans-domestication* for this process.

This lack of coincidence of areas of origin and domestication seems very unlikely if domestication was intentional or adaptive. There seems to be no reason domestication should not have taken place where the plant was originally found. Also, we must explain the movement of the plant in an undomesticated form from the region of nativity to the region of domestication.

Most students of the problem of agricultural origins would probably agree that one obvious way in which the undomesticated crop can move from one area to another is by becoming what is generally called a *"weedy camp-follower."* Yet this notion stands in contradiction to the assumption that consciousness was required on the part of people to bring about the origin of agricultural plants. In the literature, there has generally been a slight uneasiness about the role of weediness and the origin of the domesticated plant. On the one hand is the desire to recognize the naturalness of the origin of agriculture, and on the other hand is the need to maintain the paradigm in which people have a role in *consciously* choosing the plants that were to form an agricultural subsistence.

The same processes that lead to the rise of domesticated plants can be used to understand the nondevelopment of domesticated plants in certain regions and the development of domesticated plants in regions outside their original homes. The interaction of humans and plants cannot be removed from the biological environment. The sexual systems of the plants, the relationships of plants to animals other than people, and the need for isolation to occur in a gene pool before domestication becomes morphologically expressed are all significant in the evolution of domesticated plants. The weedy camp-follower is only an early cultivated plant. Most, if not all, ancient crop plants began their evolutionary history by establishing a relationship with humans in which people dispersed and protected the plants. In this relationship, both humans and plants were maximizing their fitness relative to lineages not involved in the relationship.

Although intentionality may be used as a shorthand, or literary convention, for describing certain types of changes, it tends to be taken as literal truth. To use adaptation as an explanation for cultural change is teleological and ultimately requires consciousness to ensure that the adaptations taken by a culture are indeed ''appropriate'' to the situation at hand. Thus most cultural–ecological models must be seen as Lamarckian in effect, if not in intent. Finally, intentionality as a mechanism begs the question of what stimuli and forces in the environment were significant in fostering the formation of new approaches toward the interaction of humans and the plants around them. Thus our attention is diverted from the detailed study of the ecosystem of which people are a functioning part. As I have tried to show, much of the information given by the plants is incompatible with or at least difficult to explain in an intentionalistic or adaptive framework.

I have considered the effects of an intentional decision on only two phases of a plant's life cycle—pollination and germination. Analogous arguments could well be made for other phases of the life cycle, such as seed dispersal or requirements for vegetative growth. It seems to me significant that cultivated plants exhibit all the major methods of sexual and asexual reproduction known in wild plants. It is only the fact of their domestication and presence in agricultural systems that allows us to differentiate them from wild plants.

When I claim that intentionality cannot be considered causative in the development of agriculture, I do not deny that people, like other animals, act.

There is a persistent confusion between intentionality in terms of immediate acts (and their motivating stimuli) and intentionality as the recognition of the long-term (or ultimate) effects of those acts. I do not deny that people feed, or that they would feed in the most straightforward manner possible. I doubt that humans, or any other animal, would choose a difficult or inefficient mode of subsistence over easier options open to them. This does not mean that people consciously choose the most efficient means of subsistence in any environment; they can choose only among the available options in terms of already existing modes of behavior. We may, thus, accept that people chose the best, easiest, or most attractive option available to them—but this type of historical reconstruction of human behavior must be done with great care. It is extremely easy to fall into error by attributing to people knowledge that they could not have had.

The ultimate consequence of any form of behavior is extremely difficult to predict. Existing modes of behavior may have outcomes that change the situation in which decision making occurs, thus making new choices possible. But to claim that the behavioral choice was made to permit new choices in a new environment implies a precognition of the long-term results of behavioral choices. If we claim that a change in behavior was chosen by people to ameliorate an existing problem, then we must be prepared to prove that the change in behavior would have had an immediate and perceptible effect. Such immediacy in effectiveness cannot be assumed in the development of pristine agricultural systems. Thus, intentionality as the "recognition of the long-term effects of behavior" must be abandoned in our study of the origin of agriculture. To deny intentionality, of course, is not to deny consciousness, I am not claiming that people are incapable of reflection but only that reflection and consciousness are incapable of causing the initiation of cultural changes such as agriculture.

The greatest practical problem with the use of intentionalistic adaptive models for the origin of agriculture is that they severely limit the number of possible explanations we may advance. If humans and their culture are acting in an intentionally adaptive manner, then this presumption colors our reconstruction and interpretation of the development of agricultural systems. It becomes inconceivable that people responded in an inappropriate or unadaptive manner when agriculture was being initiated, even if the data would appear to demand this interpretation. We begin by assuming that people *had* to act in a certain manner, and in the process we limit our own understanding of how people *may* have acted. The attribution of conscious adaptation by people or their culture has seriously impeded our understanding of how agriculture arose. Agriculture is not necessarily an adaptive form of behavior, if by *adaptation* is meant any more than merely what an animal does. Behaviors that merely increase the possible numbers of a species are not adaptations, per se. Agricultural modes of subsistence are highly complicated systems, some aspects of which are adaptive in that they permit continued survival of the species, whereas other aspects may threaten the survival of many members of the species.

It seems eminently reasonable that, beside the origin of agriculture as a historical process based on a materialistic phenomenon, there was also an origin for the *consciousness* of agriculture. But the latter is the recognition of an already existing mode of behavior. It is obvious that at some point agricultural humans became aware of being involved in a mode of subsistence that differed from that of other cultures, or from that which might be imagined as possible. Consciousness is the recognition of the long-term effects of behavior, but consciousness is reflective, not predictive. It is the recognition of "what we are." Although it is tempting to believe that by knowing what we are we shall know what we are to be, it is a useless form of conjecture without the knowledge of the forces controlling change.

Finally, a distinction must be made between intentionality as an explanatory device and intentionality as a literary convention. It is common in the ecological literature to use intentionalistic terms as a means of summarizing evidence. Thus we read that "organism X adapted to the increasing cold by developing a thicker coat of hair." Clearly, no one interprets this to mean that organism X decided to grow more hair. It merely summarizes in a conventional form the understanding that (1) increasing cold decreased the adaptiveness of certain progenitors of organism X and they died off; (2) progenitors with thicker coats were better adapted to cold and therefore more fit than other members of the taxon; (3) thus organism X came to have a thicker coat than its predecessors. Whether intentionality is being used as a literary convention or as an explanatory device generally can be determined from the context.

To suspend belief in intentionality as a causative mechanism is not to deny consciousness; rather, it focuses our attention on determining how much of the human way of life is biologically comprehensible. If it also earns us some new insights into problems such as agricultural productivity and instability, it may prove to have been worthwhile.

Animal–Plant Relationships

The best way to demonstrate the naturalness of domestication is to point out that it is found in widely separated taxa. From this it can be inferred that it arose independently in different groups and that its cause must be found in some process shared by these groups. I look to evolution, specifically coevolution, as the unifying concept that allows the general argument of the coevolution of domesticates to be extended to the specific case involving humans.

Coevolution is an evolutionary process in which the establishment of a symbiotic relationship between organisms, increasing the fitness of all involved, brings about changes in the traits of the organisms. Janzen (1980:611) characterizes it as "an evolutionary change in a trait in the individuals of one population in response to a trait of the individuals of a second population, followed by an evolutionary response by the second population to the change in the first.'' Coevolutionary sequences frequently may be described as cooperation, but they do not depend on any recognition by the organisms of the advantages involved. Most interspecific forms of coevolution began with a situation in which one organism was the prey of the other.

Coevolution is not a cause of agriculture in the same sense that religion, sexual division of labor, or even the well-known ''dump-heap'' theories (Anderson 1969) are postulated to be causes. It is a cause only in that it is prerequisite to the development of agricultural systems. Thus the description of the origin of agriculture presented here should not be taken as an ''evolutionary determinism.'' Evolution is not deterministic but opportunistic. The evolution of reptiles did not lead inevitably to mammals and birds; rather, these groups separated and evolved in tandem. The ancestor of the three groups most closely resembled a reptile, but calling it a reptile is merely convenient and in no way indicates that modern mammals evolved from creatures identical to modern reptiles. The study of evolution involves the reconstruction of unique events and is done only after the events have occurred. The potentials offered to a society by coevolutionary interactions with plants need not be exploited, but if they are the systems that evolve will show certain resemblances rooted in the forces controlling their common history.

Agriculture is not a particular adaptation to the environment, but a type of animal–plant relationship. The various forms of agricultural behavior are the elaborations of certain traits in people, each of which originally conferred fitness on an individual organism and, through reproduction, became visible in a population. It is not necessary to prove that a specific cultural behavior is a genetic trait transmitted solely through sexual reproduction; cultural traits are capable of transmission through learning. Yet, as is discussed later, this different mode of transmission does not free us from the consideration of differential fitness.

Thus I define *agriculture* as environmental manipulations within the context of the human coevolutionary relationship with plants. Although not perfect, this definition points to the two most important aspects of agriculture: that it is at least partially culturally transmitted and that it involves a certain type of relationship between people and their immediate environment, including the plants that populate that environment.

For the purpose of this volume, I have found it useful to make a clear distinction between domestication and agricultural origins. Although it cannot be claimed that the two processes are fundamentally different, it is useful to note that in the study of man and his relationships to the greater environment there are several foci of concern. One is the origin of certain types of behavior in people— in this case those behaviors called *agricultural*. This is the fundamental concern behind the investigation of agricultural origins in its most strict sense. Then there are the related questions concerning the development and spread of agricultural behavior through time. I treat these as a subset of agricultural origins. Finally, there is the question of the evolution of the plants used in agricultural systems. I separate this out as the distinct phenomenon of *domestication*.

The real distinction present in this separation is that of object of inquiry. In domestication studies we are concerned with the plant and with its relationship to human behaviors and to agricultural systems. We are primarily concerned with the evolution of a plant in a particular type of environment. Changes in human behavior and the effect they have on the environment and thus for further domestication underlie the origin of agriculture. Although interrelated, the two problems can, I think, be treated as distinct, Domestication, the human–plant relationship, has generally been confused with agriculture, the highly developed form of the symbiosis. Most authors have seen agriculture as characterized by such traits as protection, harvesting, and sowing, or by the development of morphologically distinguishable cultivars, or by cultivation activities such as weeding and plowing. By any reasonable judgment, it is unjustifiable to say that cultivated plants arose only in relationship to humans, for all these traits are to be found in other animal–plant relationships.[2]

Protection

The relationship of division of labor and protection of resources to the origin of agriculture and cultivated plants was initially developed by Ames and Sauer, who stressed the significance of taboo and other forms of protection behavior. Ames (1939) developed the concept of *horticulture* to describe an early stage of agriculture in which individual plants, notably trees, were domesticated by selective preservation of the plants. During the horticultural stage, people were able to experiment with plants, and these experiments eventually yielded agriculture. Ames argued that horticulture differed fundamentally from agriculture: in horticulture people cultivate individual plants, whereas in agriculture numerous plants are cultivated for their collective yield. The importance of this distinction is that it distinguishes evolution via simple protection and the more advanced type of evolution that requires mass selection and the development of techniques of environmental manipulation. Ames placed considerable emphasis on the interaction of protection modes of horticulture with the food taboos and agricultural rituals that many societies have developed. His arguments appear naive today, but his fundamental position is still sound: the protection of such reliable food resources as trees by taboos could be of advantage to a society and could ultimately affect the evolution of crop plants.

[2]The following presentation can only serve as a brief survey of a large literature. For an introduction and further examples and discussions of the ecological and evolutionary principles involved, see Ridley (1930) and van der Pijl (1972).

Protection symbiosis has developed frequently in nature. One of the most complicated elaborations of this particular coevolutionary system can be found in the myrmecophilous plants of the tropics. Here we find a highly developed mutualistic relationship between ants and plants, in which the plant provides domicile, food, or both for the ant. In "return," the ant protects the plant.

Hartzell (1967) has reported on an ant–acacia relationship in which the ants inhabit hollow enlarged thorns on the acacia and receive a sugary exudate from special foliar nectaries at the base of the leaves. They also harvest modified leaf tips, called *Beltram bodies*, which are rich in both proteins and lipids and are used to feed the ant larvae. The ants patrol the foliage and remove and feed upon herbivorous insects. This is so effective that when ants were experimentally removed from acacias, the plants were severely attacked and all died within a year. Other acacias are protected by their ants from herbivore grazing (Hocking 1975); *any* animal disturbing the plant is very painfully bitten (personal experience, 1978).

The loss of chemical defense mechanisms within species of acacia involved in a mutualistic relationship with ants is apparently common (Rehr *et al.*, 1973). In an analogous fashion, many cultivated plants have lost chemical defense systems: with humans acting as protective agents, the chemicals are no longer required for the survival of the plant. This phenomenon could frequently have the effect of increasing the vigor of the plant. Many protective substances are highly complex, "expensively produced" compounds. Thus their loss *could* allow for diverting some of the energy used for their manufacture into greater vegetative or reproductive vigor. Another effect of the loss of these compounds would be an increasing palatability of the plant to humans. Thus the loss of the defense mechanism could in these cases be seen as a way of "attracting" people.

Numerous examples of this process could be noted. The wild eggplant was highly bitter (Choudhury 1976:80). The bitterness of wild *Cucurbita* species is the reason that Whitaker and Bemis (1975) postulate an original domestication of these fruits for their seed crop. Lettuce domestication has been accompanied by both a reduction of spininess and a decrease in the latex content (Ryder 1976:39). Wild cabbage can contain as much as four times the quantity of bitter and dangerous glucosimates as cultivated strains (Josefsson, cited in Thompson 1976). Work conducted by Pimentel (1961) indicates that the increase in palatability of cultivated cucurbits is likely implicated in high levels of predation upon the plants artifically grown in wild areas. Cassava (*Manihot esculenta*) has two forms, one highly poisonous when raw, the other lacking in poisons; the presence or loss of toxins is likely correlated, at least in part, with methods of cultivation in different agricultural systems (Rogers and Appan 1973, Rogers 1965). Yet another example is the relationship between *Cucumis humifructus* and the aardvark. *Cucumis humifructus* is a cucurbit closely related to the cultivated muskmelon that grows in desertlike regions of southern Africa. After pollination of the flowers, the flower stalk elongates and buries the developing ovary in the soil, where the fruit comes to maturity. The mature fruit is large and juicy, with numerous seeds. The primary and perhaps only dispersal agent for this plant is the aardvark. The plants are commonly found growing near the tunnels inhabited by aardvarks. Meeuse (1958) has shown that this melon is the primary water source for aardvarks during the dry season and that the seeds of the plant germinate poorly unless pretreated by the digestive tract of the aardvark. In this symbiosis, the aardvark not only consumes the fruit, but also plants the seed; because the aardvarks' toilet habit are like those of the domestic cat, *C. humifructus* seeds are planted in fertilized "hills." The advantage in this relationship for the aardvark is the freedom to avoid waterholes and the predation attendant upon frequenting them. Particularly interesting for our purposes is the observation by Richard Robinson (personal communication, 1981) that *C. humifructus* is the only "noncultivated" cucurbit he knows of that lacks the numerous bitter substances characterizing the wild species of the Cucurbitaceae. Seen in terms of this relationship between plant and animal, the assumption of Whitaker and Bemis (1975:364) that "primitive man in harvesting the fruit of *Cucurbita* for seeds found fruit whose flesh lacked the bitter principle, and commenced selection which later led to the cultivated species" appears in a slightly different light.

The ant–acacia relationship described above can be very complex. Beside the direct protection that the ants give to the acacia by preventing or diminishing grazing, the ants may also indirectly protect the acacia by removing competing vegetation from around the bushes. This can place the acacias at a distinct advantage, and localized pure stands of ant–acacias (as these plants are known) are often produced by the weeding activity of ants (Janzen 1970). Another indirect protection that this gives the acacias is protection from the brush fires which are characteristic of the savannas to which they are native (Janzen 1967).

The establishment of a symbiosis between ants and acacias probably also permitted the acacia to enter a new ecological zone. Species of *Acacia* closely related to the ant–acacia of Central America are restricted to dry regions. As Stebbins (1974:48) points out, "Apparently, most species of *Acacia* pay the price for alkaloid production in terms of slower growth rate, which enables them to compete only in the drier regions where growth of all plants is slower and less vigorous." Having dispensed with the necessity for alkaloid production, it was possible for the ant-acacia to compete successfully in regions of more luxuriant plant growth. Thus the establishment of a symbiosis may indirectly permit a plant to enter and grow in a region that earlier would have been unfavorable to it. A close relationship exists between domestication and potential changes in distributions of plants and thus of their coevolved agents. Changes in the morphology and physiology of the domesticate may permit new opportunities to arise for both members of the mutualism.

Protection of the acacia by the ants is clearly advantageous to both plant and animal. The relationship, it might reasonably be conjectured, might have arisen out of a predation connection. We might imagine that the ants were originally predators either on the acacia itself or on its insect hervibores. If the ants were purely herbivores on the acacias, those plants producing "expendable" tissue would suffer less from the depredations of the ants. More likely, the ants were also preying on insect herbivores that were feeding on the acacia. In this case, the acacias producing either Beltram bodies or their precursors would have an advantage in being able to maintain a resident force of protection against possible insect herbivores; the acacia would "store" ants as a defense. It is reasonable to assume that the ants were relatively successful in reducing the numbers of their prey to the point at which the acacia could no longer support the ant population, which would have to leave the bush and forage elsewhere. Those bushes providing some sort of sustenance, such as precursors of nectaries or Beltram bodies, for the ants during periods of low insect population would not be abandoned in this manner and would again be protected when the insect population increased. A similar argument might be made for the evolution of specialized ant domiciles.

This whole process depends upon increases in the fitness of each species. The ant does not really "protect" in "return" for food—it simply feeds. The acacias evolve *in the presence of* the ant's activities, not as a *response* to the ant's activities. The acacia whose morphology encourages the presence of ants produces more offspring than related acacias lacking that morphology.[3] This does not mean that if we were to observe the relationship evolving we would necessarily be aware of the "desirability" of the relationship at each of its evolutionary stages. We might, for example, describe a certain ant as being a predator of the acacia if we were to observe the relationship at an early stage in which the Beltram body was not yet morphologically differentiated. Likewise, we might not appreciate the significance of the foliar nectary and would probably describe it as a "nonadaptive" morphological organ of the acacia. At a still earlier stage, we would simply note that the ants fed on insects that were present on the acacia. It is only the presence of a highly coevolved system that directs our attention to the role that the ant plays in the protection of the acacia. Yet at all earlier stages an overall increase in fitness was occurring that would eventually yield a striking example of mutual adaptation.

In the coevolution of ants and acacias, it is not necessary to postulate that the original acacia gene pool became extinct, leaving the successfully evolved ant-acacias as its only descendents. This may have happened, especially over

[3]Evolution need not stop with the ants and acacias. Hocking (1975) points out that there is an unusual mantid that lives on acacia bushes and feeds on ants. This particular mantid shows a remarkable mimicry of the domiciles produced by the acacia for the ant's benefit. Clearly, by mimicking the home of its prey, the mantid must have its own subsistence reasonably well assured.

long periods of time, yet it is just as likely that the possibilities exploited by certain acacias resulted in a sympatric differentiation of the original population into two distinct groups. Dispersal of the ant-acacia into new areas might have been permitted by the adaptiveness provided by the ants; thus some of the population could have become geographically isolated and allopatrically speciated. Alternatively, ancestral and ant-acacia populations could coexist sympatrically, but the hybrid offspring could be extremely unfit by lacking the adaptations of either parental type.[4]

It is difficult enough to understand the evolution of plant morphology; the understanding of the evolution of behavior is even more taxing. For example, how does one explain the evolution of weeding behavior by ants? It might have developed out of foraging by the ants in the vicinity of their residence, but how do we get to the point at which the ants remove *all* rather than *some* of the vegetation? We might claim that the ants were already clearing vegetation from around their colonies before they established a relationship with the acacias, but this merely displaces the problem into the past. We must also confront the question whether the weeding behavior is genetically or culturally determined. This question lends itself to answer by study of the mode of colony fission in the ants. It would also permit some experimental analysis.

It is tempting to assume that the ants, at some point, "invented" weeding behavior—that a "behavioral mutation" occurred that increased the fitness of colonies practicing the behavior relative to the fitness of colonies not practicing it. For example, the colonies that were not weeders might have been more likely to die off in a brush fire. This is again a conceptual possibility, but it is also possible that brush fires do not have this particular effect. Also, it is logically troubling to rely upon mass extinctions from a natural cause for the establishment of one form of behavior over another: if a new form of behavior were capable of permitting survival of the group and the lack of that behavior caused extinction, how are we to account for the survival of the group before the new behavior arose? Moreover (leaving aside for the moment the vexing problem of what constitutes an "individual" in an ant colony), we may interpret the survival of one group over another as only the statistical results of individual selection—it is not in fact one group that is being chosen as more fit over another, but one individual over another, the statistical results of which appear to be the choice of groups. The appearance of group selections in this context would then be an artifact that appeared when we made the taxonomic error of assuming that the groups really represent entities instead of being constructs of the taxonomist.

[4]Furthermore, the relationship with the ants is only one of a number of other coexisting and coevolving relationships; relationships with pollinators will also be significant in this context. The logical possibilities are legion, and only a study of the *particular* relationship under consideration can resolve the problems.

In any case, this group model does not really provide an understanding of why the behavioral mutation might have appeared. Yet because of inadequacies in our present understanding of cultural change it may be necessary to posit such cultural mutation to permit cultural change. It may be necessary to describe selection by means of selection of groups to understand why one particular type of cultural behavior might have spread. For the moment, let it be understood that this is a literary convention analogous to the use of intentionalistic terms to summarize an argument based on relative fitness.

Storage and Planting

Many scholars have used the hoarding of seeds and their subsequent planting to define agriculture. Seed saving is widespread throughout the animal kingdom and, as it relates to plant dispersal, can be of major significance in the evolutionary history of plants.

Various birds, including nutcrackers, jays, and woodpeckers, are known to store seeds of pine, beech, oak, chestnut, filbert, and various *Prunus* species. Vander Wall and Balda (1977) have provided a good example of the relationship between storage and planting in their study of seed storage by birds. Nutcrackers store the seeds of piñon pines for winter use on sheltered, south-facing slopes. These sites are ideal for retrieval of the seeds because they do not accumulate deep snow covers and they are good sites for germination of the seed and growth of the pine. In mild winters, when alternative food sources are available, as little as one-third of the stored seeds are consumed. Small mammals, such as mice, squirrels, and chipmunks, also store edible seeds in their nests and in special caches (van der Pijl 1972; Stebbins 1971). Pikas store a hay made of dried grasses and weeds in caches. These hay stores are guarded, a phenomenon also common in various bird stores (Emlen 1973:164). Janzen (1971) notes that seeds of stored species can be found growing to adult status in the nitrogen-rich soils near nest entrances. It can be assumed that these seeds were ''lost'' or ''rejected.'' These plants clearly provide a useful resource for the inhabitant of the nest.

Gunda (1981) has recently given a fascinating summary of the exploitation of these animal caches by humans. He notes that the nest of numerous species of animals in Hungary and other regions of the world have been utilized as sources of seeds, nuts, and tubers by the local residents. This pattern of utilization of animal caches is widespread and at times may even develop into a symbiosis between man and the gathering animal.

> A most efficacious symbiosis has developed between man and a mouse variety among the Itel'mens of Kamchatka, who go so far as to feed the mice in order to keep them alive and make them gather *sarana* tubers (*Lilium martagon*) in sufficient quantity. (Gunda 1981:82) . . . In Siberia it was a common practice to rob mice of their winter stores, this being a relatively easy way of obtaining a fair amount of roots. This was done by

[numerous Arctic tribes]. . . . It was commonly held that a few roots should be left for the mice themselves, and some Eskimos replaced what they took with fish in order that the mice might survive through the winter. (Eidlitz quoted in Gunda, B., Comment to Rindos 1980, *Current Anthropology*, p. 82. Published by The University of Chicago Press. Copyright © 1981 by Wenner-Gren Foundation for Anthropological Research.

Although the interaction of humans and animals is outside the scope of this volume, it is fascinating to note that existing symbioses between animals and plants may become the basis for animal–human symbioses. Also, knowledge of storage gained from the observation of animal caches might have been important as man began to develop symbiotic dispersal relationships with specific plants.

More interesting for our purposes here is the phenomenon of ant-mediated seed dispersal (*myrmechochory*). Harvesting ants have established a symbiosis with plants from well over 225 genera in 63 families of angiosperms (Nesom 1981); they collect and thus often disseminate the seeds of these plants. The ants are "apparently able to control the termination of the seeds they collect and also exert considerable influence on the vegetation in the vicinity of the nest" (Hocking 1975:83).

Harvester ants have also developed quite sophisticated forms of behavior to protect their seed stores. At least some species of ants will remove wet seeds from the nest and dry them in the sun. Many species of the ants "thresh" the seeds they gather, and mounds of chaff may be found in the vicinity of the nest. It appears that both ant and plant benefit by this association (Hartzell 1967:127):

> The ant is assured a food supply during cold periods and drought. The plant species is aided in its distribution and it is protected against excessive drought and cold by being stored in the underground galleries of the ants. Since the ants normally collect their food supply in excess of their requirements, this is an important factor in the survival of the plant species during climatic changes that would normally destroy them above ground.

Stebbins notes that myrmechochory is characterized by a whole series of morphological modifications in the plant:

> In comparison to their relatives . . . [these] plants have the following characteristics: the flower stalks are relatively low, and the peduncle becomes recurved when the seed is ripe so that the mature capsule is close to the ground. The capsule . . . does not dehisce regularly by valves or pores, but irregularly and over a relatively long period of time. . . . This makes possible repeated visits by the ants to the same plants. The seeds themselves have several modifications. Sometimes they have a special oily seed coat; the ants remove this after they have carried the seeds to their nests, and they they discard the stripped seeds. More often, each seed has a particular fat-bearing appendage, or elaisome, which the ant clips off with its mandibles and carries underground, after it has transported the seeds to the vicinity of the nest. (Stebbins 1971:243; reproduced, with permission, from the *Annual Review of Ecology and Systematics*, Volume 2. © 1971 by *Annual Reviews* Inc.)

Transportation of the seeds to the vicinity of the nest can allow for the establishment of fairly large populations of the plant in an area where it most benefits the ants. Like the establishment of food plants in the vicinity of mammalian nests, this form of ant dispersal bears a certain similarity to the "dump-heap" model for the origin of agriculture (see Anderson 1956, 1960, 1969).

Of far greater interest for an understanding of agricultural origins is the recognition that myrmechochory may allow for the survival of otherwise un-adapted organisms. Handel (1978) has studied the competitive relationships of three closely related species of sedges (*Carex*) that grow in the northeastern United States. Both fieldwork and extensive experimental investigation have shown that the one ant-dispersed species is a relatively poor competitor. It has survived because it does not have to compete directly with the other species; it is dispersed by nesting ants into a habitat (rotting logs) unavailable to the other species.

Handel (1978) also points out that the ant-dispersed sedge behaves like a typical weedy colonizing species. This species has an adaptive strategy that involved the evolution of means for rapid introduction into gaps that may appear in the environment. Most weedy colonizing plants are not climax species; they are usually displaced rather rapidly by others. For the sedge studied by Handel, ant dispersal is the means by which this rapid colonization of disturbed areas could occur. The logs provide a refuge for the ant-sedge, at least until the log rots away, allowing for colonization by the more aggressive sedges. In a sense, this particular weedy colonizing species is using the ant as its dispersal agent in a race to keep ahead of the other, more competitive species of *Carex*. The ant-dispersed species must move into new habitats as they become available, for once the log has decomposed the plant will be outstripped by its relatives.

As part of the syndrome of a weedy colonizing species, the ant-dispersed sedge has evolved several other traits. It has, on the average, twice the seed set of its relatives. Because the species is always on the move, the necessity of produc-ing large numbers of propagules should be obvious. It is also typified by immedi-ate germination of the seeds; the other two species of sedge studied require overwintering (vernalization) before they germinate. This is also probably an adaptation to greater seed production and thus the maximizes the total seed crop during the relatively short life-span of the disturbed habitat. The weedy coloniz-ing species has to devote a great deal of energy to the production of seeds; only by maximizing the potential for dispersal can this type of plant survive.

The similarities of the ant-dispersed colonizing species and the cultivated plant are rather striking, but they are not unexpected if we recognize that they are the results of a common process. The problems confronted by the cultivated plant during the early phases of domestication are essentially identical with those of any other plant maximizing its potential as a weedy colonizing species. Total crop of seeds must be maximized to allow for the transport of the plant to a newly opened habitat. This crop is augmented both by an absolute increase in yield and by such devices as precocious germination, flowering, and ripening. But the plant, if it is to survive, must be dispersed to the favorable, newly disturbed habitat. Many weedy plants do this by the production of large numbers of seeds that are dispersed passively by an agent, such as the wind. Dandelions are a perfect example of a plant using this tactic. Yet passive dispersal means the transport of many seeds to such unfavorable locations as forests, established fields, or the ocean. Another dispersal syndrome can maximize fitness by the "high-quality" transport of limited numbers of seeds to favorable habitats without the losses that passive dispersal entail. This is the tactic taken by the ant-sedge; it has tied its morphology to the behavior of an animal that transports the plant to the requisite habitat. The only "cost" to the plant has been the production of the seed appendage that is used to attract the ants. The same sort of phenomenon has occurred during the domestication of plants by humans. Finally, the ant-dispersed sedge, like the cultivated plant, is a poor competitor relative to other plants. This does not mean that the plant is therefore "unadapted." The plant has evolved in relationship to its dispersal agent; both the ant-dispersed and the human-dispersed plants are *consistently* and *predictably* dispersed to habitats in which this lack of competitiveness is irrelevant.

Cultivation

Planting, harvesting, seed storage, and the protection of stored seeds characterize certain relationships of animals to plants. Morphological modification of the "cultivated" plant is not restricted to the effects of man's interaction with plants. A fascinating range of insects has established symbiotic cultivation relationships with fungi. More than 40 species of ambrosia beetles, as well as some wood wasps, cultivate fungus as sources of food. These insects bore into, but do not directly eat, wood and other plant tissue; they inoculate the tissue with small amounts of fungus, usually carried to the host plant in specialized organs called *mycetangia*. Mycetangia are derived from different organs in various groups of ambrosia beetles, and we may thus assume that the habitat of fungus cultivation is polyphyletic in origin. In all cases, fungus serves as the chief source of food for all the developmental stages of the insect.

The fungi that are cultivated come from many different genera but can be broadly classified into two groups. Some of the ambrosial fungi are known outside of the symbiosis, but the majority are highly specialized and are not known to exist independently (J. M. Baker 1963). They are generally hard to classify because they do not have a true sexual cycle, but are adapted to and dispersed by their respective insect vectors.[5]

[5]It would be hard to hold that the lack of a sexual cycle is necessarily a result of the association; there are numerous fungi not associated with insects and that possess no sexual stage—the so-called *fungi imperfecti*. Of course, these fungi do not possess morphological adaptations, such as the kohlrabi bodies, to ant cultivation.

The fungus will not grow in the host plant unless the insect is present. The insect, in fact, maintains pure cultures of its associated fungi by removing any competing foreign fungus. If the insect is experimentally removed, the site is usually taken over by alien growth. The insects also frequently prepare special beds of wood chips and feces to enhance the growth of the fungi (Francke-Grosman 1967). Growing these fungi under laboratory conditions has proved to be extremely difficult. Complex media that include vitamins, lipids, and amino acids are required. It appears that many of the fungi are dependent upon insect secretions for maintenance of growth characteristics in the ambrosial relationship. Under laboratory culture, the fungus loses the growth pattern and morphology that are present in the ambrosial state.

Ants and termites have also developed sophisticated cultivation symbioses with fungi. For example, certain species of tropical termites maintain fungus gardens in their nests. Unlike most termites, these species lack the intestinal microorganisms necessary for the digestion of cellulose: their carefully nurtured fungus provides their food supply (Trager 1970). It is again among ants that we find the most highly developed cultivation analogues (Hartzell 1967; Hocking 1975; Trager 1970; Weber 1966, 1972; Wheeler 1973). Cultivator ants prepare special beds—generally of plant debris, cut-up leaves, flowers, and excrement—in special chambers in the ant nest. The ants are meticulous about growth conditions within the chamber; numerous ventilation passages are dug and these are opened or closed to regulate both temperature and humidity.

To construct the beds, the ants chew the substrate material to make a pulpy mass and deposit it in layers in the chamber. The bed is then planted with propagules from previously maintained beds. Constant care is given; the ants remove alien fungi and add anal and salivary secretions that apparently have a positive effect on the growth of the fungi. These cultivation activities encourage the production by the fungus of small whitish round bodies, the so-called kohlrabi bodies, which are the principal food of the ant colony. While under the care of the ants the fungus undergoes no sexual stage. If the ants are removed, the reproductive mushroom may appear. More frequently, however, the effect of removal of the ants is contamination of the beds by alien fungi and other organisms to the quick exclusion of the ant-fungus.

Relationships between the ants and the general ecology are also quite complex. The ants show remarkable care in the collection of a large variety of leaves from which they construct the substrate (Rockwood 1976). This large variety of plant leaves from differing plants may be necessary to optimal growth of the fungus on the substrate. The ants also apparently practice conservation of resources—they do not completely defoliate those plants providing the most valued sources of leaves for manufacture of the substrate (Cherret 1968). The success of this tactic is permitted by the fact that these ants are strictly territorial, giving them complete control over the utilization pattern for local resources—perhaps another parallel with human agriculture.

The evolution of agricultural behavior in the various insects considered here occurs by changes in the morphology of the coevolved plants and changes in the behavior of the animal. As in the ant-acacias, the plant has evolved in the presence of the ant's activities. But more provocative questions are posed by the evolution of the agricultural behavior itself. One theory holds that the ants originally lived in rotted wood and gradually acquired the habit of cultivating fungi that were found to inhabit their environment. Another theory considers the ants originally to have been harvest ants that learned to cultivate the fungi found infecting their grain stores. These conjectures are not very interesting because they only provide an environment in which the evolution of cultivating behavior might have occurred. They do not try to elucidate the particular behavior patterns that might have preceded the agricultural way of life. The relationship between domestication and agricultural origins is not clear at this point, even for animals other than people.

It would be highly desirable to present an analysis of the evolution of agricultural behavior in ants, based on our concept of fitness. Unfortunately, it is difficult to give a Darwian description of the appearance of the new behavior in the ant, behavior that makes sense only in light of the associated plant and its cultivation. Perhaps building the beds might originally have been related to housekeeping activities of the ants. Those colonies not practicing such sanitary measures might have been more susceptible to disease. The appearance of food on these dump heaps might have prevented the removal of the heaps from the colony or, alternatively, would have given a relative advantage to colonies not removing such a food source. This conjecture might be supported by further investigations of the ant's domestic ecology. Perhaps the planting and transport behavior might have arisen by the fungi evolving the kohlrabi bodies in imitation of larval ants. These "cryptic mushrooms" would not be consumed until they reached a critical size and could thus have been protected and transferred with the colonies during migrations. Again, as with human agricultural systems, certain experimental approaches could be useful. The rather weak arguments presented here for the origin of ant-agricultural systems points out again the need for more fieldwork and experimental investigation into the general problem of the evolution of agricultural behavior.

Perhaps it is not coincidental that highly evolved types of cultivation symbiosis have taken place in populations of social animals such as ants, termits, and people. Perhaps an understanding of the evolution of agricultural behavior is predicated upon a more complete analysis of the impressive set of intraspecific coadaptations that we call *social behavior*. It is important to recognize that the appearance and evolution of agricultural behavior in the ants could have arisen only by means of a natural evolutionary process. If the similarities of human and animal coevolutionary sequences are not coincidental, it would be sufficient reasoning for assuming that the development of cultivation behavior in people might also have arisen by means of a natural coevolutionary process.

Coevolution and Dispersal

Coevolution in plant–animal relationships is determined by various types of selective pressures, among them protection, pollination, dispersal, storage, or even simple predation.

Mediation by dispersal pressure has been fairly well explored in the ecological literature; it is also closely related to our primary interest, the evolution of domesticates. Many of our domesticated plants are cultivated for their propagules.[6] We might note that the morphological changes generally held to be typical of the domesticated state are most frequently found in the morphology of the propagule. Thus it is not unreasonable to assume that certain characteristics of the propagules might be of significance in understanding domestication. Further, agriculture itself has been dispersed widely throughout the world. Thus we might assume that a certain adaptation to easy dispersal by man might well be expected in a domesticated crop. Crops that are relatively difficult to disperse, grow, or cultivate will probably remain in their area of initial domestication. Finally, dispersal lends itself fairly easily to mechanistic interpretation.

Unlike protection, cultivation, or planting, dispersal is a fairly straightforward concept. It has been reasonably well investigated in the field (see Stebbins 1971), and it places emphasis on understanding the relationship between the plant and the animal. A discussion of various modes of dispersal can, I believe, show that the cultivated plant exhibits a highly specialized type of adaptation—adaptation to the conditions of cultivation—and that the adaptations in the cultivated plant act as a selective force that increases the fitness of the cultivating animal. By looking at coevolution in dispersal terms, I hope to establish that non-Darwinian concepts are unnecessary to account for the morphological and autecological changes that evolve under domestication. In focusing on dispersal-based coevolution, however, I do not mean to suggest that dispersal is the only or even the most important kind of relationship in the development of the domesti-

---

[6]Propagules are of two types: sexual and asexual. The first are composed of the seed(s), and any associated accessory structures; the second are vegetative structures, such as tubers, corms, and bulbs. In evolutionary terms, the important distinction between them is that sexual propagules will produce plants that differ from both parents to the extent to which cross-fertilization and combination of traits has taken place, whereas asexual propagules perfectly reproduce the characteristics of the mother plant.

cated plant. It is merely a starting place—one that seems logical to me, yet one that may eventually prove less important than other coevolutionary relationships in domestication.

Discussing the significance of dispersal in the evolution of plants, Janzen (1969, 1970) points out that there are numerous situations in which it seems logical to assume that advantage accrues to plants that disperse their propagules beyond the area occupied by the parental plant. Besides contributing to the success of the taxon as measured by range occupied, such dispersals enhance the prospect of survival for the propagules. It is obvious that the parental plant is, in a sense, a reservoir of pathogens and predators that affect the species under consideration. Although the adult plant may be able to survive despite disease and predation, the young plant, if for no other reason than its smaller size, is likely to be killed or severely weakened by the presence of disease or by even limited predation. Dispersal also prevents the offspring from competing directly with the parental plant.

The very existence of coevolved dispersal schemes indicates that, from the plant's point of view, fitness may be maximized by taking advantage of such mechanisms. Dispersal agents, like any other element in the environment, may be in relatively short supply. Were there always an excess of effective dispersal agents, plants that produced propagules especially attractive to particular dispersal agents would not thereby gain in fitness. If dispersal agents on the whole are in short supply, then there is a quantitative aspect to dispersal systems, as seen in the requirement that the plant attract enough of certain types of agents to survive.

Different potential dispersal agents will vary in the way they affect the dispersed propagule. It is obvious that a small bird eating a fruit will have a different effect on the seeds contained in that fruit than will an elephant. Various animals will be either predators or dispersal agents for the plant, depending in part upon the type of physical and chemical treatment they give a propagule. Thus not all potential dispersal agents will be capable of increasing the fitness of a plant on whose propagules they may feed. In the same way, not all agents will have behavior patterns of potential advantage for the plant; there will be a difference in the distance and in the environments to which they may transport the propagule. More complicated behavior patterns, such as hoarding, obviously will be of great potential significance for an increase in the plant's fitness. Because of the various types of treatment of plant propagules, it can be said that there is a qualitative aspect to plant dispersals.

It may be argued that, unlike insect attack or pollination, the treatment of a propagule is without direct feedback to the plant and thus has limited potential to affect a species's fitness. In evolutionary terms this is not true. Seeds have a more advantageous dispersal agent, in either qualitative or quantitative terms, will be selected for and will thus preferentially spread those aspects of the phenotype of the parental taxon that are favorable to interaction with the dispersal agent. Thus it is not necessary to posit the existence of a ''real'' high-quality dispersal agent. To describe an agent as giving high-quality dispersal is only a shorthand way of indicating a certain type of relationship in which a particular mutual advantage is found at the level of the individual organism. The basis for a high-quality dispersal relationship is merely the ex post facto recognition that the plant's fitness has been increased and that it has been increased along either quantitative or qualitative lines: the plant has found either more dispersal agents or a better dispersal agent. As I have said, the immediate outcome of such increases in fitness is not necessarily a concomitant loss in the fitness of the ancestral taxon. Indeed, at the earliest stages of the development of the coevolutionary relationship it would be very hard to recognize that an evolutionary change was occurring because only a small portion of the entire population would be slightly more effectively dispersed by its early coevolved agents.

Historically, dispersal agents are basically predators (Janzen 1971; C. C. Smith 1975). Coevolutionary dispersal relationships may arise when an agent feeds upon a propagule; they also frequently arise when animals store or hoard propagules (Woodmausse 1977). The establishment of coevolutionary dispersal relationships involves various compromises for the plant, which obviously must produce a propagule that is attractive to the dispersal agent. The propagule will have to be adapted to—or, more accurately, not be destroyed by—the dispersal agent. Because agents may vary in the effectiveness with which they disperse the propagule, the plant may increase fitness by compromising on the number of dispersal agents and by evolving a relationship with a few high-quality agents. Yet, because the agent is basically a predator, it would be difficult to establish a coevolutionary dispersal relationship without providing some benefits to the animal. This benefit, frequently food, attracts the agent to the propagule.[7] Thus it is commonly found that the plant offers some of the propagules as bait for the dispersal agent. Although some of the propagules (or parts of a given propagule) may be destroyed, dispersal is nevertheless accomplished; hence the plant arrives at a compromise between attraction and dispersal.

Animals involved in coevolutionary dispersal relationships must, of course, be able to utilize the propagule as a source of energy and must have morphological adaptations or behavioral patterns that permit them to do so without contributing to the extinction of the plant species. In other words, the animal must contribute to increased fitness of the plant. This relationship can develop a true mutualism in which plant and animal are obligately tied to each other for their survival. It is at this extreme that we most easily recognize the highly coevolved dispersal relationship, but in reality there is only a continuum based on slight increases in fitness over successive generations.

---

[7]Burs and other passive use of the animal for dispersal are coevolution in only the most broad sense—I would rather see them as the plant ''exploiting'' the morphology or behavior of the animal that is present in its environment.

Two classes of dispersal agents, based on their characteristics in relation to coevolutionary dispersals, can be identified: *Opportunistic* agents utilize the propagule but are not specialized to or dependent upon it. For example, birds may be seen as opportunistic dispersal agents for many species of soft fruits in the temperate zone. Although birds may eat blueberries, and thus disperse them, they do not require the plant for their survival at existing densities. The birds also show no particular morphological modification that can be considered adaptations solely to blueberry feeding. *Specialized* agents, in contrast, have developed a close relationship with a given plant species and depend upon it for at least seasonal subsistence. The generalized predator readily able to make substitutions in its subsistence base in response to changes in relative abundance or availability, may have specialized relationships with certain plant species. For example, collecting and storage behavior by ants is necessary for the survival of ant-dispersed plants, and the plant shows special morphological modifications that are adaptations to the ant's behavior. Nevertheless, these ants also collect seeds that are *not* specifically adapted to ant dispersal. In contrast, the obligate predator's subsistence is on relatively few sources of energy. For the agricultural ants, which have restricted energy sources, a decrease in their food supply may cause decrease in numbers or necessitate a change in their feeding patterns if this is a genetic or behavioral possibility. Just as generalized predators may have specialized dispersal relationships with one or more plants, obligate predators may opportunistically disperse other plants. The distinctions between opportunistic and specialized dispersal agents and between generalized and obligate predators are merely abstractions from nature; the vast majority of animals are far more flexible in behavior than these distinctions suggest. Nevertheless, the distinctions are useful for understanding certain coevolutionary processes.

McKey (1975) points to two extremes of adaptation to the problems presented by seed dispersal: (1) production of large numbers of propagules that are dispersed helter-skelter by large numbers of opportunistic agents and (2) coevolution with certain specialized agents. In the latter, the quality of dispersal will be higher; because the species of agents are fewer, they are more predictable. Whereas opportunistically dispersed propagules are dispersed by large numbers of unrelated generalized predators with differing modes of behavior, propagules disseminated by specialized agents are generally dispersed by only one or a few agents whose behavior will be more constant; thus there can be selection to maximize the fitness of plants to the habitats to which the animals relatively uniformly disperse them. Human-mediated plant dispersals are a good example of this behavioral constancy. On the whole, the environment of the garden is highly predictable for the plant: competition is rather unimportant; soil fertility is rather high, allowing for the expression of exceptionally vigorous genotypes (Darlington 1973); and certain early cultures show remarkably consistent patterns of transhumance. It seems obvious that man can offer high-quality forms of dispersal.

Besides the geographic dispersal pattern, the chemical and physical treatment of the propagule has been found to be more constant and predictable when the plant is tied to a specialized agent, again because there are fewer species of dispersal agents involved and the plant will be selected to optimize the survival of the propagules during and after treatment by the particular dispersal agent. Opportunistically dispersed seeds have to accept a compromise in seed morphology. Seed coats will be treated differently in the guts of different agents and will have to be thick enough not to be destroyed by the average one; many propagules will therefore fail to be dispersed. Plants dispersed by specialized agents, however, do not have these high predispersal losses. Propagules that fail to be dispersed by the specialized agent but are dispersed instead by gravity, for example, will *not* place a contravening pressure on this selective program—for had this contravening pressure been successful, for example, by overcoming dispersal problems created by proximity to the parental plant, the coevolutionary episode would never have developed and we would not be discussing it. We have to remember that fitness is being maximized at each moment; specialized dispersal schemes are not "goals" of any evolutionary system.

As we have seen, the establishment of a specialized relationship with a dispersal agent presupposes certain morphological modifications in the propagule. Because obligate predators get their energy from a limited number of sources, plants on which they feed must be able to provide the energy they require. Thus, whereas most opportunistically dispersed fruits are high in carbohydrates and water, the cheapest nutrients for the plant to produce and store, the fruits eaten by obligate fruit-feeding tropical birds are usually high in proteins and lipids. These nutrients are expensive for the plant to produce and generalize predators get them from insects and other animal foods. Producing the large number of propagules required for opportunistic dispersal allows for only the cheaply produced high-carbohydrate fruit. The high-quality dispersals provided by the specialized agent permit (and, in the case of obligate predators, demand) the production of high-quality food. This, of course, further concentrates the energy of the agent on the plant. A syndrome develops in opportunistic and specialized dispersals that causes feedbacks in the evolution of both plant and agent.

It appears that the generalized agent would do well to feed on the fruits produced for specialized dispersal agents, but this does not often occur in nature because fruits dispersed by obligate predators are usually morphologically modified to prevent opportunistic dispersal. The most common kind of exclusion is physical—the specialized propagules are too large for animals other than the obligate predators to swallow. Many fruit-eating birds of the tropics have specialized guts that are adapted to the handling of large seeds. Other specialized plants produce fruits with intoxicating seeds that are ground up in the gizzards of normal birds but pass undamaged through the digestive tract of the specialized agent. Thus the plant protects itself from being dispersed by low-quality agents.

Certain traits of cultivated plants with a symbiotic relationship with humans may have originally arisen, and are certainly in part maintained, as means to exclude nonhuman dispersal agents. The large fruits of the cultivated apple, mango, pear, watermelon, or peach effectively exclude most dispersal agents other than people. Of course birds may peck at apples or stone fruits, but they are generally incapable of carrying off the fruit and actually dispersing it. The indehiscent rachis of the small grains and the cob of maize serve an analogous function. Thus certain traits of the cultivated plant not only benefit people, but also serve the plant in its dispersals. The major evolutionary effect of this type of specialization is an intensification of the symbiotic relationship between plant and animal.

McKey (1975) notes that the obligate predator is a more reliable visitor to its specialized plant than a generalized predator could ever be. This is another reason a higher quality of dispersal is effected. Unlike a strictly opportunistic agent, the specialized agent is less likely to desert the plant for other food sources when relative abundance of the plant foods changes. Besides the nutritional tie, there is the behavioral tie that is the animal's necessary contribution to the coevolution; without it the symbiosis could not develop.

Because the evolution of the plant and animal have been tied together, there is a potential danger to any plant that establishes a specialized exclusive relationship with a given animal species. If it is dispersed, or pollinated, by one and only one agent, extinction of that agent is likely to be followed by the extinction of the plant (assuming, of course, that no other effective agent is available). This has apparently occurred in nature. Temple (1977) has reported that the extinction of the dodo (*Raphus cucullatus*) has caused the near extinction of a tree on whose fruits it fed and for which it likely acted as a specialized dispersal agent. The fruits of the *Calvaria major* are large, and their seeds have extremely hard seed coats that require mechanical abrasion for germination. Because the dodo was apparently the only, or, at least, the overwhelmingly important agent feeding upon these trees, the hard seed coat evolved as the optimum one. The extinction of the bird left the *Calvaria* in the unfortunate position of producing seeds that cannot be processed; hence there are no young *Calvaria* trees. It is fortunate for the *Calvaria* gene pool that the individual trees are long-lived; thirteen are still living, all of which are sufficiently old to have germinated before the extinction of the dodo.[8] It is striking that not one *Calvaria* tree has germinated since the

[8]Owadally (1979) has contested Temple's contention that *Calvaria* was obligately dispersed by the dodo, pointing to an interpretation of the age distribution of surviving trees that indicates at least some trees germinated after the bird went extinct. Even if this is true (tropical trees lacking growth rings are exceptionally difficult to date), obligate dispersal need not imply that an *occasional* tree might appear unaided by pretreatment in the dodo's gizzard—some variability is inevitable in seed coat thickness. Other phenomena, such as accidental abrasion on the ground, might also be responsible. Biology seldom allows us to use the word *never*. Obligate relationships are tied to the continued survival of the dispersed species, and until the relationship between abrasion and germination was

dodo became extinct. There is some hope that *Calvaria major,* unlike the dodo, may yet be saved from extinction; experimental abrasion of the seed coats, a process that stimulates the treatment of the seed in the dodo's gizzard, has proved efficicious and the first seedlings are now growing.

Species adapted to specialized dispersal will generally have long fruiting seasons. Several reasons exist for this, the relative scarcity of the dispersal agent being of major significance; a system evolves that permits the specialized agents a relatively stable subsistence base. Also, production of the fruits for a relatively long period of time by means of sequential ripening, with small quantities of fruit available at any one moment, discourages feeding on the plant by generalized agents. Conversely, mass presentation of the fruit crop encourages opportunistic dispersal by agents that tend to feed upon the largest single resource base existing at any given moment. On the whole, opportunistically dispersed species will tend to sequence their ripening periods, whereas species dispersed by specialized agents will each tend to maintain a long fruiting season, and the various species fed upon by the agent will have overlapping ripening periods.

This relationship need not hold if the animal dispersing the plants is capable of storing part of the total crop for use during periods when it is not ripe on the plant. Under these conditions, a tendency exists for synchronous ripening by the various species having a specialized relationship with the agent because, by "flooding" the agent with food, a larger portion of the crop will be buried and thus be more likely to be dispersed.

Smythe (1970) has investigated the impact of hoarding of seeds by a fundamentally opportunistic agent on seasonality of ripening of trees in Panama. The agouti, the principal dispersal agent for many species in the region in which Smythe did his research, subsists for the greater portion of the year almost entirely upon seed stored during the time of abundance. Smythe recognizes two basic syndromes relating seasonality to dispersal: Type 1 involves the association of large edible seeds, active transport and caching by the agouti, and a relative simultaneity in ripening by these species; Type 2 involves the association of small seeds, ingestive dispersal with chance dissemination by numerous agents, and a tendency toward sequential production of fruit throughout the year by the various species. Why should these two types of seeds, associated as they are with differing agents, have contrasting ripening ecologies?

The sequential ripening of species having Type 2 fruits is fairly easy to understand if we recall our discussion of opportunistic dispersal. Because opportunistic agents are generally "unreliable" in their relationship with a plant, feeding upon those species presenting the greatest abundance of fruit at a given time, we can posit interspecific competition for the attraction of agents. The species yielding the most fruit at a given moment will be preferentially fed upon.

---

discovered, *Calvaria* was fast on the way to extinction: *something* previously scarified *Calvaria* seeds.

There will be little dispersal of members of other species fruiting during the peak season for the "dominant" trees. The result will be a division of the year by means of sequential ripening patterns into peak periods of fruit production by various species.

The agouti's feeding behavior is adapted to the dispersal of nut-like and plum-like propagules. The animal usually eats the fleshy part of the fruit (if any is present) and carries the seed away and buries it for later consumption. Smythe notes that after eating any edible tissue or some part of the total number of seeds in the fruit, "the remainder will be carried away individually and buried. Agoutis usually bury seeds near some structure such as the base of a palm clump or a fallen log. Later when there is a scarcity of fresh fruit, they search near such structures" (Smythe 1970:32). Type 1 fruits are dispersed in a specialized manner by a generalized predator and, unlike fruits dispersed by a specialized agent, show few morphological adaptation to dispersal. Because the agent is not tied to any one plant, it is not necessary for a given species to provide the high-quality food necessary for the survival of the agent that feeds on only one or a few species. But most important for our purposes is the relationship between hoarding by the agent and simultaneous ripening of the dispersed species. From the point of view of the plant, the agouti is both potential predator (eating the seeds) and potential distribution agent (burying the seeds). The probability of successful dispersal is directly proportional to the number of seeds buried. The probability of a seed's being buried is likewise directly proportional to the total amount of food available at a given time—a seed that falls during the period of lowest fresh food availability is likely to be eaten rather than buried. Thus those species that ripen fruit synchronously will be dispersed with a greater frequency. Here, predator satiation (providing more individuals than the predator can consume at a given moment) interacts with hoarding-dispersal and the result is synchronous ripening in agouti-dispersed species.

Smythe (1970:33) distinguishes the two systems he studied in this manner: "those seeds that can pass through the gut of an animal unharmed (type 2) benefit from fruiting as diversely through the year as possible, while seeds that are too large to be swallowed and are dispersed by hoarding rodents (type 1) benefit from fruiting as nearly synchronously as possible." Species adapted to specialized dispersal generally have long fruiting seasons (one of the reasons for their scarcity in temperate zones). As we have noted, several reasons exist for this, the relative scarcity of dispersal agents being of major significance: a system evolves that allows the specialized agent a stable source of food. Also, production of the fruits for relatively long periods of time, with only small quantities of fruit available at any one moment, discourages feeding by generalized agents and prevents the development of selective trends leading away from specialized and toward opportunistic dispersal systems. The possibility of storage of the dispersed plant's propagules changes the pressures considerably. The agent is supplying its own constancy of availability and the propagule is also protected from dispersal by other agents. When storage is combined with planting, a further pressure contributes to simultaneity in ripening within the entire community of species dispersed by the agent: those species ripening during the period when hoarding occurs will be most likely to be dispersed; those ripening in the off-season will be more likely to be consumed.

The evolution of cultivated plants came about by the development of specialized dispersal relationships between humans and numerous previously opportunistically dispersed plants. Like the agouti, humans are not tied to one species of plant for their survival and are capable of storing quantities of plant material for off-season nutrition. The development of specialized dispersal systems by man has been encouraged by the plants already in the environment, some of which were type 1 and others type 2. These types represent preadaptations to the two major types of agricultural crops: (1) staple plants, usually grains or tubers, which may be stored and which provide the bulk of calories and (frequently) protein in the diet, and (2) variety plants, such as fruits and vegetables, which tend to form a replacement series comprising various species that ripen at different seasons of the year and provide sugars, minor nutrients, and diversity in the diet.

The evolution of the cultivated plant is a response to the selective pressures existing when humans feed upon the plant. The opportunistically dispersed colonizing plant will maximize fitness by being dispersed to disturbed habitats. Those plants producing more propagules will have a greater chance of surviving predation. The dispersal by humans of the propagules of these colonizers would be of a higher quality than the average opportunistic agent *if* people were to become specialized dispersal agents for the plant. Morphological changes in the plant could offer the option to humans of becoming specialized dispersal agents. This would occur through the competitive exclusion of other agents. For example, the initially large number of propagules borne by a plant would attract both humans and other predators. A simple specialization, such as the indehiscent rachis, could exclude other agents from dispersing the crop. Significant increase in the size of the propagule could have the same effect. Once humans began to disperse a crop, its introduction into areas where there was no native dispersal agent would permit the survival of the crop only if people successfully dispersed it. The evolution of domestication and of the domesticate has been the refinement of the coevolutionary relationship between humans and plant.

Weeds and Mimicry

Independent confirmation of the importance and effectiveness of coevolved dispersal systems can be seen in the evolution of seeds that mimic the color and shape of bird-dispersed fruits and have independently evolved in several genera of flowering plants. These plants take advantage of the high-quality dispersal offered by the specialized dispersal agents, but they do not provide the reward of high-quality food offered by the coevolved plants. Like specialized avian dispersal systems, these mimetic systems are generally confined to the tropics.

The mimetic plant can be viewed as a parasite on the coevolved dispersal relationship. It can also be seen as exploiting the behavior of the bird. The evolution of a mimetic dispersal system may be imagined as follows: while the bird and the specialized plant are both maximizing their fitness by means of a developing coevolutionary sequence, certain other plants may be able to "exploit" this growing relationship, again through maximization of fitness. Having seeds that resemble the coevolved fruits—in a sense, being preadapted to mimicry—these plants will be dispersed by the high-quality agent. The agent, of course, is incapable of knowing that it is being exploited, and the mimetic plant merely evolves a closer resemblance to the specialized plant by becoming adapted to the same conditions of dispersal. The requirements for dispersal for the mimetic plant are identical with those for the specialized plant. A further requirement for the mimic is relative abundance; it is unlikely that relationship would persist if the mimetic plant were to become common relative to the specialized one (van der Pijl 1972). Probably both plants would lose their agent if it had behavioral options. If it were an obligate predator, it might become nonadaptive because of the loss of time and energy spent feeding on the mimetic plant. More likely, however, the agent would evolve greater discrimination and continue to feed selectively on the specialized plant.

The evolution of the mimetic seed is homologous to the evolution of weeds. Weeds are mimics of crop plants, at least at certain stages of the life history of the crop. The weed may vegetatively resemble the crop, or it may be inseparable from the crop at the dispersal stage. Weeds may also, of course, have their own means of dispersal, but in this case they may be seen as parasites of the environmental niche created by agricultural behavior. Weeds have evolved as inevitably as crop plants; they have taken advantage of the behavior of people for their own protection and dispersal.

The long association of humans and weeds has produced a vast number of plants that are preadapted to survival in the greater agroecology. Weeds are not merely facultative colonizers, but are highly specialized members of the agroecology. Emlen considered the success of European weeds in America in contrast to the rather poor showing of North American weeds in Europe and explained it as follows:

> Part of the answer lies in the coevolution of European and Asian species with man. Man was farming there for a long time, on a larger scale than in America. . . . These species, then, were not invading a wild America, they were invading a human environment, something to which American species had been exposed for only a few generations. The

strangers were the American, not the European and Asian species. (Emlen 1973:371–372)

Thus the introduction of a particular type of agroecology into a new area should be expected to bring with it the weeds, as well as the cultivated plants, characteristic of that agroecology.

Again, the weeds of New Zealand are a reflection of the agricultural ecologies in which they arose and in which they currently prosper. Thomson (1922) enumerated more than 600 species that were introduced and became established in New Zealand. Practically all these species are natives of Europe and are restricted in their distribution in New Zealand to zones of human disturbance. In New Zealand, as in the rest of the world, the agricultural weed (taken in its largest sense) presents little danger to the indigenous flora. These weeds are actually highly specialized denizens of the agroecology. Discussing the predilection of certain European weeds for colonizing disturbed habitats, Good (1964:38) has suggested that weedy species of a plant have been selected for long periods of time and may actually be somewhat different from the species as it exists in the wild: "If this is so, then it might be expected that these weeds would flourish in all artificial habitats no matter where they might be." The success of European weeds in America and New Zealand was not due to their being "better" weeds; rather, they were adapted to the agroecosystem that was introduced to these regions when colonized by Europeans. We cannot separate weeds from the system in which they occur. Thus we may expect different types of agriculture to be prey to different weeds, even in the same region. One of the easiest ways to see the interrelationship of system and weed is to look at changes that occur over time in weed floras.

Weed floras, like the repertory of cultivated plants, will evolve over time in response to changes in techniques of environmental manipulation. Examination of the changes that have occurred in the weed floras of the world is a neglected subject that may eventually provide significant data on the ecological effects of environmental manipulations. One summary of the changes in the weed flora of England is provided by Salisbury (1964), who notes that some 30 species of obligate weeds have been recovered from Paleolithic and Mesolithic strata in England. It is possible that these plants existed independently of man, but Salisbury holds it more likely they were introduced to England by Paleolithic man "who must often have been the unconscious agent for dispersal of many seeds and fruits in the course of his wanderings" (1964:28). During the Neolithic, numerous new species of weeds appeared in England, probably as components of the early agricultural systems of that time. New species of weeds continued to appear during the Bronze and Iron Ages, and the rate of introduction appears to have increased at least through the Roman Occupation. However, the basic weed flora was well established at quite an early date. Salisbury notes that the cornfield weed flora of [England] probably exhibited no striking qualitative changes . . . until comparatively recent times. There have, however, doubtless been marked fluctuations in quantity accompanying the vicissitudes of agriculture. . . . When the private enclosure acts, from the time of Queen Anne, and the general Enclosure Act of 1845, brought to an end the "common fields" system, the initial stages of agricultural hygiene, with respect to weed pests, became practicable. The introduction of the Drill, the mechanical winnowing and screening of the threshing machine, and the Seeds Acts, all played their part in changing the pattern of the prevalent weed flora. The advent and continual augmentation of chemical herbicides will no doubt carry this changing pattern a stage further as the reservoirs of certain dormant seeds become depleted and suscepti-ble species give place to those which are resistant, until resistant strains become selected out. (Salisbury 1964:147)

The last point made by Salisbury has already been borne out. Contrary to the prediction of Harper (1956), who felt that evolved resistance would arise very slowly, the weed flora in Great Britain has undergone a drastic change since the introduction of herbicides (Fryer and Chancellor 1970). As Parker (1977:253) notes, "the whole history of herbicide development since 1950, particularly for small-grain cereals in the United Kingdom, has been one of adjustment to chang-ing weed floras." A similar process continues worldwide wherever chemical herbicides have been used on a relatively large scale.

Human disturbance of the soil, whether agricultural or preagricultural, has been of prime importance in the evolution of weeds. As King puts it:

The disturbed habitat—with its challenge to the invasions and establishment of new types, and its destruction of many of the barriers between populations—consistently forms one of the most important environmental causes for setting intricate and diverse genetic mechanisms into operation. This results in more aggressive and successful forms—often weeds. (1966:26; reproduced by the kind permission of The Blackie Pub-lishing Group, Scotland.)

Thus it is impossible to separate the evolution of weeds from that of crops. Both originate in a similar genetic and physical environments: "[Man] has been a breeder of weeds as well as crops!" (King 1966:117). It would seem that the only way to distinguish the weed from the cultivated plant is in terms of the relationship between plant and people: crops have a symbiotic relationship with people, whereas the relationship of weeds to people is exploitative. In other terms, weediness is the ability to fill a specific niche. No plant species is, per se, a weed, and without information about the relationship it will often be impossible to classify a particular species. One of the dangers in the reconstruction of prehistoric subsistance systems is the assumption that "weed" contaminants of a grain crop were, indeed, weeds.

Nowhere is the interrelationship of weed and cultivated plant more clear than in the case of the secondary domesticate. The Vavilovian secondary domes-ticate is a plant that moves from weed to domesticated status in conjunction with its movement to new ecological zones. It is the result of the replacement of the primary crop by one of its weeds as the primary crop spread into regions that were less favorable to its own growth and reproduction. This aspect of the interactions of weeds and cultivated plants has been well discussed in the literature (see, e.g., Darlington 1973), and little more consideration will be given it here.

The same evolutionary pressures that have permitted secondary domestication also have had great effects upon weeds that are not of present utility to man. Wickler (1968) gives several examples of the evolution of weeds under cultivation. He notes that the weed *Camelina sativa* has diverged from its putative progenitor *C. glabra* by virtue of selective processes operative in flax fields. *Camelina sativa* is larger than *C. glabra* and is generally unbranched. The leaves of *C. sativa* also bear a striking resemblance to flax leaves in both shape and color. The fruit is indehiscent and requires threshing to separate the seed from the husk. Although the seeds differ from those of flax in both size and weight, the combination of these two factors is such that they cannot be separated from flax seeds by winnowing. *Camelina sativa* is, in essence, a mimic of flax at both the vegetative and dispersal phases of the life cycle. At the same time, as Bonner (1950) reports, it produces a toxin that directly inhibits surrounding flax plants by repressing germination and growth (see also Grümmer and Beyer, 1960; Lovett and Jackson, 1980: Lovett and Sagar, 1978).

In a similar manner, the agricultural techniques of man have selected for the development of strains of dodder (*Cuscuta epilinum*) whose mimicry of flax is based on morphological mimicry occurring at the dispersal stage. The weedy strains of dodder produce a double-seeded fruit that approximates the flax seed in size and weight. ''Wild'' strains of dodder can be successfully removed from the linseed crop by winnowing, but the weedy strains are dispersed with the crop. Another species of dodder, *C. epithymum,* is a parasitic European native that can be a very injurious pest in clover fields. Like *Camelina sativa,* this weed has seeds difficult to remove from those of the host plant. Recently some success has been achieved in controlling this pest by the use of separating techniques that take advantage of the rough seed coat of the dodder seed. One technique involves passing the clover seed with its dodder contamination over felt rollers to which the rough dodder seeds become attached (King 1966). The other technique calls for the addition of fine iron filings to the seed mixture: the iron lodges in the crevices of the seed coat of the dodder seed and the seed may then be removed by the use of powerful magnets (Martin and Leonard 1967). Of course these techniques may eventually become useless: we are presenting an evolutionary opportunity to dodder—only time will tell whether this plant's genome will be capable of mutating to produce a smooth-seeded form.

Mimicry by weeds of the seeds of cultivated plants is widespread and causes major problems for farmers and seedsmen. Certain weeds are so much a part of a cultivated crop that in the agronomic literature they are called *satellite weeds.* Varieties of weeds exist that morphologically mimic the variation in seed size between cultivated strains of a crop, or different species of weeds may grow in association with the various races of the crop. The information that can be drawn from these associations can be of help in reconstructing agricultural history:

> adaptation of flax weeds to the various cultivated forms of flax is so extensive that, on the basis of a precise investigation of the present-day geographical distribution of the various weed species and races, conclusion about the native habitats and routes of dispersion of the individual forms of flax can be drawn. (Schwanitz 1966:124–125)

The study of seed-mimics of crops also shows that traits such as indehiscence, large seed size, and lack of seed dormancy are merely the inevitable outgrowth of evolutionary pressures placed on plants by methods of harvest and planting.

One of the most fascinating weed developments is teosinte; it so mimics the vegetative structure of maize that native agriculturists cannot distinguish between the plants until the flowering stage. At this point, of course, the large size of the teosinte and the late stage in the growing cycle are good reasons for not bothering to remove it. Removal would be pointless or, because of root disturbance, perhaps even injurious to adjacent maize plants; thus it has adapted in such a way that people unintentionally protect it. Many believe that maize and teosinte have a common evolutionary history (Doggett 1965; Mangelsdorf 1974; Galinat 1965, 1974, 1975; Beadle 1977; Iltis and Doebley 1980). If this is the case, we may view the history of these two plants as one of divergent evolution: two paths developed in a partially shared gene pool by adaptation to different aspects of the human–plant relationship. Both teosinte and maize have taken advantage of people as protective agents. While the crop was an early domesticate, people may have dispersed both of them, but once maize became obligately dispersed by people it was dispersed without the teosinte portion of the gene pool. Thus the maize cob allowed maize to escape from its original home and diverge from its closest relative, teosinte. In the area in which teosinte was already present, however, people continued to protect it, even though the intent of the protection, at least at a later date, was directed toward a different plant.

The important thing to recognize about weed-cultivated plant interactions of this type is that the entire complex is the domesticated species. Despite the great morphological difference existing between teosinte and maize, the two plants form one interbreeding complex in the area in which they are native. Teosinte serves as a source of variability for maize and thus is a "useful" weed. Furthermore, this introgression is occurring in both directions. It is likely that certain strains of teosinte bear a closer genetic and phylogenetic relationship to certain strains of maize than to other strains of teosinte. If this is the case, taxonomic recognition of them at the level of a variety (or even a forma!) would be consistent with modern taxonomic principles.[9] Portions of the teosinte–maize gene pool may thus be seen as adapted to nonhuman dispersals.

---

[9]Iltis and Doebley (1980) have taken the first step in this direction by recognizing certain teosintes as subspecies of maize.

Another example of a dynamic interaction between weed and crop can be found in cultivated species of *Sorghum*. Harlan notes that in the Sudan, weedy sorghums have persisted in the wild only in those areas in which wheat is the major domesticate. In areas in which domesticated sorghums are grown, wild sorghums have interacted with the cultivated forms and produced complexes of weedy "shattercanes" that mimic the domesticated crops.

> The plant type [of these shattercanes] is very much like cultivated sorghums with thick stalks, wide leaves and sometimes even crooknecked peduncles. The spikelets . . . however . . . shatter by way of abscission layers as do those of wild sorghum. [Shattercane] is a serious pest not only in the Sudan, but in other African countries where sorghum is grown. Typically a weed of the fields [they] tend to mimic the cultivated races with which they grow. If the crop has loose open heads, the shattercane will have loose open heads; if the crop has dense heads, the shattercane has dense heads. (Harlan 1976:15)

It is especially interesting that in this case the mimetic weeds are apparently able to appear with great speed when agriculture is introduced into new areas. In areas where agriculture has only recently been introduced, the shattercane is already present after merely two growing seasons. Clearly, land races of sorghum in the Sudan have a large number of genes in them that permit the rapid appearance of the weedy morphoform: the species is variable in its response to the agroecology and this brings with it a selective advantage. The agroecology is composed of many niches, one of which is that of the weed. In cases such as this, the mimetic weed is clearly at an advantage over other weedy plants. By relying upon people for protection while adopting the weedy trait of self-dispersal, the weedy shattercanes, like teosinte, acquire a far greater chance of survival in the agroecology.

In the United States, cultivated sorghums have taken a different route to colonization of the weedy niche. The shattercanes of the United States have evolved without any genetic interaction with wild species of the genus. Instead, mutations have occurred that allow for new means of dehiscence that permit self-dispersal. Numerous crops have taken advantage of the weedy niche existing in the agroecology: rye, peas, oats, radishes, and turnips all have weedy forms. Unlike the cases mentioned earlier, in which mimicry is occurring at the seeding stage, the mimicry in these cases occurs during the vegetative stage.

The close resemblances between crop plants and weeds have long been recognized in the literature of agriculture origins (Hawkes 1969, Vavilov 1926); both have been seen as *weedy heliophytes*. Opportunistically dispersed generally produce large numbers of propagules to compensate for the inevitable losses associated with low-quality dispersal. The weedy heliophytes, as colonizers of a specific niche, also need to produce large numbers of propagules in order to maximize the possibility of reaching a disturbed habitat; those not dispersed to disturbed habitats will not fit. Dispersal pressure is of great significance in the evolution of the colonizer plant. Whether a plant evolves as a weed (to fill the weed niche) or as a crop will in part depend upon whether it is utilized by people.

If the plant is not adapted to human dispersal, it cannot become a cultivated plant. Both crops and weeds may originally have been colonizers of the agroecology, but their divergence in relationship to humans was largely a function of the evolution of different types of dispersal systems. Humans could not easily propagate an annual crop that had small seeds dispersed by wind from an explosively dehiscent fruit. Thus, although the colonizing habit can be viewed as a preadaptation to domestication, the dispersal strategies employed by plants may well have been of greater importance in their ultimate domestication.

Humans and Domestication

Domestication clearly cannot be held to be an exclusively human-mediated phenomenon. It remains to be shown how this coevolutionary process has worked in the case of people. Therefore, I want to consider here what little evidence we now possess concerning potentially coevolved relationships between plants and primates. I hope to show that nonhuman primate behavior, and thus human behavior on the most elementary level, is compatible with the elaboration of coevolved relationships. These relationships would have evolved in and around the camps of hunter–gatherers where natural vegetation was enriched and disturbed; here colonization by the plants on which people were feeding could occur. I will consider both discarded or ''lost'' propagules and those that might have incidentally dispersed in human feces.

Teleki presents a well-documented and highly persuasive summary of the argument that human and higher primate subsistence patterns differ, at most, only quantitatively. He holds that the various arguments for a ''unique'' subsistence pattern for the hominids ''can now be replaced by a single, much simpler model wherein subsistence shifts among both nonhuman and human primates are perceived as smooth transitions within a graded continuum of evolution'' (Teleki 1975:125). Indeed, we are now experiencing great excitement over a new model for human speciation in which hominid relationships with plant, *not* animal, sources of food were primary. Jolly (1970) has advanced the conjecture that a relationship between early hominids and small grains was significant in the divergence and further development of the hominid line. Arguing from dentition and using analogies, he makes a logically plausible case for the notion that the human relationship with the reproductive propagules of plants is ancient indeed. Dunbar (1976) expanded this work, suggesting that the fundamental plant-food niche of *Australopithecus* likely resembled that of modern baboons, with an emphasis on seeds, fruits and subterranean storage organs. Other workers have contributed to this model for early hominid diet by noting that tooth wear and damage patterns are not those to be expected from a predominantly meat-eating species (Wallace 1972, 1975; Wolpoff 1973; Zihlman 1978; Zihlman and Tanner 1978) but more like those of a herbivore (Crompton and Hiiemäe 1969). As Zihlman and Tanner put it:

> We have stressed gathering plant foods as the critical innovation in human evolution, one that logically emerged from ape behaviors such as tool using and food sharing. . . . We believe that the picture of hunting emerging from the very beginning of human origins with meat as a major food source is no longer supported by the evidence. (1978:184)

Similarity between ape and human dietary patterns can be quite striking. After considering some of the foods consumed by chimpanzees, Teleki notes: "Chimpanzee collector–predator subsistence compares favorably with the range of food types utilized in Bushman hunter–gatherer subsistence, and even the technological innovations for collecting insects are precisely duplicated in some human populations in Africa and South America" (1975:130–131). Moreover,

> many human food-getting techniques, and especially those used in collecting wild plants and small land fauna, do not differ substantially from those performed by other primates. At what point, then, can we draw a meaningful line between collector–predator and gatherer–hunter patterns? Combination of the lower size limit in large prey killed by humans (c. 15 kg) and the upper size limit in small prey killed by baboons and chimpanzees (c. 10 kg) yields a range of 10–15 kg as a marker for this subsistence shift from predator to hunter. . . . But we must not, at the same time, disregard the fact that much of the game taken by human hunters and by nonhuman primate predators falls within the same size range, and that monkeys, apes and men are all omnivores practicing basically similar subsistence patterns overlaid by some differences in habits, skills and techniques. (Teleki 1975:159)

Similar observations reinforcing the importance of parallels between the omnivorous human and chimpanzee subsistence patterns have been given by Teleki (1973), McGrew (1975), Tanner and Zihlman (1976), Clutton-Brock (1977), Gaulin and Konner (1977), Hladik (1977, 1981), Zihlman (1978) and Harding and Teleki (1980).

Peters and O'Brien (1981) have recently summarized the literature on hominid dietary patterns and have considered in detail the similarities and overlap in plant resources consumed by *Homo, Pan,* and *Papio* in eastern and southern Africa. In their analysis of this one flora, they found that competition among the three genera is greatest for fruits and that significant potential competition would have existed for leaves and shoots as well as seed pods with chimpanzees and for underground storage organs with baboons. They conjecture that the fundamental plant-food niche for the early hominids would likely contain leaves and/or fruits as the most common food items. Although the question of niche overlap and thus potential competition between these genera is somewhat problematic (see Boaz 1977, and the comments published in Peters and O'Brien 1981), no question can exist but that many occasional and staple resources are shared by these three genera.

Similarities among primates, including humans, are also present in the "cultural" aspects of subsistence. Sharing behavior and sexual division of hunting and gathering activities show remarkable similarities in different genera (Galdikas and Teleki 1981; Goodall 1971; Harding 1975; Suzuki 1975; Teleki 1975). Teleki (1975:157) notes that "films on gathering and hunting activities among the !Kung Bushmen demonstrate that numerous parallels exist in the subsistence techniques employed by chimpanzees and bushmen, from the methods of collecting berries and eggs to ways of acquiring and killing small land fauna." Summarizing the changes in our attitude toward the subsistence of other primates, Harding (1975:256) says that it now seems, "less likely that predatory behavior was a major shaping force in human evolution. On the contrary, the characteristic human diet, a mixture of meat and vegetal foods, is probably a continuation of an ancient primate pattern." In a comparison of the diets of chimpanzees and humans, Hladik strongly supports the same view and points out that geographic variation in diet and even in modes of subsistence may be found within the range of modern chimpanzees.

In large part because of these studies of primate diet and behavior, the big-game hunter paradigm for the evolution of *Homo* has rather precipitously fallen from favor in recent years. No longer do we see the early human diet as being qualitatively different from the general primate diet. Human sexual "specializations," likewise, are no longer viewed in the same light. In place of the earlier, rather sexist, view of "man" the hunter as cause for the evolution of the species, we now see human patterns in division of labor described in a less stereotyped manner (Gale 1970; Harding and Teleki 1980; Tanaka 1976; Zihlman 1978; Zihlman and Tanner 1978), and we are beginning to appreciate that the complicated patterns evidenced in human subsistence patterns, like the diet itself, are based upon a fundamental primate model (Galdikas and Teleki 1981; Gaulin and Konner 1977; Lancaster 1975, 1976; McGrew 1975; Silk 1978; Tanner and Zihlman 1976).

The basic primate diet is characteristic for hunter–gatherers today; most being fundamentally vegetarian. According to Woodburn (1968), the !Kung receive over two-thirds of their calories and protein from vegetable foods, whereas the diet of the Hadza is over 80% vegetable. Lee (1968) notes that although the !Kung and the !Ko consume at least slightly higher amounts of meat per capita than monkeys do, this meat comes from large game that is supplied as an irregular supplement to a basic diet of small game and plant foods. Unfortunately, little seems to be known about just how much potentially coevolutionary interaction these groups have with plants. More attention must be given to how plants respond to such primitive groups—consideration of this problem so far has been dominated by the question of how humans respond to plants, and most researchers have been content merely to ennumerate a list of utilized species. The value of studies of groups like the !Kung for the study of domestication would be in the information they might provide about the *types* of interactions that humans are capable of establishing with plants in their environment. We should be concerned with understanding *how* people interact with the plants in their environment: the /Kade and !Kung may be hunter–gatherers because of limitations placed by the environment or even by the plants themselves on the development of symbioses with plants.

Although primate-mediated dispersal systems have received only limited attention to date, they do occur and coevolutionary relationships are likely present. An example may be found in the relationship between the fever tree and the vervet monkey. For example, Struhsaker (1967) reports that the fever tree (*Acacia xanthophloea*) is important in providing a sleeping place and refuge from predation for the vervet monkey, as well as serving as its staple food. The monkeys consume the young vegetative parts, the flowers, and both green and ripe pods of the acacia. Struhsaker implies that the pods are retained, at least for a while, on the tree where the monkeys can feed on them without strong competition. Brenan reports (1959:108) that, not surprisingly, the fruit of this tree is indehiscent. This allows the moneky to feed on the tree under conditions of reduced competition because it is capable of shucking the pods. Indehiscence, or delayed dehiscence, of the fruit encourage dispersals by the monkey in several ways. It discourages consumption by ground-feeding agents. Indehiscent pods retained on the tree also serve to attract the monekys to the trees: indehiscence and retention have an advertising function in addition to the direct adaptations favoring monkey consumption. This same sort of indehiscence has occurred in several woody legumes utilized by people, including the carob (*Ceratonia siliqua*), and the tamarind (*Tamarindus indica*).

The relationship between the vervet and the fever tree can profitably be compared to that between the hunting-and-gathering /Kade-San and the legume *Bauhinia micrantha*. According to Tanaka (1976), the pods are transported to camp before being shucked, even though 75% of the weight of the pod is composed of inedible tissue. The morphology of dehiscence of this species, however, is strikingly different from that of the fever tree because the pods are not indehiscent. Thus they may be profitably gathered only before the seeds are entirely ripe. Tanaka notes that when the pods dry and turn brown they open, scattering the seeds upon the ground, and are therefore no longer easily utilizable as a food source. The relationship of the people to this plant is one of strict predation. If humans were sufficiently numerous to interfere with the natural dispersal of the plant by consuming most of the unripe seeds, we would predict that over numerous generations the plant would be driven out of those areas inhabited by people. Indeed, this may have occurred with another species of *Bauhinia* in the area in which the /Kade-San live. Tanaka reports that *B. esculenta* is preferred to *B. micrantha*, but it is seldom gathered because it is only found at great distances from the traditional camping grounds. Either an incompatibility exists between the areas in which people prefer to camp and the ecological conditions required by *B. esculenta*, or humans have exterminated the species from the areas in which they normally camp. Whereas the vervet monkey has developed a domesticatory relationship with its major legume staple, the /Kade-San have not succeeded in doing so with theirs. The difficulty in establishing a domesticatory relationship with *Bauhinia* species may have acted to limit the potential population size of the /Kade-San and may also have affected the degree of sedentism they practice.

Besides feeding on the fever tree, according to Struhsaker (1967), the vervet eats the basal portions of various grasses and the leaves, flowers, and bark of several genera of dicots. It could be reasonably assumed that this is basically a predatory interaction. Like the feeding of the /Kade-San upon *Bauhinia*, there is little possibility that a coevolutionary relationship will develop. Vervets also feed on the ripe fruits of several herbs and shrubs; a dispersal function would be possible here, and we might expect appropriate coevolutionary adaptations by both plant and animal.

Another type of primate–tree interaction has been reported from the Canal Zone of Panama. Oppenheimer and Lang (1969) write that *Cebus* monkeys exert considerable influence over the growth pattern and reproductive behavior of *Gustavia* trees. By consuming the terminal buds of the branches, they remove the inhibition effect of the terminal bud on the lateral buds. Branching is increased, and thus the number of terminal buds available for food in later years is also increased. This feeding behavior also has an effect on the tree's fruiting pattern. The *Gustavia* tree bears its flowers terminally, so increase in the number of branches yields an increase in the number of sites at which fruiting may occur. Because the monkeys also feed on the large fruits of *Gustavia*, feeding on the terminal buds increases both the number of future terminal buds and the potential number of fruits. Oppenheimer and Lang feel that under higher population pressure than currently exists, feeding on the branch tips could interfere with fruit production. This is indeed possible, but it is not the only possibility. If the monkeys act as a dispersal agent for the *Gustavia* trees, any increase in the monkey population would be accompanied by a higher dispersal rate. Depending upon the growth and dispersal rates of the two organisms and the number of sites available to the *Gustavia* for colonization, an increase in the total available fruit and bud supply could be achieved by the greater distribution potential of the larger monkey population. Thus a positive correlation could develop between the size of the animal population and its effectiveness as a dispersal agent. Higher population levels could produce higher total food resources.

At times the dispersal activities of monkeys have influenced the vegetation in specific areas, causing large concentrations of preferred foods to grow in restricted areas. Jackson and Gartlan (1965) studied the flora and fauna of Lolui Island in Lake Victoria, Uganda, that was once inhabited and cleared by humans, but was vacated in 1905. During the intervening years, cleared areas were recolonized by grasslands and woody thickets. The latter are the result of the feeding and dispersal activities of vervet monkeys that spread in their feces the seeds from the fruits upon which they feed. Jackson and Gartlan analyzed the contents of monkey feces and found unfleshed fruit seeds comprised around two-thirds of the total dry weight. The thickets created by the activities of the monkeys were found to be composed exclusively of species of woody vegetation having edible fruits. Some of these *monkey gardens* may have developed around a tree originally protected by humans; *Canarium* is an important focal point in the formation of thickets and this tree is used by humans as well as monkeys. Younger thickets generally nucleate around exposed rocks or termite mounds used by the monkeys as perches. The thickets, besides providing food and shelter for the monkeys, have a microclimatic influence, slightly increasing the available precipitation in the immediate area. Jackson and Gartlan conclude that the monkey, by means of its dispersal activities, is changing the species composition of the flora of the island, and because the monkey is extending its "own habitat, some relation between monkey population and forest increase can be sought" (1965:594).

Because Jackson and Gartlan did not ignore the floristic contributions of humans, some interesting parallels may be seen between the activities of monkeys and humans. Numerous relics of cultivation were found in all the ecological zones of the island. As mentioned above, these relics at times have been utilized by the primate now dominant on the island. Other relics, such as several species of weeds normally associated with cultivation, have remained common in areas where successional processes are still incomplete. It is interesting that relic, utilized plants and weeds could survive over 50 years of abandonment—the human-mediated ecosystem need not be ephemeral. Illuvial areas were especially rich in relic species; here natural processes could be relied upon to mimic human cultivation activities.[10] Other areas with many human-associated plants were fishermen's camp sites found on the island. These camps have had temporary occupations of varying duration and some have been abandoned for substantial periods of time. Jackson and Gartlan note that "a flora distinct from the rest of the island and several cultivated plants are found where camps are still occupied or recently abandoned" (1965:582). They list numerous species of cultivated and utilized plants growing in these areas, including such

[10]The persistence of weedy human-associated plants in illuvial areas may have bearing upon Sauer's (1952) hypothesis for a riverine genesis of agriculture: favored foods transported to these regions would have a greater potential range and presistence. Transhumant populations frequenting such areas would find, over time, increasing concentrations of desirable foods. It is therefore not necessary to claim that aquatic resources were cause for settlement.

important foods as millet, fruit trees, manioc, Cucurbitaceae species, *Amaranthus* species, portulaca, and even bananas. Lolui Island may be visualized as an environment where primate activities largely determined the structure of the vegetation, and the specific flora at any time is determined by which primate is the primary occupant.

Parallels between monkey gardens and the relics of cultivation activity by humans are based on the fact that in both cases the animal transports favored articles of diet to regions that it frequents. In the process, the environment is modified in a manner that may enhance the carrying capacity of the environment for the transporting animal. Hawkes has claimed that the first stage in the development of agricultural systems involved the colonization of human habitation areas by plants with weedy tendencies, and noted that to preagricultural man "it must have seemed little short of miraculous to find that the plants needed for food sprang up by his very huts and paths" (1969:21). Jones gives an ethnographic example of just this process. In his analysis of the interaction of aboriginal Australian subsistence patterns and the distribution of food plants, he found a striking correlation between abandoned camps and groves of edible fruit trees. Proof that groves were likely the result of human dispersal, rather than camps being located near existing food sources, was found in his observations on the introduction and dispersal of an nonindigenous fruit, the watermelon. Melons that had been obtained at a European store were transported to a dry season camp where they were consumed. At the end of the following wet season, Jones was present to observe that "a crop of several hundred huge watermelons was taken from the old rubbish heaps of the camp, and many were carried . . . to other wet-season camps in the neighboring area" (1975:24). Later in the year, Jones saw seedlings of watermelons appearing on dump heaps at these other camp sites. It is the rare and lucky observer who can witness the introduction and spread of new food sources in such a clear and striking manner; the process of human-mediated dispersal is usually far more cryptic.

Vavilov (1926) discussed the importance of human-mediated plant distribution at an early date, stressing the major importance that colonization of anthropogenic areas was to have on the early evolution of agricultural plants. His pioneering work was followed up by Anderson (1956, 1960, 1969) who gave special attention to the dooryard garden as the direct lineal descendant of early agricultural systems. Anderson viewed the dooryard garden as a specialized ecology, including weeds, utilized plants, and crops associated with human habitations. Such an agroecology arises because human activities change the general environment in localized regions and people transport selected species to, or protect selected species in, the region. In the early stages of its development, agriculture may have been little more than the dooryard garden, and although this *dump-heap* model for agricultural origins has become popular and is widely accepted, at least in part, by most scholars, the details of the dooryard garden's evolutionary ecology have received scant attention.[11]

Kimber (1966, 1973, 1978) has studied species composition in contemporary dooryard gardens and has found that the spontaneous elements in the gardens may be of surprising importance. For example, in one study done in Puerto Rico (1977), she found that an average of 45% of the utilized species found in gardens were of spontaneous origin rather than being the result of conscious planting. As she points out, the dooryard garden is a highly complex ecological system that thus far has been poorly understood, and one that may be comprehended only by taking into account the numerous natural and cultural parameters affecting it. Lathrap (1977) discussed South American dooryard gardens from a different perspective. Unlike scholars such as Anderson and Kimber, he has stressed the cultural and symbolic significance of the dooryard garden as a "safe" (albeit energy-consuming) region for human endeavors. In this remarkably wide-ranging paper, he also investigated the complex ecological structure of these gardens, their utility in the reconstruction of cultural history, and the possible role on demographic structure of induced agricultural instabilities.

Anderson (1969) gave special emphasis to the evolutionary opportunities presented plants by colonization of dump heaps and the regions surrounding human habitations. This area is open, with enhanced nutrients, and it may act as a meeting place for related species with the possibility of subsequent hybridization. Anderson's concern with hybridization in the origin of evolutionary novelty, arose in part because of his taxonomic investigations of the spiderwort (*Tradescantia*): here different species evolved and adapted to specific ecological and edaphic parameters. Niche partitioning tended to preserve related species in separated locales and reduced the possibility of successful hybridization. Related species in nature frequently are specialized to differing environmental conditions, hence the hybrids that are adapted to neither of the parent's niches will have no zone to which they are adapted. Within the cleared soil surrounding habitations, however, it is possible for hybrids to grow, prosper and perhaps outcompete the parental species.

Besides acting as zones of hybridization and reduced competition, the areas surrounding human habitations direct further plant evolution because selection will favor those plants possessing parts that are both attractive to humans and easy to gather. Given the colonization of such habitats, traits such as indehiscence would be selected even in the absence of conscious planting by humans.

---

[11]Recent interest in polycultural and intercrop agricultural systems, although peripherally related to the development of the dooryard garden, has generally centered on the structural dynamics of these systems of production. See Altieri *et al.* 1978; Elton 1958; Feeny 1973; Gliessman *et al.* 1981; Ighozurike 1971; Janzen 1973; Kass 1978; Murdoch 1975; the chapters in Papendick *et al.* 1976; Perrin 1977; Pimentel 1961, 1977; Ruthenberg 1976; Risch 1980; Root 1973; Trenbath 1974; and Yen 1980.

As Hawkes (1969:22) points out, "even before planting was practiced, unconscious selection by man of the plants with less efficient dispersal methods took place since it was the seeds of these that were automatically more efficiently gathered." Dump heaps under these conditions would be (*contra* Fowler 1971; Harlan and de Wet 1965; Saucer 1952; and Struever 1962) more than mere repositories for potentially useful plants awaiting the influencing activities of people to achieve domesticated status; they would be areas of active evolutionary activity, even though people might be totally unaware of the direction that evolution was proceeding (Robert E. Dewar, in preparation, 1983).

The development of the anthropogenic environment is facilitated by human behaviors, such as clearing of the land and interfering with "natural" successional processes, that modify the local environment to the benefit of certain plants. The local environment will be relatively homogeneous and predictable, yet, seen from the perspective of the broader environment, these favorable zones are both spotty and uncommon. The spottiness of the environment will be experienced by the plant as a function of its ability to attract and be dispersed by people. Those plants that succeed in doing so will effectively experience a large, continuous niche fit for growth and reproduction. Whereas dump heaps and dooryard gardens represent one type of dispersal by people, that of inadvertent or accidental loss of seeds, another type of dispersal might well have been of far greater significance during the earliest phases of domesticate evolution. Many plants, those bearing edible fruits in particular, are inherently adapted to dispersal after having passed through the gut of an animal.

On the whole, there has been regrettably little scholarly interest in the dispersal consequences of primate toilet habits. Wickler (1968) has noted that seeds of the sausage trees (*Kieglia* spp.) and the baobab (*Adansonia digitata*) require passage through a vertebrate digestive system in order to germinate effectively. The baboon is a major feeder upon these fruits. Gartlan and Brain (1968) report that monkeys seem to be responsible for the rapid dispersal of tree and shrub species with edible fruits into disturbed and open habitats. Van der Pijl (1972) considers higher primates to be the major dispersal agents for species of such fruiting genera as *Citrus, Gardenia, Strychnos, Theobroma* (chocolate), *Cucumis* (melons), and *Durio* (durian), many of which include important domesticated species. Hladik and Hladik (1967) list several species of plants dispersed in Gabon by primates and note that the seedlings obtained from seeds processed in monkey intestinal tracts were more vigorous than seeds planted directly from the fruits. They also refer to several reports in the literature concerning the human consumption of such fruits as *Garica* and *Passiflora* preparatory to planting; intestinal treatment enhances the germination of the recovered seeds. Glander lists numerous genera of trees on which howler monkeys feed in Costa Rica and points out that howlers "act as seed dispersal agents for many of the tree species present in their range" (1975:50). Klein and Klein report that *Ateles belzebuth,* a basically frugivorous monkey, recognized ripe fruits and ate them in preference to unripe ones. They point out that most of the fruits were eaten whole and that "intact and recognizable seeds comprised the major volume of [the monkey's] feces" (1975:70–71). Ridley (1930) has pointed out that seeds of several fruits will germinate after consumption by people. These include saguaro and prickly-pear cactus (*Carnegia* and *Opuntia*), mangosteen (*Garcinia*), litchis and rambutan (*Nephelium*) and guavas (*Psidum*). He also notes that human manures are avoided by inhabitants of St. Helena because its contamination with *Opuntia* seeds would cause the fields to become overrun with this cactus. Van der Pilj (1972) reports the same to be true in Mesoamerica. I have personally been struck by the fact that, like sweet corn, the seeds of tomatoes, melons, and cucumbers pass intact, and possibly still viable, through the intestines of infant humans. The increasing importance of coprolite studies for archaeology highlights the frequency with which seeds may pass, in an at least recognizable form, through the human digestive tract.

A survey of ethnographic information on the latrine habits of 58 societies (Miller, in preparation, 1982) shows that, not surprisingly, defecation is not done indiscriminately in the habitation area, but is restricted to peripheral areas of the camp. Not only do such practices provide an opportunity for the dispersal of preferred foods in the area of camps, but also serve to alter significantly the soil chemistry of the latrine areas by increasing the fertility, especially in terms of nitrogen. Latrine areas are kept cleared by traffic of daily activities and cover was also probably removed to discourage snakes or insects. As valued plants began to grow in latrine areas, it is likely that they would be permitted to remain and the latrine areas would be moved further afield; over time, the "humanized" region of the local ecology would increase.

The creation and maintenance of the anthropogenic environment will continually interact with other cultural traits. People, like most other animals, have always been capable of destroying the resources necessary for their survival. Even before people were effective tool-using animals, they could have destroyed necessary resources, thus contributing to a lowering of subsistence base the environment could provide. With the development of more sophisticated tools and technologies, their destructive potential would have grown as a function of the increase in time any resource would be vulnerable. For example, if the seedlings of a potential plant species are preferentially spared, the total potential yield of that species will generally be increased because more will be likely to reach reproduction. Seedlings, as any gardener knows, are capable of being destroyed by an act as seemingly inconsequential as a misplaced step. With the development of sophisticated tools, the vulnerable stage of any plant's life will be greatly increased; adult trees previously immune from attack become highly vulnerable before the axe. The cutting of a particular tree is far more than the removal of one individual, and instead will have some effect on the future reproduction of that species. It will also have effects upon the growth of other species by creating a gap that may be filled by competing species. The habitual destruction or preservation of species will have major effects on the floristic structure of a region, and eventually on the directions open in plant evolution. Such habitual activities, passed as a cultural trait, are inseparable from human language.

The development of the agroecology cannot be separated from the human ability to symbol. Language in humans, as in bees or chimpanzees, may be used to convey information about the locations and desirability of food resources (Menzel 1975). In humans, unlike these other species, however, language can be used to classify resources according to the amount and type of utility, both immediate and potential. Language in humans permits the preferential preservation of resources at times before their utility is apparent. Thus a species may be protected at a vulnerable phase of its life cycle even though no immediate benefit is forthcoming from it. Such behaviors, modified though they may be by symbolic factors, will vastly increase the ability of people to affect coevolutionary relationships that already exist. Yet we must recognize that such activities are, and can only be, based upon the utility of the plant to the individual as modified by the ability of symbolic language to make the future crop apparent in the present. The decision to protect is not based upon any recognition of long-term evolutionary effects of that protection upon the plant species. The decision to cut a useless species for firewood is not perceived as having an effect on the floristic structure of the ecology in succeeding generations. The decision to step around, or on, the seedling of a useful herb or tree is not related to a recognition that the plant will eventually be dispersed by the same agent that could have destroyed it. Nevertheless, all these activities can only enhance the development of coevolutionary relationships between humans and the plants on which they feed. Indeed, we might conjecture that the most important distinction ever made in any human language was between *crop* and *weed,* for in this distinction lies the likelihood of population increase over time.

CHAPTER 4

The Evolution of Domestication

Formerly there were neither gardens nor cultivated plants. An old woman was pestered by her young nephew who was hungry and was asking for agricultural produce which did not yet exist. She had a section of forest cleared and burned and informed men of all that would grow there. . . . And she explained when each plant should be harvested and how it should be cooked and flavored. . . . She had herself buried in the garden and from her body sprang all of the plants.

MUNDURUCU MYTH

FROM C. LEVI-STRAUSS,

THE RAW AND THE COOKED

Cursed is the earth in thy work; with labor and toil shalt thou eat thereof all the days of thy life. Thorns and thistles shall it bring forth to thee; and thou shalt eat the herbs of the earth. In the sweat of thy face shalt thou eat bread till thou return to the earth.

GENESIS 3:17–19

Introduction

Domestication is a process mediated by morphological and autecological adaptations in the plant and by behavioral changes in man. Human interaction with plants introduces a new dimension into the evolution of plants and may ultimately have far-reaching effects upon human subsistence patterns. Three major modes of domestication may be distinguished: incidental, specialized, and agricultural.

Incidental domestication is the result of the relationship between a nonagricultural society and some of the plants on which it feeds. As a result of selective pressures placed upon the plants by this relationship, certain morphological traits in the plant will be at a selective advantage. That is, by virtue of

138

a coevolutionary relationship with humans, certain plants will become relatively more fit. The ultimate expression of this evolutionary process is what we today call the *domesticated* plant. These early domesticated plants, however, were not the result of agricultural techniques—agriculture could not arise without the plants necessary for its functioning. The evolution of agricultural plants was tied to, and in part a cause of, the evolution of agricultural techniques. Seen in these terms, the indehiscent rachis of the small grains is as much the cause as the result of agriculture.

The evolution of early domesticates permitted new types of interactions between people and their environment. I call these interactions *specialized domestication*. As humans came to be a dispersal agent for various species of plants, these plants were dispersed into zones of human habitation. A new type of ecological succession was initiated in which the plants on which humans feed became more common in the areas in which people resided. Differential destruction and protection of various plant species by humans intensified the effectiveness of this successional pattern. These effects upon the local environment arising from human interactions with plants set the stage for the development of complex agricultural systems.

The agroecology, however, is not merely the theater in which people act out their agricultural performances. Under incidental domestication, humans are directly *interacting* with the plants on which they feed and with which they are establishing symbiotic relationships. As the agroecology begins to develop, selective forces generated within this community of domesticates begin to gain great importance in directing the further evolution of domesticates. Seen in other terms: human activities are sufficiently all-encompassing to affect domesticates throughout their life cycles, not only at the dispersal phase; people come to establish a symbiotic relationship with a *community* of coevolved plants.

Agricultural domestication is mediated by human behaviors and evolutionary tendencies within the developed agroecology. Human behaviors initially helped establish the agroecology by means of such environmental manipulations as fires or the clearing of land surrounding areas of settlements. The introduction of weeding, irrigation, and eventually tillage within the agroecology increased the rate of evolution of the domesticate. With the establishment of the agroecology, new opportunities were created for plant evolution; it is only within the context of the agroecology that it becomes meaningful to speak of the development of weeds. Weeds are opportunists that, in essence, parasitize the relationship existing between people and their coevolved domesticates. Weeds initially may be simple colonizers of the agroecology, but over time we may expect many of them to come to mimic cultivated plants in morphology.

Perhaps the most successful weeds of agroecosystems are those that mimic cultivated plants to the extent that they too come to be valued and protected. These plants, called *secondary domesticates* (Vavilov 1926), differ from primary ones in the type of ecology in which they established mutualistic relationships with humans: whereas the primary domesticate arose in the indigenous ecology, and was involved in the founding of the agroecology, the secondary domesticate arises in the agroecology. As we shall see, primary domesticates are usually plants whose reproductive propagules are used by people, whereas secondary domesticates are frequently valued for their vegetative organs.

Incidental domestication is largely mediated by human dispersal and protective activities, which are a direct result of feeding behaviors. Specialized domestication is largely a demographic and distributional result of incidental domestication that occurs as people begin to affect the environment in a manner that indirectly benefits the domesticated plant. The most important effect of specialized domestication is the establishment of the agroecology, a process that ultimately permits the initiation of agricultural domestication. The selection of those characteristics in the plant that allow for the development of obligate symbioses between humans and plants provides the foundation upon which agricultural systems are built and is domestication in its broadest sense.

Morphological change in the plant is the most important indicator of domesticatory pressures and serves to tie the survival of the plant to the activities of humans. The most highly developed cultivated plants are incapable of survival in the wild, yet this does not mean that they are not the products of natural selection: the ant-acacias are incapable of survival without the ant, yet no one would hold that they are therefore unnatural. The morphological changes occurring during domestication tie humans and plant together. As people disperse the plant they also provide a greater potential source of food for their own use. Unlike the situation that exists under "pure" predation, in coevolutionary relationships the success of the predator does not depend on the possible ultimate destruction of the prey. Rather, as people increase in numbers, their quantitative dispersal activities increase. Thus, increasing numbers of people may yield increasing quantities of food (at least during the earliest phases of the development of agricultural systems). The plant is increasingly tied to the quantity dispersal provided by humans. Competition will exist among plants for human dispersal abilities, and successful dispersal will be based on the successful attraction of humans. Local dispersals may occur because of planting behavior; dispersal may occur through the dissemination of propagules by transhumant humans. That the coevolutionary sequence has been successful for the plant needs no demonstration—one need only consider the present distribution of a crop like wheat or maize and compare it with that of a putative ancestral taxon for the groups.

Except during the earliest phases of domestication, analysis of the evolution of domesticated plants cannot proceed without a consideration of the general environment and local ecologies in which the evolution is occurring. The evolution of the agroecology depends, inlarge part, upon human behavior, but the environment places certain constraints upon that development. Beyond this, although the early domestication of plants can be understood in terms of simple coevolutionary processes, understanding the development of agricultural behavior in almost all human cultures calls for a more ecologically oriented approach.

It will also be useful to remember that we have no evidence that people have had anything other than a generalized diet. Unlike some other primates, humans did not evolve as obligate predators dependent upon only a few plant or animal species for their sustenance.[1] On the whole, specialized relationships were developed with plants that were primarily carbohydrate, rather than protein, sources. The human relationship to plants is more like that of the agouti than like that of the specialized fruit-eating bird of the tropics. Yet this does not reduce the nutritional dependency created by the relationship. The vegetable sources of fats and carbohydrates are reliably high yielding for the time invested, especially during the early phases of the developing relationship, and could increase the efficiency of utilization of proteins from sources such as hunting. Also, readily storable vegetable food could reduce seasonal limitations of total population. The development of a mutualistic relationship with plants would increase the carrying capacity of a region for humans. Thus coevolutionary interactions including mutualism and storage may directly increase the fecundity of cultures practicing them.

Why humans began to establish coevolutionary relationships with plants is a question without real meaning. We might as well ask why certain ants established coevolutionary relationships with fungi or certain birds with specific fruits. The relationships were established as a result of the maximization of fitness in given situations in time and space; they were neither inevitable nor desirable, but merely happened. Again, to ponder *why* a plant should enter into a relationship in which its survival depends upon a single agent is misleading and inappropriate. We have for far too long perceived the morphology of domesticated plants as somehow "inefficient" or "abnormal." Like the expensively produced fruit of the obligate-dispersed plant, the cultivated plant has been seen as "unlikely," therefore "unadapted," and we have had to introduce intent to account for its origin and development. There are clear advantages that can account for the development of plants in a coevolutionary scheme with humans, and it is to these advantages that we must look. We must stop looking at the indehiscent rachis of the small grains, the maize cob, and the indehiscent pod of the pea as morphological changes that "prevent natural dispersal" of the plant. Rather, they are morphological adaptations that permit high-quality dispersal and generally limit these dispersals to people. We are, in the final analysis, not interested in asking, Why did people establish relationships with plants? but, Why

---

[1]Yet we should note that with the development of agriculturally dependent societies, human survival (at least at present densities), like that of all other obligate predators, is now tied to the survival of the few specialized food resources of humans.

did the establishment of these relationship have such far-reaching effects and how did people come to exchange one mode of subsistance for a radically differing one?

A description of the history of agricultural systems from their beginnings would show a tendency toward specialization in the morphology and breeding systems of the plants as a response to growing human control over the physical and biotic environment. The major problem with many models for the origin of agriculture is that they take the word *response* too literally and thus tend to see the plant as dragged along an evolutionary path by the actions of people. Too little consideration has been given the important role plants have played in establishing and maintaining the dynamics of agricultural systems. Plants have contributed as much to the evolution of the agroecology as have humans, for it was plants that created the situation in which certain types of human behaviors became selectively advantageous. We are, indeed, dealing with a system, and change in the system may be initiated by destabilizing influences generated by any component.

The relationship that humans established with plants was by and large that of an obligate-dispersal agent. The plants involved, previously opportunistically dispersed, had already evolved to maximize propagule production, and this helps explain the relatively high productivity inherent in the agroecology. One of the effects of the development of the agroecology on plants was a reversal of the selective tendencies that had guided the previous evolution of these plants in the wild. The protodomesticate had been selected for a suite of characteristics based on the demands of survival in an extremely stressful and unpredictable niche, that of the colonizer. The high-quality dispersal, protection, and human control of the environment accompanying the rise of the agroecology created a highly predictable environment for plant growth and reproduction. This permitted new directions in plant evolution to emerge, and was to increase further the total recoverable yield available to humans.

People benefited in several ways from establishing coevolutionary relationships with opportunistically dispersed plants. Because they are relatively common in most regions of the world, plants with similar strategies of dispersal were encountered by people as they moved from one region to another; people were thus preadapted to the colonization of new regions because of an existing behavioral pattern. Thus domesticatory humans could move from one ecological zone to another and continue to interact with similar types of plants. This potential for the establishment of symbiotic relationships with new species of plant added new and reliable sources of food to existing subsistance patterns. Of course, humans as a dispersal agent would also introduce into these new regions plants with which they already had dispersal relationships (at least to the climatic limits for any given species).

Through their dispersal activities, domesticatory humans created a new niche for plants in the world's ecology. Within this new niche, the agroecology evolved in a way that was to have major effects upon the spread of domesticatory and, more significantly, agricultural systems.

The introduction and spread of agroecologies was to have important effects upon the regions in which they were located, and upon their own ultimate stability of yield. Agriculture, like other mutualisms, may have a destabilizing effect upon community structure (May 1974): while peopel were experiencing increases in fitness by means of coevolved relationships with plants, they were also initiating irreversible changes in their own subsistence systems and in the world in which they lived. They were changing from generalized predators into obligate ones, and because of the interaction of their changing subsistence pattern with changes in the stability of the ecology from which the subsistence was extracted, humans found themselves, perhaps for the first time in their evolutionary career, confronted by a relatively "difficult," "stressful," and "unpredictable" ecology.

Utilization and Domestication

Domestication is a coevolutionary process in which any given taxon diverges from an original gene pool and establishes a symbiotic protection and dispersal relationship with the animal feeding upon it. This symbiosis is facilitated by adaptations (changes in the morphology, physiology, or autecology) within the plant population and by changes in behavior by the animal. Domesticatory behaviors need not evolve into agricultural relationships, and domestication does not evolve orthogenetically or inevitably into agriculture. Indeed, the degree of development of the domestication relationship will often vary among the various plants with wich an animal may be interacting.

Because domestication is a natural evolutionary process, this is not surprising. First, the plant is experiencing selective pressure not only from its relationship with humans, but from the whole environment; thus the development of the symbiosis may be interrupted by disease, predation, and competition for the plant between humans and other potential agents. Furthermore, the development of the symbiosis requires the appearance of "useful" mutations in the plant and "appropriate" behavioral changes in humans. Again, there is competition among potential domesticates, with those best adapted to the relationship at any given point tending to exclude others. Finally, domestication seldom affects the plant at all stages of its life cycle. Inherent characteristics of the plant (e.g., requirements for germination) may dictate limits to the symbiosis, and preexisting relationships with animals other than humans (e.g., pollination syndromes dictating outcrossing between the potential domesticate and wild populations) may interfere with the development of another relationship. One of the major tasks of the evolutionary view of plant domestication is to explain how a particular domesticate developed to the extent and in the direction that it did. It is necessary to investigate the whole complex of relationships existing between a given plant and the environment. No longer can we be content with assuming that a particular plant was domesticated because it was "adapted" to the conditions of cultivation; rather, we must begin to explain the biological basis of this adaptation.

One way to understand the great variation in degree of domestication among cultivated plants is to look to the evolutionary tendencies that are favored by interactions between humans and plants, focusing on the particular parts of the plant that are used by people.[2] Three major classes of plant parts may be recognized: reproductive propagules, vegetative organs, and asexual propagules (vegetative structures possessing secondary adaptations as dispersal propagules).

The Use of Reproductive Propagules

When humans utilize the reproductive propagules of a plant, the potential for the development of a domestication symbiosis is great, for the plant may be able to employ people as disperal agents. We have seen that the development and elaboration of dispersal symbioses is extensive throughout nature. Further confirmation of the important role of the reproductive propagule in the development of domestication symbioses may be found in the types of crops people generally grow. There are numerous plants whose vegetative parts are highly defended, either mechanically or chemically, but whose reproductive propagules are edible. This edibility is frequently advertised by the development of conspicuous coloration when the propagule ripens. Of course, as would be expected, this advertising and defense syndrome is not restricted to cultivated plants; it is extremely widespread in truly wild plants (including those that have established relationships with agents other than humans).

There are indications that many plants valued for their vegetative parts were originally domesticated as seed plants. Keimer (cited by Ryder 1976) suggests that lettuce was originally cultivated by the Egyptians as an oil-seed crop; Simmonds (1976:29) points to the possibility that the use of chenopods as "greens" in Mesoamerica is a secondary domestication for a crop originally dispersed into the area as a seed crop; Vavilov (1926) was among the first to argue that the turnip and its relatives were first utilized for their edible oils. This process

---

[2]Referring to *utilization* rather than to *consumption* here is not simply academic rigor. The use of a plant for any purpose is sufficient to initiate selective pressure; I do not believe we can uncritically accept the idea that domestication had to begin with or be restricted to food plants. Indeed, a medicinal, religious, or even ornamental plant that was utilized by people might have been as likely a source for symbiosis as a food plant. Only a few workers have given any consideration to this problem. Notable among them are Sauer (1969) and Anderson (1960). Other workers, such as D. Harris (1972), have mentioned only in passing the possibility of a nonfood domestication of plants.

continues today. Herklots (1972:177) reports the recent appearance of peas (*Pisum sativum*) used as a green vegetable:

> Prior to 1945 this [vegetable] was never seen in Hong Kong though shoots were commonly sold in the Shanghai markets. Pea plants are now grown, prostrate on the ground, for their shoots; they are not allowed to develop pods and their flowers, white or purple-red, are removed as they appear. Young shoots . . . are sold in bunches.

There are two basic modes by which plants adapt to the problems and opportunities of an animal-mediated dispersal system. The first and most common involves the elaboration of an accessory structure of expendable tissue that serves to attract the dispersal agent and is consumed by it. The actual propagule itself is variously protected and is carried passively with the accessory structure. Many of the propagules we call *fruits* have this mode of adaptation: the flesh of the fruit serves to attract and feed the animal dispersal agent, whereas the actual reproductive units are inedible and are discarded, and thus dispersed, after the fruit is consumed. The seeds in stone fruits, such as cherries and plums, and the core that contains the seed in apples and pears are examples of this type of adaptation by the plant to dispersal by animal agents. The abundance of feral apple trees growing in hedgerows throughout the northern United States and the feral peaches of the south are evidence of the effectiveness of this mode of dispersal.

The other basic mode of plant response to potential animal distribution agents may be described as aggregation, which involves the collection of propagules into a larger structure. The propagules themselves are consumed, and dispersal occurs despite partial losses from the total propagule population. The indehiscent rachis of the small grains, the ear of maize, and the heads produced by sorghums may all be seen as examples of this type of attractive aggregation, as may the large seed masses of the New World cucurbits. Selection for high seed number, and thus for larger fruits, allows for a secondary adaptation; an accessory structure—the flesh of the gourd itself—becomes the means by which the plant attracts people. Even the indehiscent legumes of the cultivated bean species may be seen as examples of this tendency toward aggregation in cultivated plants.

Aggregation and the development of accessory structures may be described as responses to the predatory behavior of animals. Either can occur without the presence of developed agricultural behaviors. Both are frequent occurrences throughout the animal world.

The Use of Vegetative Organs

When people utilize strictly vegetative structures of the plant—roots, stems, petioles, leaves, seedless fruits, and flowers[3]—the tendency is toward the evolu-

---

[3]Flowers, although botanically a sexual organ, are vegetative organs in relation to the dispersal of the plant; seedless fruits, because they contain no propagule, are likewise vegetative organs.

tion of inedibility. Typical evolutionary sequences include the development of chemical defenses, such as poisons and tannins, or physical deterrents to predation, such as tirchomes, sclerids, and thorns. Among annual plants, there is a pressure toward the evolution of precocious flowering, which decreases the life span of the plant and thus its susceptibility to consumption. The potentials for the development of cultivation symbiosis are initially rather small when vegetative structures are consumed. It seems likely, therefore, that plants whose vegetative organs are utilized by people were domesticated only after people had begun protecting plants within the agroecology. Under these conditions, plants that lacked defense mechanisms might be spared because of their desirability, and it would be possible for them to pass the unprotected state to succeeding generations.

Although humans are acting as predators in consuming the vegetative organs of plants, numerous plants have been domesticated that are now utilized for their vegetative organs. Propagation of these plants, however, must be assured if the plant is to remain available to people. Propagation may occur through either vegetative or sexual reproductive means. The early evolution of vegetables originally domesticated for their seeds may be understood in terms of the relationships outlined above and below. Another way in which plants whose vegetative organs are utilized may enter cultivation is illustrated by Bye's (1979) work on *Brassica campestris* in Chihuahua, Mexico. *Brassica campestris* is an Old World species with numerous cultivated and wild forms, including turnip, rape, the Chinese cabbages, oil-seed crops, and various weedy races common in grain fields. It was apparently introduced into the New World both as a weed and in its various cultivated forms, but only the weedy form reached Chihuahua. The development of a symbiosis between humans and this leafy green vegetable depended, Bye found, on site preference, alleopathic capabilities, and the variable phenotypic responsiveness of the plant.

Bye found the plant to be restricted in its distribution to areas of human-mediated disturbance, such as empty corrals, recently abandoned fields, and the disturbed soil near dwellings. Although much of the *Brassica* population is spontaneous, planting of seeds does occur to a limited extent during the autumn months. Bye believes that *Brassica campestris* is, at best, only a "protodomesticate" at this time:

> The plant has advanced to a weed-crop stage of domestication in which the seeds from weedy plants are consciously sown during a specific period of the year in prepared habitats so that a greater proportion of the biomass of the plant is allocated to the preferred edible tissue and yield is increased. (1979:253)

This conscious planting has not brought about any change in the morphology of the plant. Rather, it is only a method by which humans may exploit a response pattern already existing in the species. During the short days of the fall, the plants tend to become larger and more robust before producing flowers. Thus the recoverable yield is higher. Seeds that germinate in the spring produce plants with such a small recoverable yield that they are generally ignored. However, as Bye points out, these plants may serve as a source of seeds for the fall crop. The response of the species to varying day lengths permits two generations of plants, one of which is utilized by humans whereas the other serves to reproduce the species. Bye's preliminary investigations indicate that the seeds have not yet lost variable dormancy characteristics and therefore planting is not a necessity. Some of the seeds could germinate in the fall, producing a vegetable crop, whereas other seeds would overwinter and germinate in the spring to produce a seed crop. The tactic utilized by the plant in this instance is avoidance of predation during the reproductive phase by having an inconspicuous morphology.

Protection of the seed crop in *B. campestris* is likely enhanced by another aspect of the plant's physiology. Bye indicates that the plant is not gathered once it comes into flower—informants report that "at anthesis . . . the yellow flowers impart a bitter flavor . . . and that the flower stalks become too stringy" (1979:243). Thus, changes in the morphology occurring during the development of plant to maturity serve to tie the plant to humans while also ensuring the continued survival of the plant within the symbiosis. The plant adjusts to the protective and dispersal activities of humans while still protecting its own reproductive propagules from consumption.

Many plants now used as vegetables show modifications similar to those exhibited by *B. campestris* in its relationship with people in Mexico. Lettuce, spinach, radishes, kale, and mustard greens all have a day-length response, a temperature response, or a combination of the two that permits the separation of the life cycle of the plant into vegetative and reproductive generations (at least in certain climates). Also, all these crops tend to be bitter and woody or otherwise unpalatable upon reaching anthesis. Most plants grown for their vegetative organs have a highly distinctive leaf coloration or texture. Species of plants used as leafy vegetables usually have cultivated varieties with pale green, yellowish, reddish, or purple leaves, or are variegated with a dark color. Likewise, most leafy vegetables have cultivated varieties with rugose (wrinkly) or serrated foliage. These plants are recognizable immediately after germination, and colors are often strongest at this time. For several years, I have been attempting to fill the weed niche in my garden with cultivated plants by allowing several species of vegetables to ripen and scatter their seeds. Highly distinctive foliage in the seedling stage acts as a means of drawing my attention to the fact that a useful plant has germinated. Protection in the seedling stage is a prerequisite for successful colonization of the agroecology. Likewise, the development of bitter substances or a fibrous texture as the plant is beginning to reproduce facilitates the successful maintainence of the plant in the agroecology by preventing human consumption at a crucial stage of growth. One of the most interesting vegetables

I grow, the edible-podded radish, is used for both its vegetative and immature sexual propagules, and it exhibits adaptations to the agroecology in several ways. The seed-leaves of this plant are exceptionally large permitting early identification, and thus protection, of the plant. The plant exhibits a series of edible and protected stages. Plants germinate erratically for several weeks in the spring and may be used as a leaf or a root crop while young, but they get bitter as they begin to flower. Flowering plants produce large, edible, fleshy pods for several weeks. These pods, if unharvested, eventually become fibrous and the crop is therefore protected from total consumption. Plants growing in the fall under short day-lengths do not go to flower quickly, and they produce large fleshy roots in several shapes and colors (likely due to introgression from other varieties of radish I used to grow—this variety has now replaced all others). These roots are usually totally harvested, a matter of little consequence to the plant here because it will not survive severe winters. The subsequent spring's crop germinates from dormant seeds. This plant is utilized throughout the growing season and has not only survived in my garden but has actually spread despite the fact that no more than 1 out of every 100 seeds that germinate eventually reaches sexual maturity. The greatest mortality occurs in the seedling stage when plants come up in what is for me an inconvenient location. These seedlings are never consumed in any form but are merely destroyed. In a similar manner, I have found that recolonization of the garden by parsnips (one of the few root crops that can survive our winters) may be accomplished by leaving two or three plants unharvested each fall—a task more likely accomplished by accident than design.[4]

In this context it is important to note that only a very small percentage of the utilized plant population needs to survive for the relationship to be successful. Plants growing in the agroecology are generally characterized by the production of large numbers of propagules. This is, in part, related to the opportunistic

---

[4]The reader might guess by this point that I have a rather untidy garden. This is true, but as the years go by, the garden becomes increasingly self-sustaining and the amount of work I have to put in decreases—I can only imagine how the agroecology of my garden might function if I had inherited it rather than having to create it. Besides the radishes and parsnips already mentioned, I also have been allowing other vegetables to naturalize: cherry tomatoes; a red-leafed lettuce; mustard and turnip greens; herbs, such as dill, cilantro (coriander), and fennel; and several species of flowers. Other vegetables, including late varieties of tomatoes, squash, peppers, potatoes (both from seed and tubers), peas, lagenaria, and spinach have occasionally appeared but thus far they have not proved adapted to my climate and garden. An interesting tale of intraspecific competition based on differential yields also bears mention: I originally grew an early variety of the tomatillo (*Physalis ixocarpa*) that naturalized well. I introduced another variety that was much more vigorous and potentially higher yielding, but it did not begin to ripen its fruit until September and the total recoverable yield before our usual late-September frost was small. Frosted fruit, nevertheless, apparently contained viable seed and the contribution to the next year's crop of tomatillo plants largely came from the late variety. Because I received little benefit from these plants and I could not distinguish the two varieties until they began to fruit, *Physalis* moved from crop to weed and was eradicated from the garden. I recently located a source for the early variety and am allowing it to reestablish in the garden.

dispersal strategy that is the evolutionary heritage of many cultivated plants. Yet development of a symbiosis with humans may tend to increase seed production per plant, even for those plants whose vegetative organs are consumed. In nature, a large percentage of any plant's seed crop is wasted in that it is not successfully dispersed to an area in which growth of the plant may occur. As we have seen, one of the advantages for the plant in establishing a symbiosis with an animal is the higher quality dispersal the relationship affords. Successful colonization of the agroecology allows the opportunistically dispersed plant a relatively predictable environment that is analogous to the one offered to the obligately dispersed plant—a smaller number of propagules need to be produced for survival of the species within the agroecology. Overall seed production may be reduced either by a decrease in per capita productivity or by a decrease in the number of successfully reproducing individuals; in most plants whose vegetative organs are utilized, it is the latter that occurs. At least in part, this is an outgrowth of the relationship with humans. The number of seeds that will be successfully dispersed by any given plant in the agroecology is increased by human protective activities. More important, many of the traits that make a vegetable crop valuable to people (e.g., large leaves, hyperdevelopment of storage organs) tend to bring with them a general increase in the vigor (as expressed within the agroecology) of the plant: more energy is stored in vegetative organs and this energy is available for the production of a greater number of seeds per plant. Thus even when a symbiosis with humans develops on the basis of the utilization of vegetative organs, there are clear benefits for the plant. The symbiosis permits the elaboration of tendencies present in nature that are inhibited by the compromises necessary for survival of the plant outside the agroecology (see Stebbins 1971, 1974).

Increases in vigor and the competitive advantage it brings may also help to explain the evolution of crops no longer capable of sexual reproduction. Numerous "root crops" (e.g., sweet potatoes, potatoes, manioc) are propagated by asexual means, and lost the ability to set seeds. The strategy of increasing vigor at the expense of sexual reproduction has also been noted in wild plant communities (See Harper, et al., 1961; Stebbins 1971, 1974). Although loss of the long-term evolutionary possibilities associated with sexual reproduction would seem costly for the plant, evolution is indifferent to such notions. The increased vigor accompanying the loss of sexual reproduction tends to replace the sexually reproducing plants by nonsexual ones as long as efficient means of vegetative reproduction are present. Many crops that have lost the ability to reproduce sexually are propagated by means of organs that are not palatable to or easily utilized by people. Crops such as manioc and sweet potato are reproduced by means of stem cutting, bananas by offshoots, and horseradish by means of slender, runner-like roots produced near the top of the consumed portion of the plant.

A final method of domestication for plants whose vegetative organs are utilized is simple colonization of the agroecology by perennials whose vegetative organs are palatable to people. Here the plant depends on people for protection within the agroecology, but there is no striking modification of methods of seed dispersal—if anything, there is selective pressure to maintain or intensify opportunistic-dispersal mechanisms. We would, however, expect selection for such traits as gigantism and reduction in chemical defenses to occur because of human selection. Crops that may have been domesticated in this way include rhubarb and sorrel (Schwanitz 1966).

The Use of Asexual Propagules

Human utilization of an asexual propagule allows for an adaptive response on the part of the plant; but analysis of this type of symbiosis presents a number of problems. The first issue that must be clarified is the distinction between reproductive and storage organs. The domestication of crops such as manioc, turnips, and carrots are best considered within the scheme just presented for the domestication of plants, in which the vegetative structure is consumed. These plants are not propagagted by the structures that are generally consumed. However, the potential of the top of the carrot or turnip to continue growth and eventually produce a seed stalk cannot be overlooked. The apical growing point of these roots is generally discarded (at least today) in preparation, and the discard of these vegetable tops on a rubbish heap *might* allow for seed production and thus colonization of the area by the plant.

Plants such as the potato, in which the consumed organ has a natural dispersal function, are capable of developing a symbiotic dispersal relationship. The gathering of the tubers, with subsequent inadvertant losses, will permit those plants producing food most acceptable to humans to spread preferentially. Because the propagules are asexual and thus reproduce the parental phenotype, desired characteristics of the plants will be established in the crop much more rapidly than when sexual propagules are being dispersed. Plants whose asexual propagules are utilized have an advantage that is only approached by plants whose sexual propagules are apomictic: any particular morphological form is immediately fixed if it is successful in the symbiosis because variation from sexual recombination is eliminated.

In general evolutionary terms, the most likely response of a plant like the potato to the predatory activities of an animal is the development of small tubers widely spaced from the parental plant. Individuals possessing large tubers close to the mother plant would be under negative selection pressure because a non-dispersing predator could easily retrieve them. Indeed, Lemmon (1885) noted just such an evolutionary tendency in North American species of potatolike *Solanum*. Species growing in areas where they were subject to sustained and intensive harvesting by people tended to produce numerous small tubers on long, running rhizomes; populations not subject to this pressure and related species not consumed by people tended to have larger tubers more closely clustered around the mother plant.

If we accept that no planting behavior by humans was occurring during the early stages of tuber utilization, it would seem that the "desirable" types would experience a lowered chance of survival. An understanding of the natural dispersal mechanisms of wild plants with vegetative propagules could help us understand the evolution of the cultivated plant. It is theoretically possible that potatoes or other tubers that are cultivated today were already coevolved with another dispersal agent and that people were able to "steal" the crop from another hoarding animal (see Gunda 1981).

The domestication of the potato is further complicated by the fact that sexual reproduction of potato plants also occurs with great frequency in the region to which they are native. Most of the potatoes growing in the Andes set fruits abundantly. This may be more than a climatic response. Brush *et al.* (1981) report that the presence of volunteer and feral potatoes contributes greatly to the diversity of the potato gene-pool in its native habitat. Fruiting ability may be a result of unintentional selection for the trait because fruiting potatoes may be more likely to colonize fields. Although a seed will almost certainly be overlooked, a tuber will probably be harvested and thus be unavailable for growth the next season.

Fruiting ability may also be a relic of the original domestication of the crop. It is possible that those potatoes bearing desirable tubers were originally carefully dug. Reproduction of the desirable types occurred by the inadvertent planting of the fruits in ground disturbed by harvesting activity. Because considerable disturbance of the soil occurs during this process, the fruit would be situated in an ideal position for future growth. Competition from other plants would also be reduced because of this digging activity.

Preliminary data presented by Brush *et al.* (1981) indicate to me that the role of sexual reproduction of the potato in indigenous agricultural systems of the Andes has been little appreciated. The evidence comes from two sources. First, native cultivars are less affected by viral disease than are introduced cultivars. Because viral diseases are frequently eliminated in the offspring resulting from sexual reproduction and are maintained when asexual means of reproduction are employed, we may assume a heightened incidence of sexual reproduction in the native cultivars. Second, studies of the relationship between native classification systems for potato cultivars and the gene pool (as indicated by isozyme study) of these cultivars indicates a moderate level of diversity at the local level, with higher levels of diversity as the size of the sampling region is increased. Native agriculturalists of the Andes recognize numerous classes of the common potato: as many as 100 distinct classes (although frequently only half this number) occur within a single locality. These classes show a superficial morphological similarity, but investigation of the genetic similarity of the classes indicates they are relatively polyphyletic. The preliminary investigations of Brush and his colleagues have a significance, however, that extends far beyond the role of sexual reproduction for enhanced diversity in the potato. Perhaps their greatest interest for the ethnoecologist is the insight they provide into the interaction between native classificatory systems and the evolution of crop plants. The native agriculturalists of the Andes have developed a system for the naming of potatoes that may encourage diversity in the cultivated potato population. The existence of a large number of classes based primarily on tuber characteristics could permit the integration into the cultivated flora of practically any tuber arising as a volunteer seedling. Rather than true phylogenetic clones, morphotypes, similar forms arising from different parents, are lumped together under one name. Although conscious propagation is vegetative, the classificatory system encourages the unconscious assimilation of tubers resulting from sexual recombination.

The importance of this system of classification to the successful cultivation of potatoes can well be appreciated given the large number of diseases and pests to which the crop is subject. Maintenance of a high rate of evolutionary change in the crop, especially in the region to which it is native, permits resistance to various pathogens to ''track'' the evolution of the pathogens. Sexual reproduction also encourages hybridization between cultivated, wild, and feral populations of *Solanum,* further enhancing the probability of assimilation of protective genes into the cultivated gene pool. Although much of this investigation has, of necessity, focused on morphological change in plants in the development of domestication symbiosis, the work of Brush *et al.* highlights the contribution that cultural traits such as taxonomic systems may make in the development of a given domestication interaction. However, the study of these cultural aspects of domestication must be integrated with an understanding of the biology of the organism under investigation if we are not to fall into the errors of mentalism or cultural determinism.

A Taxonomy for Domestication from an Evolutionary Perspective

Viewing domestication and the rise of agricultural systems from an evolutionary perspective requires a vocabulary that permits us to make distinctions in what is, in fact, a continuous process. To make these distinctions is not to claim that ''stages'' exist in the development of all agricultural systems. Rather, the purpose of such a taxonomy is to facilitate accurate descriptions of the relationships between humans and plants—in particular, descriptions independent of the various types of ecological processes affecting the rise and development of the agroecology.

General application of the terms I introduce here is subject to a number of reservations. First, morphological evidence provided by the plant will have to be carefully interpreted if it is to be used as an indicator for the type of relationship that existed between people and plant. Unfortunately, much of the paleoethnobotanical record is inadequate for precise identification of selective forces existing at a given point in time. Thus, although I hope my taxonomy will serve to broaden the concepts used in investigations of human–plant interactions, it cannot reduce the importance of careful case-studies of specific cultivated crops. Second, although the development of pristine agricultural systems depends on the preexistence of nonagricultural relationships with plants, not all cultures need have developed agricultural subsistence patterns autochthonously; in seeking to understand independent invention, one does not thereby deny diffusion. Third, the three types of human–plant relationships described here are not mutually exclusive: although humans may have highly developed agricultural relationships with certain plants, they may still maintain and even initiate simple incidental relationships with other species. Thus one of the most felicitous implications of this approach is that the processes that were the initiating cause for the development of agricultural systems may be studied even today in the field. Fourth, continuity of species between the three types of relationships does not need to be assumed (although in certain cases it seems likely): the plants of highly developed agricultural systems need not have been involved in relationship with humans until they began the environmental manipulations that were to bring about the rise of agricultural systems. Thus the plants need not go through a series of stages in their development either. Finally, although the distinction between the three types of domestication described here is not as clear as might be wished, this is only to be expected because I am drawing artificial boundaries around components of an integrated, natural process.

In the simplest terms, *incidental domestication* includes simple dispersal and protection actions by people that create and maintain coevolutionary interactions outside the agroecology; *specialized domestication* focuses on the forces initiating and maintaining the agroecology; and *agricultural domestication* is largely concerned with the forces controlling the function, evolution, and spread of developed agricultural systems—here the important variables arise from within the agroecology. Although I believe these distinctions serve a useful end for arranging and describing the behavioral, ecological, and evolutionary variables impinging on plants and developing agricultural systems under various conditions, I hope the reader will bear with me at those times when the distinction between them appears less than absolute: continuity in descent is ultimately irreconcilable with clear, unambiguous classification.

Incidental Domestication

The simplest and most common type of domestication relationship found in nature is incidental domestication. Animals involved in incidental domestication relationships with plants need not exhibit specialized behaviors; they need only harvest the reproductive or vegetative propagules of the plant and effectively disperse it. Plants involved in incidental domestication relationships may be involved with several agents; they may also employ an opportunistic strategy for their own dispersal. A possible example is given by Maurizio:

> The Chukchi [hunter–gatherers of northeastern Asia] gather large supplies of vegetable foodstuffs for the winter. They are, so to speak, unintentional plant breeders. Plants grow densely almost everywhere around their tents; some of them by chance, while others owe their location to the Chukchi, having been gathered far away and come to grow in the refuse individually. Among these, a cineraria, a composite plant, deserves particular mention, for it is found only around the tents, where it contributes its annual share to the support of the Chukchi. (quoted in Schwanitz 1966:12)

We may suspect an incidental domestication relationship between humans and plant when we have information indicating that humans are an effective, although not necessarily the *only* agent for a plant species' dispersal. Clear evidence, especially paleoethnobotanical evidence, of such a relationship is frequently difficult to obtain and depends on careful analysis of patterns of plant distribution, routes of human migration, and the biology of the plant under consideration. Thus, for example, in the domestication of the tomato, an early incidental domestication is a likely explanation for both the patterns of distribution of the crop and the timing of the domestication. In this case, incidental domestication permitted allopatric differentiation to occur within the gene pool of the crop, and this differentiation was enhanced and fixed by the evolution of morphological traits such as self-fertilization. This change in the reproductive strategy of the plant encouraged the fixation of traits such as indehiscence and large fruit size, traits that served as means by which the plant became obligately dispersed by humans.

Given the far-reaching changes agricultural people have made in the environment, few "pure" incidental domesticates will be found in modern societies; rather, we will expect to find them in "primitive" settings, where human environmental manipulations have been less far-reaching. Harlan, de Wet and Stempler, in a review of the useful plants of Africa, stress that these plants must be viewed as representing a continuous series from wild to domesticated. They point to a close relationship between human settlement patterns and the distribution of certain useful trees, even though we have no information indicating that these trees are consciously planted:

> In many parts of the African savanna there is a close correlation between village sites past and present and the occurrence of the baobab (*Adansonia*). To what extent villages

are located near baobabs by design or how frequently baobabs become established after the village is founded we do not know. The tree has multiple uses. . . . We have seen villages in Sudan that could not be occupied year around without baobab cisterns, since there is no surface water in the dry season. The baobab is not man dependent, but man can sometimes be baobab dependent.

Among some peoples *Moringa* sp. is an important part of the diet. In Konsoland, southern Ethopia, the trees are grown in the villages near the houses. Again to what extent they may be actually planted we do not know, but in that region they are confined to the villages. . . .

In the West African savanna the karite or shea buter tree (*Butyrospermum*) is protected and encouraged. The original arborescent flora probably contained a number of species, but selective cutting has been practiced for a long time. The Karite is protected and every other species is cut for firewood, construction or other uses. As a result, vast areas are covered with a uniform orchard-like savanna of nearly pure stands of karite. (Harlan *et al.* 1976:11)

One need not go so far afield to find incidental dispersal and protection relationships between humans and trees. According to Schwanitz,

Acorns are an ancient gathered fruit; they have not been used for human consumption for centuries, but they were still used in the Middle ages as a kind of fruit for breadmaking, and for a much longer time have been used to eke out bread-flour supplies, at least in times of famine. . . . According to Konstantin vonRegel, the oak . . . was planted near dwellings to make it available in sufficient quantities at any time. The large oak groves that we find today around many farm buildings in Lower Saxony may be left from those ancient times when acorns still served as human nutrient. (Schwanitz 1966:10, 12–13)

The mere existence of a dispersal relationship may offer unique evolutionary possibilities to the plant. We have seen that people and other primates have commonly established coevolved dispersal relationships with legumes. Many species of legumes are highly susceptible to attack by seed predators, and numerous morphological and physiological defenses have evolved in legumes to reduce the impact of this form of predation (Janzen 1969); bean-seed parasites have evolved rather impressive oviposition tactics for dealing with such variables as seed size (Mitchell and Williams 1979; Mitchell 1975).

As Flannery (1969:n. 71) notes in a different context, insect infestations of seeds may additively and synergistically increase the value of vegetable sources of protein. Chronically infected legume seeds may provide a better diet than uninfected ones. Again, one of the common ways in which legumes attempt to reduce predation is by the development of a hard seed coat that reduces the digestibility of the bean, especially for animals without a dentition specialized for grinding. Injury to the seed coat by the insect parasite may make the seed (and its inhabitant) easier for the animal to digest. Here the animal would receive a highly nutritious source of food at little or no cost to the plant because the seeds that are most easily digested would generally be incapable of germination. Also, besides dispersing the viable seeds of the plant, the dispersal agent would be acting as a predator on the insect. The fitness of the plant might even be maximized by a reduction of its chemical defenses against insect predation (if the loss of defenses were more than compensated for by increases in feeding and subsequent dispersals by the agent). Thus more effective dispersal could permit higher levels of insect infestation because the plant would need fewer viable propagules in order to maintain its fitness; it would also benefit the insect parasite in that greater numbers of the staple plant permit greater numbers of the insects. Thus a simple dispersal relationship might greatly benefit all the organisms involved.

Incidental domestication differs from both specialized and agricultural domestication in that the relationship between humans and plants is played out within the nonagricultural environment. The most important effects of this relationship for human subsistence will arise from the fact that the amount of niche space available for any given incidental domesticate is predetermined by intrinsic environmental parameters. Thus the potential population of the incidental domesticate (and thus the total additive yield from these domesticates) is not directly under human control. Nevertheless, as Harlan *et al.* (1976) point out human activities may indirectly cause major floristic changes that bring with them modification of the local environment and make abundant useful plants. Thus, at the most general level, it is merely a matter of convenience to distinguish incidental from specialized domestication.

The concept of incidental domestication may be mre readily understood as describing a type of relationship between people and plant that is, to a large extent, independent of the subsistence strategy taken by a society. Incidental domestication encompasses both situations in which the morphological features usually associated with domestication are apparently lacking (wild plants) and cases in which they have been, to a greater or lesser degree, secondarily lost (feral plants and weedy morphoforms of crops). This joining of seemingly disparate phenomena under one rubric serves to direct our attention to an important fact: domestication relationships established with feral and weedy forms of domesticates are for all intents and purposes indistinguishable from those established with wild plants. In both cases, the plant has taken advantage of human dispersal and protective behaviors and thereby increased its fitness. Under incidental domestication, changes in the morphology and distribution of a species are a simple and direct result of human feeding behavior.

Incidental domestication is an ongoing process and occurs even when agriculture provides the primary mode of subsistence for a society. It has been easy to overlook its significance in the origin of agricultural systems because, compared with the importance of the agricultural domesticate, the caloric (although not necessarily the nutritional) contribution of incidental domesticates to the diet is frequently small. In addition, it has been hard to see how such an interaction between humans and plants could have any quick effect on the development of agricultural systems. However, this attitude grows out of an assumption that agricultural systems arose all at once. This assumption is probably mistaken:

agriculture is but an intensification of a preexisting model of subsistence, and its development was likely the result of a slow process.

Incidental domestication is a realtionship that preserves and promotes a conservative, traditional relationship between humans and the environment. Rather than create positive feedbacks that bring about changes in subsistence patterns, incidental domestication tends to reinforce the negative feedbacks that maintain existing subsistence strategies. The small potential for change inherent in incidental domestication relationships is rooted in ecological and evolutionary processes. Prime among these is one factor already mentioned: the niche space for the incidental domesticate and thus its potential yield are a function of preexisting ecological parameters. The yield obtainable from the environment places an upper limit on the carrying capacity of the environment and thus on human populations. The intensity of the human–plant relationship is in part a function of human population size; thus the rate of evolutionary change in morphology and autecology of the incidental domesticate will be low compared to that existing under specialized or agricultural domestication.

Several other evolutionary processes will act in concert with the ecologically imposed limitations on the potential importance of the incidental domestication relationship to human diets. First the spatial distribution of the incidental domesticates will reduce the speed with which human selection will effect the evolution of the domesticate. The rate of evolutionary change in a taxon is, at least in part, a function of the probability that an adaptive mutation will be positively selected by the agent. A ''spotty'' distribution of incidental domesticates reduces the chance of encountering a desirable mutation. Thus the diffuse distribution pattern of incidental domesticates will encourage a slower rate of evolutionary change than the concentrated distributions that occur under specialized and agricultural domestication. Second, the growth and reproduction of the incidental domesticate in the nonagricultural environment call for adaptive mechanisms that may be successfully lost once the agroecology arises and the plant's distribution is restricted to this special ecological zone. One of the ways in which the domesticate can increase its yield is by partitioning into yield the energy that may be gained from the loss of these mechanisms. Third, incidental domesticates are likely objects for coevolutionary interactions with agents other than people. Thus the incidental domesticate is likely to oscillate (both over time and across its distributional range) between various forms in response to changing selective pressures. All in all, the rate of change in subsistence patterns caused by incidental domesticates is likely to be extremely low.

The very ambiguity of the incidental domestication relationship is proof of its importance an an evolutionary process: it shows how easily plants may adapt to human activities. The details of any particular incidental domestication relationship however, are not subject to a generalized exposition but must be approached on a case study basis. Although we may state the general directions and principles underlying the initial domestication of plants, the specific changes occurring in a species must be explained with full knowledge of the biology of the plant within the context of the environment in which evolution has occurred. We may nevertheless recognize two basic types of incidental domesticates. The first is the plant that, although seemingly wild, has a distribution that appears to be the result of human feeding behavior. The second type of incidental domesticate includes those cultivated plants that have portions of their gene pools incidentally dispersed or protected. Although human relationships with these plants cannot usefully be considered as precursors of agriculture, they remain useful exemplars of the coevolution of humans and plants.

Although incidental domestication is interesting in itself and deserves far more study, its most important aspect from the perspective of agricultural origins is its potential for the development of more specialized relationships. Although the animal agent is behaving in a relatively unspecialized fashion, there will be a certain amount of consistency in its behavior: it is likely to value some morphs of the species more than others, and in preferentially feeding on these, it is more likely to disperse and protect them. In its extreme form, this tendency will permit the exclusion of alternative dispersal agents and an intensification of the symbiosis, the ultimate development being an abligate relationship between humans and plants. Of course, evolutionary changes of this type will be dependent upon the appearance of the "appropriate" mutations in the plant.

It is of extreme importance to recognize that the processes and relationships of incidental domestication are of far more than peripheral concern to the study of the origin of agriculture: they are the very basis, the ultimate cause, of agriculture's origin, elaboration, and spread. The appearance and gradual development of the incidental domesticate creates the feedbacks that ultimately change the values humans place on the plants on which they feed. Furthermore, it is the demographic effects of the elaboration of incidental relationships that bring about the transition to specialized and then agricultural domestication.

Specialized Domestication

Specialized domestication involves an intensification of certain tendencies present in incidental domestication and the rise of new phenomena affecting the evolution of domesticates. The most important aspect of specialized domestication for humans is that they are becoming more or less obligate their agents in relationship to domesticated plants. The relationship, in terms of the concepts described in the last chapter, is a specialized one—hence the term *specialized domestication.*

Perhaps the most important new variable in specialized domestication is the change in the behavior of the agent. No longer is the animal acting as an opportunistic agent for the plant; instead, it is using specific behaviors that enhance the success of the plant, especially within a particular region. When an animal feeds opportunistically, its choice of any given plant does not necessarily require that the plant provide a reasonably balanced diet. Other measures of food choice, such as obviousness or convenience, may be of greater importance than nutrition. As the animal becomes increasingly obligate in its subsistence patterns, nutritional concerns become increasingly important. Opportunistic animals tend to have both a diversified diet and behaviors requiring fewer specializations. The development of specialized domestication relationships will change scheduling patterns and the basis of the subsistence strategy of the animal species.

The origin of certain behaviors and of techniques of environmental manipulation may well be tied to the selective pressures placed on humans and their plants as incidental relationships intensify. Increasing success in coevolutionary relationships permits specialized relationships to arise between humans and their incidental domesticates; we may then describe humans as becoming obligate agents in their relationship with the plant. Humans become sufficiently dependent upon certain plants so that their survival, at new densities, is dependent on the survival of the plants. Of course, at the same time, the plants become dependent on people for their survival at higher densities in particular local regions. Concomitant with, or perhaps a cause for, the historical development of an increasingly obligate relationship between humans and plants was the increasing success of primitive people in affecting their environment. The firing of grassland, disturbance of the ground cover, the cutting of trees, and other activities that influence the local environment would increase the total amount of potentially habitable area for certain plants coevolved with humans. These changes in behavior share a very important corollary: no longer is the plant population limited by intrinsic environmental parameters; instead effective niche space has been expanded as a result of the habits of the agent. As the productivity (here, localized biomass) of the coevolving plant increases, the potential also arises for increase in the agents' population.

Evolution of the plant with in the context of the symbiosis is often necessary for the change from incidental to specialized domestication to occur. Morphological changes may be required to attract successfully the agent and to permit successful dispersal; autecological changes may be selected for as the plant adapts to environmental changes brought about by new behaviors of humans. Plant adaptations that occur under specialized domestication are thus responses to human effects on the general ecology and especially on the local environment. The greatest benefit to plants from people as obligate dispersal agents is the human ability to alter consistently local environments in such a way as to place the coevolved plant at a distinct advantage.

Protection is another simple form of behavior that encourages specialized domestication. Protection behaviors are, as we have seen, common in nature and serve to enhance the survival of the plant involved in a relationship with an animal. For people as well as for other animals, the line between protection (the nondestruction of a plant) and weeding (the removal of competing vegetation) is very fine. If, for example, people remove a certain species of otherwise non-utilitarian tree for firewood or construction purposes, while permitting the fruit or nut tree to survive, they are both protecting and weeding. Simple forms of specialized domestication based on protection are common in all societies: examples that come immediately to mind are preferential sparing of fruit, nut, sugar, or starch sources in the clearing of forests and hedgerows and the toleration of useful wild plants, such as berries or medicinal herbs, around settlements. Because people are not only effective dispersal agents for many of these plants, but are also using a specific behavior to encourage their survival in limited areas, these are examples of specialized domestication. The probability that a specific environmental patch will be subject to this type of behavior is a direct function of the frequency of human habitation of that patch. Therefore, specialized domestication will increase with sedentism. Although certain plants involved in specialized relationships are dispersed by agents other than humans (sugar maples, for example, are wind dispersed), these plants may be considered specialized domesticates in that protection behavior helps develop the agroecology and thereby permits a whole new mode of domestication to come into play.

Two other forms of behavior that begin to become important under conditions of specialized domestication are storing and planting. Again, as we have seen, storing is common in the animal world and often is indistinguishable from planting. Subterranean caching of foods is one possible route to the development of planting behavior. The importance of planting in the domestication of crops has long been recognized, but storing has received little attention. Kaplan (1981) cogently argues that an increase in the size of the common bean (*Phaseolus vulgaris*) is not the result of agricultural selection but of the interaction of bruchid weevil predation, storage techniques, and human-aided dispersal of the species out of the range of the weevil. Human dispersal of the "wild" bean into regions that lack bruchid populations would remove an evolutionary pressure for the maintenance of small seed size in the species; many nonpoisonous legumes minimize seed size to prevent loss of the seed crop to seed predators such as the bruchid weevil. Thus (an incidental?) dispersal of the plant into higher altitudes would permit the evolution of large seed size, which as we have seen, is generally advantageous for the plant. This process would be aided by human behaviors—exposure of the seed to frost, mild heating of the seed, sifting of the seed—to reduce storage losses from bruchid infestation.

Vegetatively propagated tropical root and tuber corps provide examples of the close relationship between storing and planting. Manioc, many species of yams (*Dioscorea*), and even sweet potatoes (*Ipomaea batatas*) decay very rapidly once removed from the ground, especially in the moist tropical areas to which they are best adapted. Colonization of the agroecology by plants such as these would be an inevitable outcome of subterranean storage.

Planting encompasses two radically different forms of behavior that have very different effects on domestication and the development of agricultural systems. The first type I call *replacement planting*, which involves maintenance of the plant species in a preexisting niche. An example is the planting done by many ginseng (*Panax quinquefolium*) gatherers, who harvest the root of the plant when the seeds are ripe and plant the seeds in the hole made by the extraction of the root. Again, the loss of ripe seed that inevitably occurs when a dehiscent grain like wild rice (*Zizania aquatica*) is harvested constitutes a replacement planting for this crop. Replacement planting is likely to have little effect upon the evolution of the plant; however, it may encourage the development of the agroecology and thus an agricultural mode of subsistence. This type of planting is, in effect, an incidental type of domestication interaction and in and of itself does not increase the niche space available to the plant.

Replacement planting may be contrasted with *agricultural planting*, characterized by colonization by the plant of more or less well-defined areas created by human disturbance of the existing ecology. The simplest form of agriculural planting is the colonization by plants of areas of human-mediated disturbance— as in the "dump heap" theory of the origin of agricultural systems. In its most familiar form, agricultural planting involves extensive preparation of the land and maintenance of ideal conditions for seedling emergence through such techniques as irrigation and herbicide application. The significance of the simpler forms of agricultural planting is that they help initiate the agroecology and the various evolutionary pressures toward the development of agricultural systems.

Replacement planting and protection may interact to extend a species' range. This has apparently occurred for many tree species utilized by humans Harlan *et al.* (1976) point out that range has been extended and density increased not only for the three species mentioned earlier, but also for *Acacia albida, Parkia* spp., *Tamarindus,* and the oil palm, *Elaeis guineensis.* The development of the agroecology from such processes has been summarized by Nicholas David:

> Any human settlement has effects on its immediate environment. The present distribution of species on the quaternary terrace of the Mayo Kebi [of North Cameroon] is a human artifact. . . . There are few trees that are not encouraged or protected for their economic value, scarcely a shrub that is not is some way exploited, and the herbs are for the most part either weedy colonizers of the fields or used for thatch, matting or fodder. (David 1976:259)

Similar descriptions of complex agroecologies that arise by protection planting are given by Anderson (1969), Bye (1979, 1981) Harris (1972, 1973, 1977a), Kimber (1966, 1973, 1978), Iverson (1949), Janzen (1973), Lathrap (1977), and Sauer (1969).

At times, rather simple forms of environmental manipulation may have great effects on the total yield realizable from a given plant. An example of the effectiveness of simple techniques of water control can be found in Australia, a continent long known for its lack of an indigenous agricultural people. According to Tindale,

> Examination of activities association with . . . foods such as wild rice, water chestnuts (*Eleocharis*), and *Dioscorea* yams encourage a suggestion that many of the activities of northern Australian people were already akin to those associated with the earliest gardening cultures lacking principally the idea of the deliberate preservation and sowing of new seed. (Tindale 1977:345)

> My continuing interest in the Iliaura led to the information that the most desirable grasses were most commonly found on mulga plains that became flooded for limited periods after heavy summer rains, and that it was proper to fill the runoff channels of creeks so that larger areas of ground would be flooded when the rains came. For many years there was no substantiation, but in 1963 a Wanji man from the Nicholson River country indicated that his people knew it was an advantage to get as large an area as possible flooded . . . at a certain place where the country was suitable, they choked up the channels with stones, earth and other debris. Areas such as these were well known as grainfields and were visited at the proper times to gather the harvest. (Tindale 1977:347)

The effect of this simple subsistence strategy upon the demographics of practicing peopels was great: Tindale believes that the carrying capacity was at least doubled.

Jett reports that peach trees were introduced into what is now the southwestern United States by the Spanish and that their cultivation was taken up by the Hopi at an early date. During the early eighteenth century, the cultivation of the fruit apparently diffused to the Navajo of Arizona, especially the Canyon de Chelly Navajo. These people propagate the peach exclusively by means of seeds and "most trees start as volunteers from seeds discarded at fruit-drying sites. . . . When intentional seed-planting is undertaken seeds are selected from the largest fruits and 3–6 whole seeds are placed in a hole . . . all the seedlings other than the most vigorous are removed from the cluster" (Jett 1979:299). Although the peach trees receive some care, especially protection against pests, little further attention is given them. The establishment of peach orchards resembles, if anything, that of the planting of a field of maize. Most traditional Spanish fruticultural methods, such as pruning, thinning of fruit, and propagation by means of budding and grafting, did not diffuse with the fruit. This is not a case of lack of awareness: government agents have attempted to introduce European methods but with little success. Jett claims that Navajo practices increase the work involved in preparing the fruit for drying and also decrease the total yield. Cultural inertia created by a conceptualization of what the agroecology "is" will ensure that changes in the direction of evolution within the developing agroecology must come slowly.

Unfortunately, we have at the present time little information about all the techniques involved in specialized domestication. There is no doubt that some were radically different from the techniques characteristic of developed Western agriculture. Many authors have distinguished monoculture from the diversified-field cultivation found in such agricultural systems as the milpa, tropical vege-culture, and European herb and vegetable gardens. Most accept that the diversi-fied-field system is ecologically more stable than any monocultural one, at least within certain ecozones (see Harris 1969, 1972). It is possible that many cultures developed agriculture within the confines of the techniques dictated by diversi-fied-field agriculture. Feeny (1973:14) finds it "interesting to speculate as to whether the vulnerability of [most] crops to insect herbivores . . . may result from planting in monoculture species which have evolved chemical defenses appropriate to communities in which the optimum strategy is being hard to find." This is a provocative line of evidence that seems to lend support to the notion that even within Western agriculture specialized domestications might well have been elaborated within the confines of a generalized agroecology.

As the specialized domestication relationship between humans and plants intensifies, and especially as it is increasingly mediated by the ecological param-eters dictated by the developing agroecology, we expect to find the first agri-cultural domesticates evolving. Primary agricultural domesticates are the early *obligate* members of the agroecology. Some of these may be incidental domesti-cates capable of colonizing the developing agroecology, whereas others may be wild plants that successfully enter the agroecology and maintain themselves in it as they become adapted to a symbiosis with humans. The early primary agri-cultural domesticate, evolving under conditions of specialized domestication, need not have all the morphological characteristics of the modern agricultural plant. Instead, many of these characteristics may arise as the agricultural rela-tionship between humans and plants is elaborated over time. Thus, under condi-tions of specialized domestication, perennial plants that now reproduce only vegetatively may not yet have lost their fertility; the rachis of small grains or the fruiting structures of other crops will have dehiscence mechanisms reflecting harvest techniques, not the ecological setting in which they are growing; and plants that are now annuals may not yet have lost their original biennial or perennial duration. Interactions occurring within the developing agroecology will have effects upon the plants with which people are establishing relation-ships. For example, the importance of *Brassica campestris* in Chihuahua, to the humans with whom it is establishing a relationship, is reinforced not only by its physiology and phenotypic responsiveness, but also by its chemistry. Within the regions colonized by this species grow a variety of other specialized domesticates that are also exploited as vegetables, including members of the genera *Amaranthus, Chenopodium, Bidens,* and *Cosmos.* Apparently these species are simple colonizers of the agroecology, and no supplemental seeding of these crops is occurring at the present time. Experimental work conducted by Bye (1979) shows that extracts of the roots and foliage of *B. campestris* act as an effective alleopathic agent against the two species of these other vegetables tested. Chemicals given off by the unharvested *B. campestris* plants during the fall and winter reduce the number of other annual plants that will appear the following spring. Apparently the introduction of this species into the agroecology serves to reduce the diversity in sources of food. The human relationship with *B. campestris* is thus intensified by chemical effects of the plant on other members of the agroecology.

Agricultural Domestication

Specialized domestication creates an increasingly obligate relationship between humans and plants and is mediated by specific human behaviors and also by evolutionary tendencies found in the developing agroecology. Agricultural domestication is the culmination of these processes: the evolution of domesticated plants now proceeds exclusively within the agroecology and is consequently subject to new potentials and limitations. Agricultural domestication is the establishment and refinement of the systems of agricultural production, but it does not cause the end of other modes of domestication.

Agricultural domestication is mediated by strictly agricultural forms of behavior such as harvesting, seed selection and storage, weeding, and conditions of tillage. Frequently the effect of these behaviors is control over the plant throughout all stages of its life cycle. An integrated series of human behaviors taking place within a specific environmental setting encourages the evolution of new varieties, new techniques of environmental manipulation, new types of domesticates, and weeds. Agricultural domestication is thus very close in concept to *domestication* as it has been used in most of the literature of agricultural origins: it differs in that incidental and specialized domestication are seen as having provided the raw material on which these more sophisticated forms of human interaction with plants take place and in the emphasis placed on the systemic aspects of the agroecology in the development of agricultural systems. Under conditions of agricultural domestication, weeds and secondary domesticates become very important. Like incidental domestication, agricultural domestication is an ongoing process and is occurring today. The plant breeding that is being done today is the most modern form of agricultural domestication. "Scientific" domestication differs from what has come earlier only (although this is a big only) in that people are aware of at least some of the processes that control the evolution of plants and attempt to take advantage of them. Thus modern plant-breeding is a very special type of domestication—in essence it is an attempt to be intentional about domesticate evolution. How effective this attempt will ultimately prove to be in an issue with potentially grave consequences, but one that is beyond the scope of this volume.

The error of earlier treatments of the origin of agriculture has been a confusion of the origin of domesticated plants with the origin and development of the agroecology. During incidental domestication, the relationship between people and plant is fairly straightforward and human behaviors directly affect the evolution of the domesticate. One of the outgrowths of incidental domestication is the origin and subsequent development of the agroecology, which I have called *specialized domestication.* The relationship between people and plant undergoes a fundamental change. No longer do people *directly* affect the evolution of the plant; rather the direction and force of selective processes are increasingly mediated by the agroecology.

Davis and Bye give an example of this in their discussion of the evolution of *Jaltomata,* a "wild" nightshade valued for its edible fruits. In several regions of Mexico and Central America, this plant exhibits a syndrome of morphological variation that may be the result of the plant's growth in cultivated fields and its association with humans. The authors note that human activities have encouraged certain populations of *Jaltomata* by "1) the creation and maintenance of disturbed habitats . . . 2) dispersal and establishment of the seeds and seedlings through defecation and of the rootstock through plowing, thus expanding the populations, and 3) perpetuation of these solanaceous plants in cultivated fields by retaining them during the removal of other weeds from the crop fields" (Davis and Bye 1982:234). Pressures exerted by life in the agroecology have brought about several changes in the plant, of which the most striking is its synchronous ripening with the maize crop. Plants from fields also have more flowers per inflorescence than do wild plants; the extra fruits in the clusters are another example of the domesticate's tendency toward aggregation in edible parts. Finally, most of the variant populations of *Jaltomata* have been found in cultivated fields rather than in forest openings, the original home of the plant. It is possible that these variant populations have moved between the several areas of Mexico and Central America in which they are found by means of unintentional human transport.

We have already noted that the establishment of an obligate dispersal relationship between a plant and an animal is based upon greater reliability in dispersal: morphological change in the plant both reinforces the relationship and is its result. In an analogous manner, agricultural behaviors carried out within the agroecology permit new opportunities for plant evolution. The environment inhabited by the domesticate becomes more predictable, and many of the demands placed on the plant for survival outside the agroecology are redirected. Competition is not eliminated—indeed, it may even be increased—but is placed in a different context. Many morphological and physiological expressions of the plant genome become unnecessary or even maladaptive within the agroecology. Natural selection occurring within the agroecology thus tends to eliminate such traits, while at the same time encouraging new traits as a response to conditions existing within the new environment.

One of the most important tendencies under agricultural domestication is toward selection for greater productivity. Competition among and between crops will select for those individuals and species that have the highest yield. This will occur because of simple statistical sampling and need not be the result of the human desire for higher-yielding varieties. The principle involved is a simple one: if the selection of reproductive propagules for the next generation is a random sample of all seeds produced by the parental generation, individual plants producing more than the mean number of propagules will have more than the average number of offspring in the next generation. Thus each successive generation will consist of increasingly fecund plants. An experimental demonstration of unconscious selection for fecundity was carried out by Suneson (1949), who grew four varieties of barley in a mixed field for 16 generations. Starting with equal numbers of seeds from each of the four varieties, Suneson harvested the crop and randomly selected seed for each subsequent generation. By the conclusion of the experiment, one variety had risen from its initial 25% to 88% of the crop. Subsequent work Lee (1960) proved that the change in composition of the field was due not to variation in seedling mortality but to differential productivity of the varieties. Although in these experiments variability in yield was inherent in the performance of different varieties in a given environmental setting, the same type of selection will occur within one variety of a crop or within the domesticated species as a whole. Mutations permitting higher yield under conditions existing within the agroecology will accumulate over generations, and these higher-yielding types will come to dominate the field.

As noted by Stebbins, selection for fecundity is to be expected in all environments, but the effectiveness of this selection is usually inhibited by intrinsic environmental parameters:

> Selection for maximum fecundity will be exerted on all populations, whether they live in favorable or in unfavorable environments. In unfavorable environments, however, limits are often set upon the extent to which the plant can grow and the length of time over which it can flower and produce seeds. These limits require a compromise between the genetic advantage of fecundity and the limits on seed production imposed by the environment. (Stebbins 1974:174)

Thus this particular mode of increasing fecundity is not to be confused with selection for a higher intrinsic rate of increase because it is occurring within a favorable, predictable environment with relatively high competition.

Complexity and Stability

Agriculture is based on environmental manipulation and every form of manipulation occurs within a particular setting: the agroecology develops within a larger ecology, and this ecology places limits upon the form the agroecology may take. I have thus far ignored the question of environmental limitation on biological, and thus on cultural, phenomena. This has been a deliberate decision; environmental determinism is little better than demographic or cultural determinism for developing an understanding of the factors underlying the human interaction with nature. Although environmental analysis provides a poor vantage point for the initial analysis of agricultural systems and their pristine development, it can be extremely useful for summarizing and describing the final form these systems take. It should not be surprising that tropical vegecultures mimic in certain ways the local nonagricultural ecology. The constraints of predation and species diversity placed on the wild flora of the tropics are also placed on the agricultural systems developing in those regions. These agricultural systems are neither primitive nor necessarily predecessors of monocultural systems. Similarly, monocultures are not inherently unstable or ill-adapted in all environmental settings: they are the outgrowths of coevolutionary relationships within other types of environments.

We can expect that specialized domestication will produce agroecologies that are well fitted to the larger environment: the interaction between the developing agricultural community and the external ecology is significant and involves merely a rise in the carrying capacity of the local environment for the cultivated plant. Under conditions of agricultural domestication, however, the further evolution of the domesticated plant has been, to a great extent, restricted to conditions existing within the agroecology. This restriction upon the parameters affecting the evolution of the agricultural domesticate has two major consequences: first, it permits the survival of the system in ecological zones different from those in which it originated, and second, it enhances the potential long-term instability of the system.

David R. Harris (1969, 1972, 1973, 1976, 1977a, 1977b) has done much to contribute to our understanding of the importance of the ecological setting for the development of agricultural systems. In one of his earlier papers (1972) he first proposed the idea that seed-based paleotechnic (folk) agriculture tends to be less stable than the generally more complex vegecultural systems, a position that he has maintained and amplified throughout his subsequent works. Although agriculture may bring about the "transformation of a natural into a largely artificial system . . . [it] may also proceed by a process of manipulation which involves the alteration of selected components of the natural system rather than its wholesale replacement . . . [this method] simulates the structure and functional dynamics of the natural ecosystem" (Harris 1972:183). He goes on to explore the effects of the vegecultural and seed-culture systems on the stability of the resultant systems and on their resultant tendencies to expand. He concludes (p. 188):

> Both systems widely employ the techniques of shifting or swidden cultivation, but in vegecultural plots plant diversity tends to be greater, plant stratification more intricate and the canopy of vegetation more nearly closed; in other words they represent floristically, structurally and functionally more complex ecosystems than do seed-culture

plants. Vegeculture thus has greater inherent ecological stability than seed culture. . .
Ecologically, therefore, we can postulate that seed-culture, particularly when practiced
as swidden cultivation, should exhibit a greater tendency than vegeculture to expand into
new areas. (D. Harris 1972:188)

Although Harris deserves praise for helping to redirect the study of both agricultural origins and the function of agricultural systems in a direction more in keeping with the natural sciences, he has here made the traditional error of confusing domestication with the processes acting on the developing agroecology and thus on the eventual form the systems will attain. To accept the idea of a difference in stability between grain and vegecultural systems is to overlook environmental and evolutionary aspects of their origins.

To understand better the difference between nonseasonal vegecultural systems and seasonal grain crop systems,[5] it is necessary to consider briefly two pairs of interrelated concepts: complexity and diversity, and stability and resilience. The hypothesis linking stability to complexity is far from proven, and uncritical application of simplicity–instability arguments to the description of agricultural systems without detailing the nature of the instabilities and the means by which they are introduced into the system may frequently reduce rather than increase our understanding. Furthermore, agricultural domestication, in large part, describes the linkages that exist between ecological setting and the form that developed agricultural systems may attain: it is thus the ecological setting and not the evolutionary history of vegecultural and grain-crop systems that best serves to differentiate them.

Complexity is generally used to refer to the "number and nature of the individual links in the food web" (May 1974:3). Complexity is thus a measure of what is often called *trophic structure*—the degree of interrelationship within a community. All agricultural systems have relatively simple trophic structures, especially when compared with some "natural" ecosystems. Agricultural systems at their most complicated contain only three trophic levels: plants, herbivorous animals, and people. Most nonagricultural ecosystems are considerably more complex, with primary producers fed on by secondary, tertiary, and often higher-level predators before reaching the ultimate predator.

Diversity is fundamentally a measure of two descriptive statistics: the number of species present in the community and their relative abundances (Pielou 1977). Diversity is thus a way to describe the inhabitants of an ecological community, whereas complexity describes their interrelationships.[6] Vegecultural

---

[5]Although many vegecultural systems are indeed tropical, it is incorrect to hold that grain-crop systems are temperate in their origins: three of the major grain crops—maize, sorghum, and rice—originated in more-or-less seasonal tropical climates.

[6]The discussion of diversity is complicated by the concept of *pattern diversity* (Pielou 1966), which is very close in meaning to *complexity;* for our purposes it will prove heuristically useful to maintain a clear distinction between the two concepts.

systems are frequently more diverse (they include more species) than are grain-based agricultural systems, but at the same time they often are trophically simpler (they lack the trophic level of herbivorous domesticated animal).

Stability is a central and exceptionally complicated and confusing concept in both the ecological and the anthropological literatures (Lewontin 1969, Maynard Smith 1974). Its base meaning is "unmoving, stable," but it is seldom used in this way. Instead, the meaning that the word has received in the physical sciences has been appropriated by the natural sciences to describe a dynamic equilibrium in which the response to a disturbing force is designed to maintain or reestablish a preexisting position, form, or structure (that is, the equilibrium is homeostatic, not static). Thus negative feedbacks are stabilizing whereas positive feedbacks are destabilizing. Perhaps the simplest meaning of stability is the propensity of a system to return to an equilibrium point following a disturbance. This definition does not pass judgment on the relative probability of a disturbance's occurring, but focuses on the consequences of the disturbance. Stability in this sense is compatible with dynamic change: a system that cycles through a set series of changes may be as stable as one that does not change at all—although the type of stability is different (Maynard Smith 1974; May 1974; Pianka 1978).

Closely related to and frequently confused with stability is the concept of *resilience*. Resilience in essence measures the likelihood that a given "random" disturbance will destabilize a system: it is the region of parameter space over which the system is stable—its "stable domain" (May 1974). In these terms, resilience may be imagined as a measure of the quantitative and qualitative amounts of "battering" a system can endure without having a breakdown. Thus it is at least conceptually possible for a system to be both stable and fluctuating (indicating a type of stability), while at the same time showing very little resilience.

The relationship between complexity, diversity, stability, and resilience is obscure and open to numerous interpretations. The conventional wisdom of the day is that increasing complexity begets increasing stability (Connell 1978; Connell and Orians 1964; Elton 1958; Hutchinson 1959; Leigh 1965, 1968; MacArthur 1955; MacArthur and Connell 1966; MacFayden 1963; Margelef 1968; Odum 1959). The logic behind this is rather straightforward and intuitively acceptable: as the food web becomes increasingly complex, population changes in one species are less likely to induce corresponding perturbations in the populations of other species because these other species' foraging patterns are buffered by the presence of alternative sources of food. The work of Connell and Orians, Margalef, and Leigh, among others, has sought to integrate environmental productivity into stability. The underlying thesis in these cases is that high productivity in an environment correlates with species diversity and thus with stability. Inherent in this viewpoint, however, is a slightly different concept of stability than the one I have just advanced. Attention in this case is focused on productivity, the complexity of the food web, and on oscillations in populations[7] and their systemic results. The theory is also based on several assumptions that may limit its applicability to the real world (Maynard Smith 1973), and, despite the general belief in the stability of trophically complex communities, data indicating their relative fragility are accumulating (May 1974, Colinvaugh 1978).

At the present time, then, the best we can say about the relationship between stability and a complex community structure is that the correlation, although intuitively sound, remains unproven. The growing list of dissenters from this position, however, would probably find the words of Colinvaux more to the point:

> The claim that complex communities are more stable than simple communities . . . is invalid. It is an echo of the wishful thinking of naturalists, amplified by mathematics they did not understand. It has done mischief by distracting people from real problems. . . . Many species in an ecosystem do not, of themselves, lead to population stability. Stability of climate, on the other hand, leads to the collection of many species [by reducing the probability of extinction for any given species]. This seems to be the essential truth of the matter. (Colinvaux 1978:208–209)

Although uncritical application of such a train of logic might easily degenerate into environmental determinism, we must appreciate that what we are seeing is merely an attempt to simplify the methods used to understand community stability. Climate in this case is given great importance because it can, in its extremes, bring about the death of many individual organisms. Instead of looking to (often teleological) internal controls on the structure and function of the ecosystem, a new view is arising that seeks to integrate the structure and function of the community with the individual organism's reaction to the immediate environment: rather than the "balance of nature," this new model for community ecology adopts a possibilist approach. Thus the new view goes hand in hand with the increasing emphasis being placed on the role of individual survival and reproduction rather than adaptation in the explanation of evolutionary change.

Domestication must proceed along differing lines in seasonal and nonseasonal environments, and this will bring with it differing potentials for movement of the systems outside their region of origin. The vegecultures of the tropics mimic the tropical ecology not because they arose by means of a different type of domestication process—"substituting preferred domesticated species for wild species" (Harris 1972:183)—but because they are in fact part of the flora of the nonseasonal tropics. One of the benefits of recognizing incidental and specialized domestication as precedents for agricultural domestication is that our attention is focused on the evolutionary continuities rather than the cultural

---

[7]It is not suprising that Hollings (1973) and Leigh (1975) define resiliance solely in terms of perturbations in population values; parameters such as climate, carrying capacity, and fertility rates are ignored (see Oster 1975).

discontinuities involved in the origin of agriculture. The domesticates comprising any agricultural flora are first members in good standing of the local ecology and bring with them fundamental adaptations to that ecology. Thus it is not surprising that many tropical vegecultural plants display morphological modifications adapted to mechanical protection or that they possess well-developed chemical defense systems. Such traits are not evidence for the "primitiveness" of these cultivated plants but are reflections of the ecological setting in which they evolved. As incidental domestication begins to phase into specialized domestication, the structure of the developing agroecology will differ in various ecological zones. The effects of insect predation, soil structure and fertility, and interspecific competition will act on the developing agroecology in the nonseasonal tropics with the same vigor as on the native vegetation, and it is not surprising that multispecied, highly diversified agroecologies will arise in these areas. Thus it is not necessary to hold that grain-based systems originally arose by means of domesticatory interactions resembling those found in extant vegecultural systems.

The constraints placed on the agroecology by its area of genesis will affect the structure of the system, and this in turn will influence its stability and potential dispersability. The vegetational diversity found in nonseasonal vegecultural systems is best viewed as permitted by the relatively constant and favorable climate. Diversity brings with it *under these conditions* a reduction in the parameter space in which the system may exist—that is, a reduction in its resilience. This low level of resilience places constraints on the dispersibility of the system. Because a large set of conditions must coexist for the system to function, the diffusion of the system will be limited to more or less contiguous locales with roughly similar climate and soils. Notions of differential inherent stability are not necessary to explain the observation that grain cultures have spread more successfully than have vegecultural ones: the potential universe inhabitable by the vegacultural systems is simply smaller.

The relationship between seasonality and the dispersibility of grain-cultural systems cannot be overemphasized. Grain-cultural systems, although frequently less diverse, are generally more resilient. This resilience is imparted largely by the opportunistic and colonizing evolutionary heritage of the plants that comprise it. In simple terms, most grain crops require only cleared land and a growing season long enough to mature the crop. It matters little that the dormant season—the time during which the grain is present only in the form of seeds for the next generation—is a time when growth is prohibited by lack of water or lack of heat. Thus crops such as maize or sorghum are preadapted to growth in temperate regions even though they originate in seasonally dry regions of the tropics. The transport of these systems across latitudinal or altitudinal gradients is limited by fewer factors than in the case of nonseasonal vegecultural systems.

Under conditions of agricultural domestication, the important selective forces are those arising from conditions within the agroecology. The effect this has on the overall stability of the system is a complicated matter that frustrates any attempt to arrive at a simple answer, and approaching it piecemeal seems the best policy: numerous factors will eventually have to be quantified and investigated in depth before any firm conclusion may be reached. Certain tentative generalizations, however, may be advanced. First, although agricultural systems are not inherently trophically complex, the introduction of agriculture may well alter the structural complexity of the entire ecosystem. Interactions between the general ecology and the agroecology are poorly understood. In terms of diversity, agricultural systems are highly variable, but no more so than natural ecosystems. Moreover, many agricultural systems may have developed in such a way as to decrease the diversity of crops that are important parts of the system. It is likely that the development of the agroecology was accompanied by an increase in productivity. It is also likely that many of the agricultural techniques increased the resilience of the system by permitting it to survive within a greater range of climatic zones. The evolution of domesticates within the agroecology, however, probably was working at cross purposes to the development of agricultural techniques: selection for increase in yield has generally been accompanied by a decrease in the range of conditions (*parameter space*) under which yield is maximized.

If we must advance a tentative generalization about the stability of agricultural systems, it would be that on the whole they are generally stable, but this must be understood in context. If what we mean by *stability* is persistence, then agricultural systems taken as a groups have persisted and spread from the time of their origin. If by stability we mean a decrease in effective environmental variability then, from the point of view of the plant, agricultural systems have become increasingly stable. If, however, we are defining stability from the point of view of a particular group of humans subsisting upon agricultural produce, then the issue is more complicated. Agricultural dispersal has been accompanied by a selection for optimally unstable systems, selection being mediated by the effect of agriculturally induced instabilities on the rate of spread of a given system (Rindos 1980). In brief, agricultural systems (at least thus far) have been globally persistent but locally unstable; although the general tactic of agricultural subsistence has been successful, persistent, and productive for people, it has been subject to recurrent episodes of local and often catastrophic, albeit transitory, collapses that are inherent in the means by which the agroecology evolved.

Domestication and Sedentism

Barring compelling evidence to the contrary, archaeologists tend to connect sedentism with agriculture and to accept evidence of agriculture as presumptive proof of a relatively sedentary life.[8] It does seem reasonable to assume that the establishment of the agroecology requires relatively permanent settlement or, at the very least, regular seasonal transhumance. Although I am not prepared at this time to give this difficult subject an extended treatment, three factors may be abstracted as of central importance. First, the common-sense view is that sedentism is correlated with, if not causal of, (at least locally) larger populations. This is congruent with the model presented here because, as domestication proceeds, the carrying capacity of the environment increases and thus the population may rise. Second, sedentism is not restricted to agricultural peoples: fishing and foraging peoples have frequently achieved a settled way of life. This means not only that agricultural practices need not be presumed to be the only route to sedentism, but also that the processes leading to agricultural settlement may be initiated and modified by extrinsic factors. Finally, full sedentism has apparently preceded developed agricultural systems in certain parts of the world, whereas in other places agricultural systems have become well established long before the advent of settled village life. Besides disproving any immediate causal relationship between sedentism and agriculture, these cases are good evidence for the importance of nonagricultural factors in our understanding of settlement.

Several recent authors have attempted to tie the origin of agriculture directly to increasing sedentism. David Harris (1977a, 1977b) has carefully examined and explicated the destabilizing factors that might have caused people to adopt an agricultural subsistence pattern. His approach is predicated on the popular belief in homeostatic regulation of population below the carrying capacity (1977a:180): "The salient point is that regardless of habitat differences populations stabilize at levels well below carrying capacity, and there is therefore little incentive for hunter–gatherers to develop and adopt technological innovations that intensify food procurement." Harris proceeds to investigate possible variables that may induce the "stress" necessary to condition a movement to a new mode of subsistence:

> A shift to a broader-spectrum pattern of procurement leads to reductions in logistic and/or residential mobility. These in turn trigger "abnormal" increases in population which cultural controls fail to regulate. A situation of population pressure on resources is thus created, which results in an intensification of labor input into the food quest and leads, via improved seasonal scheduling of procurement of wild foods, to increasingly specialized exploitation of agricultural resources. (1977a:192)

This change in subsistence pattern is not, for Harris, sufficient to cause adoption of agriculture: it merely permits the sedentism necessary for agriculture to exist. The transition to agriculture is dependent on a final factor, *cultural invention*—"internally generated or externally introduced changes in techniques

---

[8]Higgs and Jarman (1972) have done much to point out the circularity of this reasoning, and Bronson (1977) has provided an excellent analysis of the numerous logical and factual problems confronting the use of demography and settlement as explanations for agriculture. Consequently, these general issues will not be considered here.

of exploitation and the cultural selection of genetically responsive populations of plants and animals.'' (Harris 1977a:192) Reed (1977a, 1977b, 1977c) is in apparent agreement with Harris, although he places greater emphasis on the effects of sedentism on fertility rates. A similar approach is taken by Marvin Harris (1979:88), who emphasizes the importance of child labor to the developing agricultural economy and the effect this will have on population size. Redman (1977) offers an equally detailed, although slightly more comprehensive and descriptive model for the possible forces impinging on the development of settlement and agriculture. Like David Harris, Redman is uncertain about the importance of climatic events in the origin of agriculture, but Harris and Redman differ on the causal importance of population pressure, Redman holding that needs arising from a sparsity of population are as likely to encourage innovation as are high levels of population. Redman examines the social factors influencing subsistence change both directly and indirectly through the effects of prolonged sedentism, but he does not consider the correlation between sedentism and agriculture as necessarily causal in either direction (1977:532). Instead, a mutually reinforcing system is envisioned. Bray (1974, 1976, 1977), approaching the problem from an entirely different perspective, arrives at a conclusion somewhat similar to that of Redman. Despite Bray's strong commitment to an ''evolutionary-ecological'' explanation of cultural change, he, as a cultural ecologist, ultimately falls into adaptationism and thus into ultimate acceptance of population pressure as the prime mover in cultural evolution.

Although it seems reasonable to assume that external factors may have facilitated the transition to a more sedentary way of life, it is unnecessary to reject out of hand the idea that the growth of the agroecology itself may have conditioned the growth of sedentism under conditions of pristine agricultural origins by increasing yields and restricting these yields to highly localized regions. Thus we are speaking not of causality in any sense but of a dynamic interaction between population size and local productivity that brings about radical changes in the carrying capacity of regions or subregions and thus in human foraging patterns. The dynamic inherent in the interrelationship of agricultural and domesticatory yields with population growth is a highly complex one that guarantees neither stability nor progress.

Population levels depend on the interaction of two sets of variables: an external one that includes food supply, predation, and disease, and an internal one summarized by the concept of fertility. The distribution of a population in space involves numerous parameters, including distribution of food resources, interspecific and intraspecific interactions, modes of dispersal, and foraging patterns. Analysis of demographic change is thus a highly complex issue touching on practically every aspect of the species' biology. In the case of people, the issue is complicated even further by the necessity to consider traditions, trading and war interactions, and a plethora of other social and cultural factors. It is little wonder that the study of settlement changes at times appears intractable. Nevertheless, two variables—foraging patterns and distribution of resources—may be extracted from this long list of potential influences as of likely importance in the development of agriculturally related settlement.

The origin of agriculture may be described as the result of changes in feeding behavior. These changes were mediated by the evolution of domesticates and has major effects on foraging patterns, the most important of which, for our present purposes, was the early encouragement of the consumption of low-valued resources. This process occurred *before* domesticates had begun to make any substantial contribution to the diet. Thus the development of sedentism is tied to the development of the agroecology. Settlement amplified and enhanced the feeding behavior changes that were to bring about the eventual evolution of agriculture.

A rapidly growing body of literature, in both theoretical ecology and the social sciences, is devoted to the exploration of the relationship between foraging activities and prey characteristics (Covich 1976, Emlen 1966b, Orians and Pearson 1979, Pearson 1976, Pulliam 1974, Schoener 1971, and the papers in Winterhalder and Smith 1981). The fundamental theorem of this school is that it is possible to create an objective ranking of the value of potential prey items by means of the relationship C/t_h, where C (calories) is the expected energy content of a prey item and t_h is its total handling time. Handling time includes all investments of time connected with pursuit, capture, processing, and eating of the prey. The theory of optimal foraging holds that diets are constructed to maximize the value of $\Sigma\, C_i/t_{hi} = d$, where d is actual diet.[9] Although most of this theory is exceptionally sound when used for descriptive or analytical ends, much of it unfortunately rests on the use of optimality arguments for causal explanation (however, see Schoener 1971 for a different opinion).

Central-place foraging theory is a branch of optimal-foraging theory in which major emphasis is placed on the energetic consequences of an animal's returning to a fixed location with its prey. Under these conditions, a major cost of foraging will be the time spent transporting the prey from its point of capture back to the "home" locale. Thus central-place foraging theory hold that diets are constructed to maximize the value of $C/(t_s + t_t)$ where t_t is the cost of transport incurred during the round trip and t_s includes all other costs of feeding on a particular prey object. Clearly, $t_s + t_t = t_h$ and therefore central-place foraging may be read to include noncentral-place foraging (no home locale), because under this latter condition $t_t = 0$. In central-place foraging theory the focus is on the interaction of t_t and C because t_s is invariant (Orians and Pearson 1979).

Because specialized domestication implies settlement, or at least regular

---

[9]This deals only with yield per unit of time; we are describing optimal *foraging*, not the creation of an optimal *diet* ($\Sigma[C_i -$ cost of $i]$).

transhumance, we may refer to a culture undergoing specialized domestication as *agrilocal* in that to a greater or lesser degree it shows regular spatiotemporal patterning that permits the evolution of the agroecology. (The patterning does not cause domestication, nor is the reverse the case; agrilocality is a consequence of an interaction between people and plants based on the evolutionary and ecological parameters discussed earlier.) We have every reason to believe that the early specialized domesticates were relatively "low-value" food sources, especially when compared with the high-value sources of protein from animals. The major factor contributing to their low valuation is the substantial preparation (t_s) most of them require (harvesting, threshing, leaching, parching, grinding, etc.). Indeed, given the likely relatively low value of C/t_s for domesticates, we might wonder why humans would have chosen to feed upon them at all (a problem we will defer until the next chapter). Our task now is to understand the possible consequences for settlement patterns of this change in foraging behavior.

A transhumant population arriving at a certain season in an environmental "patch" in the early stages of developing into an agroecology will be confronted with an abundance of plants whose processing investment (t_s) is relatively high and whose caloric content (C) is relatively low, but whose transport investment (t_t) is zero. (Indeed, the very reason for the apparent primacy of plant domestication in the archaeological record may be traced to the nonmobility of the plant as prey.) The presence of $t_t = 0$ resources will undoubtedly encourage the establishment of camps where they grow, especially given their long preparation times and low caloric value. (The establishment of camps in a region of a resource abundance is not a response peculiar to domestication, but one that occurs whenever camps are located in proximity to resources such as water that have a very high cost of transport—forays for game and wild plants, and incidental domesticates may be staged with comparable probabilities of success from numerous bases.)

Increasing agrilocality will reinforce evolutionary pressures favoring specialized domestication: parts of the home camp will have a high probability of being colonized by incidental domesticates or utilized wild plants, and as people return to the locale over generations and centuries the yield generated by domesticates in it will increase. Because energy is saved by not having to make foraging expeditions to gather these plants, they can require fairly elaborate processing and still be maintained within the diet. Also—and this issue is not considered by central-place foraging theory—the preparation of such plants need not compete directly for time with other foraging activities: grinding, parching, and other processing activities may occur during bad weather and in the evenings; they may be done by members of the community whose contribution to other types of subsistence activities would otherwise be minimal (e.g., postpartum women, children, and the aged). A positive feedback system initiated by early specialized domestication will, because of increasing agrilocality, further change the local environment to the benefit of the domesticate.

A human group that moves in a basically random manner, utilizing resources as they are encountered, would show no potential for agrilocality if all resources were uniformly distributed. Most resources, however, are not uniformly distributed. Depending on the environment, certain resources will be limited in abundance and restricted in location. Two resources that are frequently limited in this manner are water and shelter. Thus, although incidental domesticates need not be located within any particular patch in a region, they will tend to be transported to, and thus have a probability of colonizing, regions containing already limiting resources such as water. An increasing abundance of early specialized domesticates over long periods of time may permit camps to be occupied for longer periods of time.[10] If transhumance occurs over several ecological zones, specific early domesticates are likely to evolve in camps and even to be transported between them (Lynch 1973, 1980). In the latter case, allopatric differentiation within the gene pool will be encouraged, and this may increase the domesticate's rate of evolution. Thus we may conjecture that the major effect of domestication on transhumant behavior will be an increase in agrilocality: in terms of the foraging pattern, more time will be spent in the camps and less traveling between them. Over long periods of time, this will be expressed as a transition from wandering to inhabiting seasonal camps, and eventually to sedentism. If the domesticates in seasonal camps have developed morphological traits (e.g., indehiscence, large propagule size) that reduce or prevent dispersal by nonhuman agents, we will find a seemingly anomalous combination of morphologically domesticated plants and nonagricultural settlement patterns.[11]

Here sedentism is encouraged, not caused by domesticatory or nondomesticatory interactions with the food supply. Furthermore, the effects of sedentism upon humans and their environment are not restricted to interactions with the food supply (see Ucko *et al.* 1972, Spooner 1972, Lee 1972a, 1972b), and a multitude of other processes has doubtless had great impact upon the development of settlements. I have attempted to demonstrate only that a coevolutionary model for the origin of agricultural systems is congruent with present understanding of the processes important in the origin of sedentism and that this approach may offer new insights into settlement in general.

[10]However, as we shall see, the yield from the evolving domesticates will be relatively small at this time. Thus, ceteris paribus, specialized domestication will encourage agrilocality. Increases in agrilocality will encourage further specialized domestication, eventually leading to agricultural domestication. But the exact effect of the process on sedentism per se is likely to be extremely complicated—after all, how often have all other things been the same?

[11]One of the clearest examples of this phenomenon is given by Lynch (1980), but other work, such as that of MacNeish (1964b, 1967) supports the same model.

Specialized domestication and the development of the agroecology stand central-place foraging theory on its head. Given the fact that $t_t = 0$ when foraging is done in the immediate locale of the home camp, the consumption of relatively low-value resources is encouraged under sedentary conditions because the relationship $\Sigma\ C/(t_s + t_t)$ may be maximized by including low-value (low C or high t_s) foods. Thus sedentism brings about conditions favoring a foraging pattern mathematically equivalent to that optimal for an animal with no home base—an animal that consumes its food as food is encountered in the course of the animal's random travels. Here we can begin to see that a broad spectrum diet would be optimal if the animal were somewhat sedentary (thus sedentism of the type that includes agrilocality is not identical in all respects to pure central-place foraging) and if food were available at a very low cost in foraging energy (resources are present in high densities in the immediate locale of the central place).

This analysis raises an interesting observation that runs counter to intuition—it is possible that the most innovative aspect of sedentism was not the feeding on local, abundant resources (shellfish, specialized domesticates), but the staging of long-distance transport-mediated foraging expeditions for large game. Seen in terms of the sedentism-modified view of central-place foraging, the $t_t = 0$ component of the diet (lack of transport investment) is conservative in that it reflects the behavior of the animal when it was nonsedentary (and had no home camp and thus no transport investment for any food). This continuity and conservatism in behavior, when applied to pristine agricultural development, may cause us to wonder why settlement and agricultural origins did not occur with far greater speed—what processes acted in a negative feedback manner to reduce the rate of development of specialized domestication. This question must also await the extended analysis of feeding behavior in the next chapter.

Domestication and Changing Selective Pressures

We have considered in some detail the effects of the dispersal relationship on plant, morphology, and we have seen how the development of the agroecology altered the setting in which domesticate evolution occurred. I would now like to consider these two phenomena in an integrated manner, looking at the more subtle effects of domestication on the life-history strategies of both domesticates and people.

The theory of r selection and K selection provides one of the most popular and fruitful means available for studying the effects of differing ecological settings on the evolution of strategies of reproduction and dispersal. Numerous workers have contributed to the development of this theory; several papers, including those of Gadgil and Bossert (1970), Harper (1967), King and Anderson (1971), MacArthur and Wilson (1967), and Pianka (1970 and 1972), have been especially influential. Described in simple terms, this theory attempts to identify the environmental conditions that tend to bring about adaptations that maximize one of two fundamental variables, r and K—r being the maximal potential reproductive rate (the "intrinsic rate of natural increase") obtainable by a given species and K, the carrying capacity of the environment, the inherent limitation placed by resources on the size of a population may achieve (see Table 4.1). It would seem that these two concepts are so different that it would be difficult to relate them to each other: r describes the intrinsic and potential aspects of rate of population growth, whereas K defines the upper limit placed on the population a sepcies may actually achieve. Although r and K do have a clear and unambiguous mathematical relationship defining them and thus linking them, such a relationship provides little intuitive "feel" for the concepts. Thus I approach them by showing how they are used to describe the interaction of selection and environment.

Let us consider the type of environment that would be likely to favor selection for an extremely high reproductive rate. In the most abstract of terms, this environment would be a "perfect ecological vacuum with . . . no competition. In the ecological void the optimal adaptive strategy channels all possible resources into progeny, thereby maximizing the rate at which resources are colonized" (McNaughton 1975:251). This hypothetical environment is one in which resources are unlimited in relation to the ability of the organism to use them. Thus the population can expand indefinitely. In the real world, such an environment is impossible, but a relatively good approximation of these conditions exists during the early phases of colonization of a superabundant resource by an organism. Thus the colonization of a large vat of grape juice by yeast and a decay bacterium encountering a dead elephant present analogues of the nonlimiting environment in the real world.

These real-world analogues, however, point out another important aspect of the r-selecting environment: rarity and transience. Large caches of grape juice or dead elephants are rather uncommon, especially from the perspective of the yeast cell or fly bacterium that must find them in order to survive and reproduce. Given the distribution of these superabundant resources in time and space, colonization will be a difficult task. Plants that colonize a rockfall or the newly opened land that appears behind a retreating glacier are also faced with a superabundance of transient resources. It is obvious that the most successful organisms in these environments are those with the highest intrinsic rates of increase. They will simply swamp other organisms by means of getting more progeny around the environment in less time (pure Darwinian fitness).

Another situation in which a superabundance of resources is available is that in which recolonization of an environment occurs following precipitous declines

TABLE 4.1

Comparison of r Selection and K Selection

Important parameters	r-selecting	K-selecting
External		
Resources	Abundant and available[a]	Mostly in use[c]
Setting	Variable, transitory, uncertain[b]	Constant, stable, predictable[c]
Community structure	Empty[a]	Crowded[c]
Species diversity	Low[a]	High[c]
Filled niches	Few[a]	Many[c]
Competition	Little[a]	Much[c]
Environmental selective forces	Physical[a]	Biotic[c]
Internal		
Population size	Variable is space and time	More constant[c]
Mortality pattern	Catastrophic,[a] density-independent,[b] directed to young[b]	Predictable, density-independent, age-independent[d]
Importance of random colonization	High[b]	Low[c]
Selection favors a tendency toward the following:		
Strategy	Maximize reproduction[b]	Maximize competitive abilities[c]
Body size	Small[b]	Large[c]
Development	Rapid[b]	Slow[d]
Reproductive tactic	Single (semelparous)[b]	Repeated (iteroparous)[d]
Reproduction	Precocious[b]	Delayed[d]
Number of offspring	Many[a]	Few[d]
Size of offspring	Small[b]	Large[c]
Parental investment	Low[b]	High[d]
Advantage of dormancy	High	Low[c]
Dispersal	Random[b]	Controlled[c]
Dispersal episodes	Many[b]	Few[c]

Source: Harper (1967), Gadgil and Bossert (1970), and Pianka (1970).
[a] Applies to humans.
[b] May be predicted from theory to apply to humans.
[c] Generally applies to domesticates.
[d] Possibly applies to domesticates.

in population resulting from a natural disaster. This disaster need not be a catastrophic event; it may be a regular or infrequent extreme in the physical world—for example, frost in subtropical regions or drastic change in salinity caused by tides or storms in marine or estuarial communities. Catastrophic mortality will result in the opportunity for recolonization under conditions in which resources are essentially unlimited. What is occurring over time under these conditions is very similar to what happens in both time and space in the examples mentioned earlier: the yeast eventually consumes the sugar in the grape juice, the elephant rots or is consumed, the plants successfully colonize the glacial till only to be displaced by a later phase in the succession (or the return of the glacier).

These environments are highly transitory and unstable—they are not predictable in time and space; the r-selected organism finds itself in a world that is variable and uncertain even though resources are extremely abundant. Simple competition does not limit the growth and survival of organisms as much as does their ability to reproduce quickly and disperse as the environment disintegrates. The environments necessary for survival (vacated or newly disturbed land, elephant corpses) are frequently located at great distances from each other, and the organism must find some means of locating or chancing on the appropriate locale for its breeding and survival; or the conditions during which explosive growth rates are possible may be widely separated in time and some means must be found to survive between them. Given these conditions, it is easy to see why the r-selected organism must evolve a series of traits that tend to maximize the intrinsic rate of natural increase. They might include the production of numerous small offspring in which little parental investment is made, a rapid rate of development with precocious reproduction, traits permitting survival during unfavorable environmental conditions, and a dispersal strategy that will maximize the probability of encountering the rare favorable environment.

K selection will occur under conditions that are the antithesis of those favoring r selection. Thus the abstract K-selecting environment is a "completely saturated ecosystem where density is high and competition for resources is intense . . . the optimal strategy channels all possible resources into survival and production of a few offspring of extremely high competitive ability" (McNaughton 1975:251). The classic example of the K-selecting environment is the tropical forest. Few would disagree with MacArthur's and Connell's description of this community:

> the tropical rain forest represents the most complex, diverse and richest example of life on land. . . . Because of the high organic productivity and the small number of physical stresses, the challenges offered to an organism are mainly biological in origin. Predators and parasites of all kinds are abundant, plants perch on plants and competition for light among the plants is intense. In these forests the number of species is astounding. (MacArthur and Connell 1966:36–37)

In contrast to the r-selecting environment, the K-selecting environment is crowded, stable, predictable, and diverse. Competition among and between organisms is the mode of life. The resources, rather than being abundant and available, are tied up by existing species. Thus, rather than the physical forces that dictate the life-history strategy in the r-selecting environment, the major selective forces in the K-selecting environment are biological in origin, and successful competition rather than rapid reproduction is the optimal tactic. This emphasis on competition rather than speed of reproduction brings with it a series of adaptations in the life history of K-selected organisms. The most important of these is a morphology and autecology that maximizes the competitive abilities of the adult plant; the reproductive tactic is thus oriented toward producing offspring with a high probability of surviving in a highly competitive environment. Frequently this means that reproduction of the parental plant is delayed until it is well established within the ecology to prevent loss of competitive energy. Fewer offspring are produced in each reproductive episode, permitting greater parental investment in each of them. This investment allows the offspring to be more competitive and, frequently, larger. K-selected organisms, like r-selected ones, face the problem of getting the next generation into a position favorable for its growth, but the means differ. Whereas the r-selected organisms may rely on a more-or-less random dispersal of offspring, the K-selected ones have to maximize the probability of successful dispersal—dispersal itself being competitive. The opportunistic r-selected organism can afford to lose most of the crop of the next generation because colonization (and thus survival of the species) can be expected to occur on the basis of a very few successful dispersal episodes. The high levels of predation in K-selected environments, however, prohibit this "lottery" tactic from being widely used: indeed, one of the major tasks confronting a K-selected parent organisms is to move its more vulnerable offspring out of its own predation "shadow" (Janzen 1970). This discussion is summarized in Table 4.1.

Before I describe domestication in terms of this theory, five important reservations and restrictions must be made. First, r- and K-selection theory, although of great utility in many situations, does not account for more than one particular type of life-history strategy (see the review by Stearns 1976). Many situations exist in which this theory is of only peripheral interest. Second, the theory is a highly simplified model of the numerous interactions that may occur in any ecological situation and at times may yield an incorrect analysis of the tactic actually used by an organism (see Demetrius 1975; Hairston, et al. 1970; Horn and Dowell 1979). This problem, however, is of little importance if we recognize that the theory is best used to structure data and conjectures concerning the evolutionary tendencies existing within a community or species without claiming thereby to have exhausted the totality of evolutionary pressures acting in the situation. Third, r- and K-selection are not mutually exclusive alternatives; rather, they are but highly abstract extremes, the vast majority of organisms inhabiting environments that are neither totally r-selecting nor totally K-selecting. Fourth, r and K selection may only be used in a comparative manner. Thus if we are considering contemporaneous communities or species we may say that X is r-selected relative to Y, whereas Y may itself be r-selected in relation to Z.

The theory may also be used to compare the types of selective pressures that are placed on a species or community as it evolves through time. Here we are concerned with changes in the direction of selective pressures and we may thus argue that the species or community is increasingly K-selected, even though in comparison with many other communities or species it is highly r-selected. Finally, r- and K-selection theory is descriptive rather than analytic. Although it attempts to identify selective pressures in relation to environmental parameters and even provides "optimal" solutions to the demands of the environment, it is obvious that evolution is not actually being guided by a specific need for optimization. The theory is merely a convenient way of discussing large masses of information in a relatively efficient manner.

We may begin to relate domestication to the theory of r and K selection by considering some of the tendencies in morphological change that have accompanied the process. Schwanitz (1966: Chapter 1) has identified the following:

1. General increase in size (gigantism), frequently accompanied by polyploidy (increase in chromosome number).
2. Reduction or loss of natural means of dissemination.
3. Decrease in the number of propagules produced, with a concomitant increase in the size of individual propagules.
4. Disproportionate increase in size of organ utilized by people.
5. Loss of delayed seed germination.
6. Loss of protective mechanisms.
7. Simultaneous ripening of the seed crop.
8. Increasing diversity of form.
9. Change in duration (annual to perennial or *vice versa*).

Although this list might be criticized, or certain items added to it, on the whole it provides a reasonable approximation of the changes characterizing plant domestication. If we take into account the fact that domestication is a symbiosis, the list speaks rather eloquently of a change to a more K-selected environment. As I have stressed, the ancestors of many domesticates are opportunistic plants, colonizers of open, ephemeral habitats. Thus their basic evolutionary heritage is one dominated by r-selective forces. This observation, however, is trivial in that it tells us nothing about changes in selective forces that may have accompanied domestication. To understand how domestication might have affected the life-history tactics of the domesticates, we must not compare them to plants in general but must look instead to the changes that have occurred within domesticates as a class.

We have already discussed many of the competitive advantages accruing to the domesticate when it is selected over evolutionary time for increases in size. The developing agroecology is a rich environment, as many authors have noted, and it is similar to the tropical forest in this respect. As I have pointed out, r-

selection occurs when the resources are both abundant and available. The developing agroecology, however, is not an open habitat like a rockfall or a glacial till. Instead, continuous competition exists for occupancy, both within the agricultural community (among the developing domesticates) and between the agricultural community and the indigenous flora. One of the bases for the relatively high level of competition found in the agroecology is its spatial and temporal stability, a result of human interference with the natural succession in the region. This stability reduces the availability of resources by permitting long-term occupation of the area by the agricultural climax. This greater locational stability leads to a *reduction in the selective importance of mechanisms for long-distance dispersal:* the human ability to act as a long distance dispersal agent reinforces the value of attempting to remain within the agroecology. As we have noted, competitive advantage is received by plants that adopt humans as dispersal agents.

Furthermore, the vast majority of animal-mediated dispersal relationships have evolved within relatively K-selecting environments, for example, in the tropics and with inhabitants of the climax flora of temperate regions. We have seen that the adoption of this tactic is expensive for the plant—it requires a large parental investment in each propagule that will be successfully dispersed—and that the establishment of the specialized dispersal relationship is dependent on the exclusion of competing nonspecialized agents. This leads us directly to *decrease in number and increase in the size of propagules.* One means by which nonhuman dispersal agents may be excluded is by the evolution of a larger propagule. Because this generally brings with it an increase in the size of the seed and thus of the potential vigor of the seedling, the very mechanism used to attract humans as dispersal agents works synergistically with the enhancement of competitive ability. We have seen that *increases in size of utilized organs* works in a similar way, increasing the ability of plants possessing these structures to compete successfully within the agroecology. *Loss of delayed seed germination* is a clear indication of the nontransient nature of the agroecology, seen from the perspective of the plant. Under any other set of conditions, the lack of mechanisms enhancing dormancy would be evidence that the ecology was stable and generally favorable, and it is therefore unnecessary to maintain morphological and autecological characteristics that would permit survival during unfavorable environmental conditions. Although loss of protective mechanisms is most often viewed as an artifact of human protective behaviors, we should remember that it has two important effects upon the general competitive ability of the domesticate. First, it permits the partitioning of energy into competitively advantageous increases in rate of growth; second, it may make the plant more palatable to people, thus giving the plant a greater chance of survival within the agroecology. *Simultaneous ripening,* although probably having evolved as a device to attract people during the early-phases of domestication, may act at later times to reduce the length of time the crop is vulnerable to attack by pathogens and predators. Thus it also may be viewed as evidence that the development of the agroecology was accompanied by increases in complexity and in biotic stress. The *increased diversity in form* characteristic of domesticates has many causes, but probably the most important is simple allopatric differentiation caused by either the gradual spread of domesticates over the globe or their independent domestication from different genetic sources in more than one region. But, as we have seen, differentiation may also arise as a result of adaptations to specific local competitive situations. *Modification in duration* of crops is applicable only to certain types of crops in specific ecological settings: many of the tropical herbaceous pulse, root, and tuber crops have experienced no change in duration; neither have the tree crops of temperate zones. Cause for the change from a perennial to an annual duration in the seed crops originating in seasonal climates has numerous components; among them is partitioning to permit higher yield and thus attract people. A change in duration may also result in an increased probability of survival within the agroecology (as the agroecology evolves and becomes more localized and temporally stable, predation will increase on those opportunistic plants that have avoided predation by being ''rare''; increase in predation may decrease the adaptiveness of perenniating structures that can serve as a reservoir of infection). Finally, changes in duration may result from an adaptation to the means used by people to prevent the agricultural climax from being overtaken by the climax evolved in the local ecology. (Thus, plowing may be seen as the mechanism reducing competition from the native flora, but it does not thereby reduce competition within the agricultural flora itself.)

If we concentrate on changes in the yield of domesticated plants, we are likely to fall into believing that the agroecology is a predominantly *r*-selecting environment for the agricultural plant. This error has two sources: the *r*-selected heritage of the domesticate and a failure to recognize the role of the symbiosis in effecting the morphology of the domesticate. Many domesticated plants, notably the grain crops have used aggregation of propagules into larger units as their prime means of ensuring quality dispersals. Rather than elaborating accessory tissues, they have sacrificed the greater part of the propagule population for attraction and rely on only a small portion of the crop for reproduction. Moreover, in the absence of planting, plants with the greatest yield will have a competitive advantage—an advantage that will increase as planting is incorporated into the behavior of humans.

The development of the agroecology is thus the evolution of an increasingly stable and predictable environment for the survival of the domesticate. The entry of relatively opportunistic plants into this environment reduces the importance of *r*-selective forces on their evolution and increases the importance of competitive interactions. The incidental domesticate, seen in these terms, is a plant that, although establishing a dispersal relationship with humans is still part of the nonagricultural ecology—an ecology that is r-selecting relative to the agroecology. Thus, in speaking earlier of the new direction and potentials for the evolution of plants that accompanied specialized domestication, one of the major phenomena I referred to was this fundamental change in the direction of selection.

It is not logically necessary that the change in the direction of selection that occurs for domesticates with the development of the agroecology be the same one that has occurred in the direction of selection upon humans. Indeed, it is possible that selective pressue on humans went in the opposite direction: the development of the agroecology was likely accompanied by a more r-selecting regime for people.[12] In essence, the difference in time and response to the developing agroecology by the two types of organisms involved in the symbiosis, people and plant, is such that whereas the system is increasingly stable for the plant partner, it is increasingly unstable and simplified for humans (see Table 4.1).

The major contributor to this development is the interaction of increasing instability in productivity with higher population levels. The competition occurring among domesticates has major effects on both yield and adaptiveness of the domesticate to local edaphic conditions: as the human population increased in response to the higher and more readily available yields of domesticates, its diet was increasingly composed of a class of resources with similar requirements for growth. Because of this, numerous factors, such as a previously insignificant variation in the climate, could have major effects upon the human subsistence base. For people, the development of the agroecology involved an increase in the availability of utilizable resources and an accompanying increase in the unpredictability of their resource base. In a sense, it became possible for humans to "use up" their environment in a way impossible in a preagricultural economy. Human demographic dynamics also underwent a fundamental shift to greater variation in local population over time (see Figure 4.1) along with a tremendous increase in total human population. As a result of the establishment of a symbiosis between humans and plants, humans were competing less with other animals for food and were also in a situation in which the total potential yield available in any environment was dramatically increased.

Although it seems apparent that domesticates were experiencing an increasingly K-selecting regime, the change to an environment dominated by r selection for human society is less easily defended. Approaching agricultural origins using the conceptual framework of r- and K-selection, however, suggests avenues of approach and poses questions worthy of investigation. Several non-

---

[12]Again, I caution the reader that I am not claiming that people are r-selected organisms; rather, I am conjecturing that selection under agricultural conditions brought about a tendency toward an increasingly r-selective regime than is found under nonagricultural conditions.

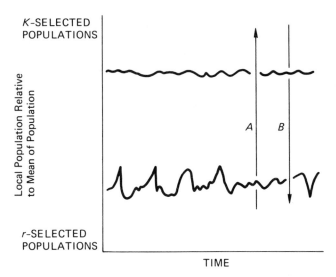

Figure 4.1 Population dynamics for *r*- and *K*-selected organisms. *A* indicates the direction of selection for agricultural plants (increasingly *K*-selected), whereas *B* gives the direction in changing selective pressures for humans (toward a more *r*-selected regime). (After Pianka 1978:121.)

trivial hypotheses concerning the effects of agricultural subsistence patterns on human culture may be generated. Many of these are not only counterintuitive, but were also potentially testable using archaeological, ethnographic, and historical sources. Thus we might expect to find that as the agroecology becomes the predominant selective force on human culture, there will be a tendency toward decrease in human stature, earlier reproduction, production of greater number of offspring, and less parental investment (e.g., training, age of independence) in these offspring. Of course, validation of these predictions would not explain how the changes occurred. As I have said, *r*- and *K*-selection theory is a descriptive tool and not an analytic one: understanding of *how* such changes occurred in human societies will require a different approach.

I have argued that vegecultural and grain-based agricultural systems have differing potentials for dispersibility based on differences in resilience, attributing the high resilience of grain-based systems in part to the (*r*-selected) opportunistic heritage of the component plants, which reduced the number of environmental parameters necessary for the system to function.

I have now offered the hypothesis that agricultural systems will increase the *K*-selective forces acting on the domesticate as a result of greater environmental predictability, and I have suggested that the advent of agriculture for people was accompanied by an increase in unpredictability in yield and thus an increase in *r*-selective pressures. How are we to reconcile these accounts?

It is really not all that difficult. The difference in dispersibility of different agricultural systems is merely a reflection of their relative resilience. Although part of the difference is attributable to their evolution within different ecological regimes, this is only of interest when we are comparing two systems to understand their relative abundances in the world. Grain crops are generally more opportunistic, but this is also merely a reflection of the environment in which they first established coevolutionary relationships with people. Increasing K-selective pressures *are* being placed upon them, but only during the growing season. The fact that they are crops that are stored as seed for the next year prohibits any tendency toward change in duration, especially when planting is combined with techniques, such as tilling or burning of the ground, that would strongly select against the perennial habit. Of course, in areas with extreme seasonality, tendencies toward change in duration would be prohibited by intrinsic environmental factors.

Reconciliation of increasing stability of the agroecology for the plant with decreasing stability for people is also a relatively easy matter. First, agriculture can be a persistant (stable) form of subsistence for our species as a whole, even though it may be accompanied by an increase in the amount of local instability. Agriculturally induced instability is basically a local phenomenon, occurring at relatively infrequent intervals. This instability is not inconsistent with increases in yield, but rather, as will be shown later, is an integral part of increasing yield. Neither is instability opposed to increases in competitive, K-selected behavior on the part of the plant; instead, it is an outgrowth of these very competitive processes. Second, there is no conflict between a simultaneous change in the direction of selection toward an r-selected regime for people with a K-selected regime for plants. We must remember that people are "buffering" the environment for plants, even during periods when environmental conditions may be inimical to their further K selection. This is clearest when we consider highly developed agricultural behaviors such as irrigation. Under conditions of great water stress, irrigation will usually go not to those fields or plants most likely to survive the drought, but to those that are least likely. Thus no selective pressures will be initiated to reduce the water requirements of the crops that are the least xerophytic. Also, even if a sudden and precipitous decline in yield occurs and a portion of the population from a settlement is forced to leave to establish residence elsewhere, no selective pressures will be placed on the crops or techniques. These people will not take with them techniques and crops that have somehow been "improved" by the experience of stress, but instead will reestablish the very system that was brought to collapse by its inherent, agriculturally induced instabilities. Thus, even though plants and people are involved in the same symbiosis, the effects of the symbiosis on the partners is not the same. Even in times of extreme stress, sufficient *unselected* seed is maintained to reestablish the agroecology without a fundamental change in its adaptation to extremes of the local environment. Thus K-selection is enhanced by people's prevention of the deleterious extremes of the local environment from having a selective effect on the plant gene pool.

Thus far, I have been considering the fundamental evolutionary processes underlying domestication and the development of the agroecology, and the static and homeostatic processes creating, limiting, and defining the form these systems may attain. In the chapters that follow, I look at how the various processes thus far described are played out through time. In Chapter 5, I show how a mechanistic model for change in feeding strategies can bring about the transition from incidental to specialized and finally to agricultural modes of domestication. In Chapter 6, I explore the role of local, agriculturally induced instabilities in the development and eventual spread of agricultural systems.

CHAPTER 5

Feeding Behavior and Change in Diet

> In the social production which men carry on, they enter into definite
> relations that are indispensable and independent of their will. . . . The
> mode of production in material life determines the general character of
> the social, political and spiritual processes of life. It is not the conscious-
> ness of men that determines their existence, but, on the contrary, their
> social existence determines their consciousness.
>
> KARL MARX, *A CONTRIBUTION TO THE CRITIQUE OF POLITICAL ECONOMY*

Introduction

A different approach may be used to elucidate the transition from incidental to
specialized and finally to agricultural domestication. This approach, drawn in
large part from theoretical ecology, benefits from its strengths and suffers from
its weaknesses. As Pileou (1977:2–3) has pointed out,

> Ecological model building takes many different forms, depending on the purpose for
> which it is done. At one extreme are so called "explanatory" models or systems models,
> whose behavior is thought to duplicate, at least approximately, the true behavior of the
> populations being modeled. . . . Their conclusions, though undeniably interesting, and
> thought provoking for ecologists, are no more than that; they relate to simple mathemati-
> cal systems, not to complex ecosystems. . . . But properly applied, they do contribute to
> our understanding of ecological processes. It is a mistake to describe these models as
> "useful" if they fit actual observations, since . . . a good fit may imply no more than
> that the observations and the model reproducing them are not capable of discriminating
> among competing theories.

Much of the impetus for the general models developed here comes from the
traditional literature of genetic fitness (e.g., Lotka 1925) and from its modern
commentators (e.g., Pileou 1977). Special emphasis has been given to the work
of Emlen on general feeding strategies (Emlen 1966b, 1973) and also to May's
(1974) comments on niche theory. In its particulars, however, the model is, to
the best of my knowledge, original (insofar as any thought in science may be so
described).

The model includes three major variables and their interrelationships: (1) the resources that comprise diets; (2) change in the abundance of these resources; and (3) the absolute and relative amounts of food consumed. These variables are used to describe alternative feeding strategies and the effects of such strategies on domestication. Our major concerns will be the nature of the change in diet dictated by the interaction of these three variables and the change in the rate of domesticates' evolution as domestication progresses from incidental to agricultural.

Perhaps the greatest pleasure provided to an author by a model occurs when the model serves to explicate a well-known phenomenon in a manner that runs counter to his intuition while preserving the integrity of his fundamental prejudices. The nicest implication of this model, therefore, is the recognition that the maintenance and even the intensification, of traditional modes of foraging behavior will ultimately serve to maximize the potential growth of agricultural systems. Put in other terms, by "not choosing" to become agricultural, humans maximized their effect on the evolution of the systems. The model also has the advantage of explaining how so gradualistic a process as domestication can be reconciled with the "sudden" appearance of agriculture. It makes it clear that the contribution of domesticates over time to the diet does not increase linearly. Instead, as the availability of domesticate sources of food increases, a "takeoff" point is reached after which, instead of continuing to increase slowly, the relative contribution of domesticates to the diet skyrockets. Further, it suggests that domestication will not have major effects on a culture's subsistence strategy for an exceedingly long time. Until the takeoff point is reached, interactions with domesticates are only another mode of foraging (albeit in a unique and evolving environment). Thus it is conceivable that domesticates were evolving 15, 25, or even 50 or more thousand years ago. We must admit that, at least in part, the age assigned to domestication has been primarily set by the biases, preconceptions, assumptions, and the excavation aims and techniques of archaeologists; compounding the problem is the fact that with increasing age comes a decreasing probability of preservation. Thus, the discovery of evidence of very early domestication will require both an adventurous mind and meticulous excavation. As long as we believed in "man the hunter," little evidence was forthcoming to question this belief. It was only after we began to see humans as primates with a typical primate diet that evidence of their basically vegetarian, gathered, diet began to accumulate. Science always involves interaction between preconception and data.

Two basic models are discussed here, a general one and a graphic one. The former deals, in a rigorous mathematical manner, with implications derived from the realization that the evolution of domestication proceeds by means of increasing abundances of domesticates in the environment and thus in the diet. Domesticates, because of their relationship with humans, have an inherent rate of increase that is larger than wild resources. I demonstrate that change in domesticate contribution to the diet will initially be quite slow, but that it will increase over time. At any given time the contribution will be a function of the abundance of domesticates in the environment. This permits us to express the relationship between domestication and the rise of agricultural systems without having to specify the exact rate of evolutionary change in domesticates or agricultural traditions. I show that the only logically possible way to affect intentionally the contribution of domesticates to the diet is by reducing the contribution of wild resources to that diet. In this manner it becomes possible to model intentionalistic thoeries for the origin of agriculture. Insights derived from these models, however, show the logical flaws in any intentionalistic explanation for the origin of agriculture. Finally, I consider the relationship between domestication and the fitness that accrues to individuals having different types of agricultural traditions. I show that cultural lineages possessing behaviors that maximize the rate of extraction of *wild* resources will, over time, have the greatest fecundity. This tactic, however, will *minimize* the rate of growth of agricultural systems.

The second part of this chapter takes the equations of the general model and applies them to a less rigorous, but more easily visualized, graphic model for the origin of agriculture. Here domesticates are derived from less valued portions of the wild plant resource set. Although this section predicts the same changes in the utilization of resources predicted in earlier chapters on the basis of ecological and evolutionary theory, and although it is also in concordance with the general mathematical model, it also clarifies some new aspects of the dynamics of domestication as follows:

1. Early domesticatory interactions indirectly increase the total potential yield obtainable from an environment. At the same time, a phenomenon I call *regression in valuation* depresses the rate of growth of domestication by reducing the evolutionary pressures placed upon domesticates. Thus the early evolution of domesticated plants proceeds very slowly. The major effect of domestication during these early periods comes from the increases in *wild* yield that it encourages. Increase in total available yield permits human population growth. Population growth, however, brings with it increased environmental disturbance and therefore increased potential for the initiation of specialized domestication. Thus, the same process provides the negative feedbacks that depress the rate of evolution under incidental domestication and also lays the demographic foundation for the conditions that facilitate the transition to specialized domestication.

2. In moving from incidental to agricultural domestication, humans experience a radical shift in feeding strategy. Rather than construct diets from the most highly valued foods in the environment, humans begin to feed on all resources in direct proportion to their perceived abundance; all available resources as perceived by the culture are utilized, or diet breadth is increased. The interaction of

shift toward higher levels of utilization of resources and the increasing abundance and yield of domesticates provides the positive feedback system that brings about the transition to full agricultural subsistence. This transition is rapid relative to the processes occurring during incidental and specialized domestication.

3. Human population growth is permitted by the increases in yield brought about by domestication. Although population growth increases the intensity of human domesticatory interactions, it is not the "cause" of domestication. Any reasonable definition of population pressure must assume that all resources are being consumed at as high a rate as is possible—to assume otherwise is to deny the very concept of population *pressure* upon food resources. However, if the wild resources are subject to high absolute levels of consumption, then the relative contribution of domesticates (or protodomesticates) to the diet will have to be correspondingly low. This could imply a very slow rate of evolution for the domesticates because they are not fed upon *preferentially,* and under these conditions the rise of agricultural systems is inhibited. Conversely, a rapid evolution of protodomesticates would require a sudden increase in their consumption (if this were not the case, they would have already been domesticated). Because feeding upon protodomesticates was always an option, however, we must wonder how population pressures could ever arise. The idea that the shift to protodomesticates arose from a change in eating patterns that reduced the contribution of wild sources of food implies that total available food was not a limiting resource and thus denies the existence of population pressure. Given that the highest yield from domesticates is obtained from *minimal* reliance upon them and this entails a very slow rate of evolution, the quick adoption of agriculture would prove ineffectual in solving the problem of lack of food in that it would require abandonment of utilizable wild resources. The tactic of minimizing the relative contribution of domesticates to the diet, however, not only contributes to the evolution of the agroecology, but also provides the highest yields possible.

In a brief application of the model to one data set, we find not only that the fit between predicted and observed data is quite good, but also that the discrepancies are capable of explanation in terms of the differing rates of evolution under the different modes of domestication. As we begin to approach the agricultural takeoff point, predicted and actual data points are in complete harmony. That the fit to the earlier periods of time is not as good is a reflection of the difficulty of finding a temporal correction to account for the slow rate of evolution under incidental and specialized domestication. Approximate temporal adjustments in the rate of domesticate evolution produce a good fit between all predicted and observed data points. Thus the model both amplifies and compliments the general coevolutionary approach taken earlier.

The General Model

We know that the interaction of humans and domesticates has produced an increase over time in the abundance of domesticates. The previous two chapters of this volume sketched out how coevolution permitted this increase in domesticate abundance and therefore led to change in human diets. In attempting to model this change we may claim that, at any given time t, the contribution of domesticates to the diet D_t was greater than at an earlier point in time. Let us call the earlier time $t - 1$. Put in other terms, $D_t > D_{t-1}$. This simple recognition permits us to represent change in the contribution of domesticates to the diet over time in following manner:

$$D_{t+1} = k_d D_t,$$

where k is the *rate* of increase in the contribution of domesticates to the diet. For example, if the contribution of domesticates to the diet at time t had been 100 units and the rate of increase in contribution per unit of time had been 101%, then the contribution of domesticates to the diet at time $t + 1$ would be 101 units.

We know far less about the type of change that has occurred in wild resources over time. The only fact of which we may be certain is that, on the average, the contribution of wild resources to the diet has not grown as quickly as that of domesticates. We know this because if it were not true then the diet of agriculturalists would not be largely composed of domesticated resources. We also have reasons to believe that feeding on wild resources had to increase, at least to some degree, as human population densities increased. For now, I therefore assume that the dietary contribution derivable from domesticates increased like that of domesticates, but at a lower rate. (Later in the chapter this assumption is relaxed to permit more realism; for the present the simple depiction of change over time in contributions derivable from wild resources has advantages in the creation of the model). Hence, let us describe the change in contribution of wild resources to the diet (W) in terms analogous to that of domesticated resources:

$$W_{t+1} = k_w W_t.$$

As before, k is the rate of increase in contribution of a class of resources. In this case it is subscripted w to indicate that we are speaking of the rate of increase in wild, as opposed to domesticated, resources. Because we know that in fact domesticated resources increased faster than wild one, we may recognize that $k_d > k_w$. Note that, in this context, we are discussing the rate of change in contribution to the diet, *not* the absolute amount of resources of either class that are actually being consumed.

To gain a better understanding of the kind of dietary change that has occurred over time, consider both the beginnings and the end of the process. First let D_0 and W_0 indicate the initial contribution of domesticates and wild resources to a diet. D_0, of course, cannot exist until the *first* domesticates appear in the diet. Domesticates, it must be recalled, are to be defined in the same terms used throughout this volume : plants involved in coevolutionary relationships with humans. Having defined D_0 in this manner we have also defined W_0 as the amount of wild resources being consumed when domesticates first enter the diet. Note that D_0 is here assumed to be very small, especially in relationship to W_0.

Having defined D_0 and W_0, we have also defined t_0 as the time at which domesticates first enter the diet; in a sense, we have started our clock. Recalling that D and W have associated rates of increase k_d and k_w, the total change in contribution of domesticates to the diet may be expressed. We will begin with the end of the process, which we call time $t + 1$.

$$D_{t+1} = k_d D_t = k_d^2 D_{t-1} = k_d^3 D_{t-2} = \cdots = k_d^{t+1} D_0. \tag{1}$$

The total change in contribution to the diet derived from wild resources may be expressed in exactly the same manner (recalling that $k_w < k_d$ and that $W_0 >> D_0$):

$$W_{t+1} = \cdots = k_w^{t+1} W_0.$$

In considering the rise of agricultural systems, our major concern is not the absolute amount of domesticates in the diet but rather their relative contribution to the diet. We must define *agricultural subsistence* as *substantial dependence upon domesticates*. No one would claim that a culture is agricultural if it is merely cultivating a crop that provides, for example, 1% of the total subsistence. Indeed, in the ethnographic literature, peoples deriving as much as 50% of their diet from agricultural activities are frequently described by terms such as *horticultural* to differentiate them from "true" agriculturalists. Likewise, as is seen at the end of this chapter, permanent settlement based on maize agriculture did not occur, in a given archaeological context, until over one-third of the diet was comprised of domesticates. The easiest manner to express the relative contribution of domesticates to the diet is to define a parameter r_t that gives the proportion of domesticates in the diet at any moment in time. Hence:

$$r_{t+1} = \frac{D_{t+1}}{W_{t+1}},$$

or, the relative contribution of domesticates to the diet at any given time is the amount of domesticates in the diet at that time divided by the amount of wild resources in the diet at the same time.

Two additional reasons exist for expressing the general model in terms of the relative contribution of domesticates to the diet. First is the practical matter of testing any of the predictions derived from the model. Here we are forced to utilize archaeological data. Within this set of data, quantitative contributions of domesticated or wild resources will be extremely difficult to determine. However, relative contributions may be reasonably approximated (subject to the standard caveats concerning relative probabilities of preservation and recovery). Second, expressing the growth of agricultural subsistence in these terms permits us to model the development of agricultural systems in terms totally congruent with an evolutionary view of cultural change. Here, the quantity k_d is a direct measure of the coevolutionary processes existing between humans and plants. Throughout the earlier discussion of domestication, I have stressed that the major effect of this coevolutionary relationship is an increase in the amount of domesticated resources in the environment. This increase in domesticates is, of course, a reliable indicator of the increasing importance of domesticates in the diet: because of the nature of coevolutionary interactions, humans must utilize domesticates or the domesticates will not increase in abundance. Also, as has already been discussed, we know that the rate of increase for wild resources has not been as great as that of domesticates. Therefore, we may use the relationship k_d/ k_w as the functional equivalent of Darwinian fitness in our analysis of the rise of agricultural subsistence patterns. We may here speak of the differential fitness of domesticates, and hence domestication, over time to gain new insights into the reasons why domestication as an evolutionary strategy came to replace gathering and hunting subsistence patterns. Likewise, we may eventually gain insights into the reasons why specialized domestication replaced incidental domestication and eventually was itself replaced by agricultural domestication.

Although it may seem a bit strange to speak of fitness in regard to rates of increase in resources, in fact a good reason exists for doing so. As is later discussed, all increases in domesticates in the diet are, essentially, additive to the diet. We have no reason to believe that humans differ from any other animal in their method of utilizing resources: resources are consumed in order of preference to limits dictated by need. Resources are not consumed randomly; rather, preferences exist in the order in which resources are consumed. However, because of the additive nature of domestication, the presence of this behavior within a cultural setting permits additional humans to exist in that area. Therefore, we may speak in totally realistic terms of domestication's increasing the relative fitness of humans practicing the behavior: cultures with agriculture grow at a faster rate than do comparable nonagricultural ones. The specific rate at which cultures grow will, itself, be a function of the rate of increase in domesticated in the diet (k_d). However, as we now see, the method by which agricultural subsistence becomes established is not simply dependent upon k_d but rather upon the interaction of k_d and k_w. Therefore. we may speak of this relationship as expressing and controlling any enhanced fitness enjoyed by cultures practicing domesticatory behaviors.

We have already noted that $r_{t+1} = (D_{t+1})/(W_{t+1})$. However, following Equations 1 and 2, we may reexpress D_t and W_t in terms of D_0, k_d, W_0, and k_w:

$$r_{t+1} = \frac{D_{t+1}}{W_{t+1}} = \frac{k_d^{t+1} D_0}{k_w^{t+1} W_0} = \left(\frac{k_d}{k_w}\right)^{t+1} \frac{D_0}{W_0}. \tag{3}$$

For simplicity we assume that k_d / k_w is a constant; this, of course, is a total fabriciation made in developing the model. We have very good reasons to expect that k_d will change radically under differing modes of domestication. Nevertheless, for the purposes of creating the model this simplification is desirable.

Within the model, k_d and k_w may be viewed as expressing the average rates of change over time. If this averaging process causes the reader discomfort, he or she should realize that, in fact, changes in these rates at different times will be largely irrelevant to the manner in which the model is developed. Changes in the rate of increase in domesticates in the diet will affect the *absolute time* at which events occur. However, as the model is developed, I standardize for time by putting all relationships in terms of the *result* of the domestication process— namely, the abundance of domesticates in the environment. Because domesticates are present in the environment only as a function of the human–plant coevolutionary interaction, the relative amount of domesticates present at any given time is an expression of both the rate of increase and the duration of the interaction. Humans cannot increase the amount of wild resources in the environment in the same way that they increase the amount of domesticates. Given the manner in which domestication has been defined throughout this volume, feeding upon wild resources must be viewed solely as predation. As we discover, the only way in which humans can influence the rate of increase in contribution of domesticates to the diet is by means of a preferential reduction in wild resource contributions. The amount of domesticates in the diet, however, is by definition a direct reflection of their abundance. Expressing changes in human diets over time in terms of the relative abundance of domesticates, therefore, allows us to compensate for any variation in the rate of k_d and therefore to treat k_d/k_w effectively as a constant.

This insight may be expressed in slightly different terms. The absolute contribution of domesticates to the diet (D_t) has been defined as immune to any intentional increase because change in this variable is dependent upon only two factors that are, themselves, merely the definition of a coevolutionary relationship. We may recognize that increasing domestication is guided by (1) natural processes, such as genetic mutation, recombination, and hybridization occurring within domesticates; and (2) cultural processes, such as innovation, intensification, and selection. Clearly, the first process is totally immune to direct cultural control (especially if we consider the means by which these processes occur and if we recognize that humans, even today, can only increase the number of such events and cannot control their outcome). In developing a general model for the origin of agricultural systems, we must assume that all cultures are equally competent in performing the cultural processes inherent in the second group of influences. Of course, accidents of insight and discovery occur in various historical settings, but these are the formal equivalents of similar accidental processes occuring at the genetic level in domesticates. Evolutionary processes cannot exist without variations to work upon. However, if we are interested in the *causality* underlying cultural change, these accidents cannot be presumed as having been primary in the development of agricultural systems. An insight, an intensification, or a discovery is not in and of itself guaranteed success. The success of these events may only be expressed over time. Expression over time is dependent upon selection: the traits must either spread throughout the culture, or they must be capable of increasing the fitness of their bearers such that, over time, the total population is increasingly composed of individuals possessing that trait. The quantity k_d must, in a general model, include any and all events, be they cultural or genetic, that have, in fact, affected the rate of increase of domesticates.

Given these observations, we may return to the development of the general model. The relationship expressed in Equation 3 may be given a more simple form. We may write k_d/k_w as $e^{\lambda}{}_t$, where λ is the logarithim of the relative increase in domesticates. Note that $k_d > k_w$ and hence that $\lambda > 0$. Therefore (substituting into Equation 3):

$$r_{t+1} = \frac{D_0}{W_0} e^{\lambda t + 1};$$

or more generally, because $r_0 = D_0/W_0$,

$$r_t = r_0 e^{\lambda t}. \tag{4}$$

This equation expresses the change in relative contribution of domesticates to the diet as a function of time. The most important thing to be noticed about the equation is that, irrespective of the value of λ a definite prediction is made concerning change in the system. Domesticate contribution to the diet does not increase in a linear fashion; instead the process increases in intensity over time. Ignoring for the moment any differences induced by different modes of domestication, we may graph the average rise in domesticate contribution to the diet as a simple "log e" curve. This has been done in Figure 5.1. The curve has been drawn as arising very near $r_t = 0$ at time $t = 0$ because we know that, in any

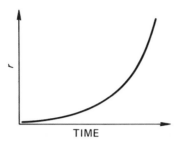

Figure 5.1   The relative contribution of domesticates r to the diet as a function of time (the time-axis is nonlinear).

pristine agricultural origin, domesticates, when they first enter the diet, constitute only a very small proportion of that diet.

Figure 5.1 shows the way in which domesticate contribution to the diet has increased over time. The nonlinearity expressed in the figure allows us to view, in a new manner, our earlier discussions of differing rates of domestication under incidental, specialized, and agricultural models of domestication: evolutionary processes enhance tendencies that would exist even if no change in the speed of domestication were to occur. The figure may also be read to express the probability of finding domesticates in archaeological deposits of varying ages. During the earliest phases of domestication, domesticates make only a small contribution to the diet and the probability of finding them preserved is likewise small. However, as time and domestication progress, the probability of finding a domesticated plant remain increases faster than the elapsed time—their contribution to the diet is increasingly large.

The major problem with Figure 5.1, as already has been stressed, lies in the observation that we have no means to compensate for the temporal changes that have occurred as a result of differing modes of domestication. Although the figure does give us some insight into the evolution of domestication, it must nevertheless be modified. The best manner in which to compensate for changes in the rate of domestication over time is by expressing r in terms of the abundance of domesticates at any given time. Define the variable μ_t to express the relative abundance of domesticates in the environment (at a given moment, i), such that

$$\mu_{t_i} = \frac{D_{t_i}}{D_{t_i} + W_{t_i}}.$$

Expressed in words, the relative abundance of domesticates is the amount of domesticates divided by the total of consumed resources. Now, (dropping the subscript t_i for notational simplicity)

$$\mu = \frac{D}{D + W}$$

$$= \frac{D/W}{(D + W)/W}$$

$$= \frac{D/W}{(D/W) + 1} \cdot$$

Now, recalling that $r = D/W$, we may express this as,

$$\mu = \frac{r}{r + 1} \cdot \qquad (5)$$

This equation makes sense; we have already accepted that domestication interactions will increase the availability of domesticates, and it is congruent with this that their abundance and contribution to the diet will increase. The limits of μ_t are also correct in that they lie at 0 and 1. Nevertheless our interest lies in expressing the contribution to the diet r_t as an expression of the abundance of domesticates. Therefore, the last equation may be manipulated (again, dropping subscripted t_i):

$$\mu = \frac{r}{r + 1}$$

$$\mu (1 + r) = r$$

$$\mu + \mu r = r$$

$$\mu/r + \mu = 1 \qquad (6)$$

$$\mu/r = 1 - \mu$$

$$r = \frac{\mu}{1 - \mu} \cdot$$

We may now graph the contribution of domesticates to diet over time as a function of the relative abundance of domesticates. This has been done in Figure 5.2. Expressing r_t in terms of μ_t satisfies our need to correct for variable rates of k_d in our analysis of change, and at the same time produces the same general type of curve found in the purely temporal model.

Comparing Figure 5.2 with Figure 5.1 we may see that the greatest difference lies in the rate of change at various points along the curve. Reexpressing the general form of the model in terms of domesticate abundances produces just the effect anticipated: during the early phases of domestication the contribution of domesticates to the diet rises exceedingly slowly until a takeoff point is reached, at which time the contribution of domesticates to the diet rises dramatically. This takeoff point is likely the ''neolithic revolution,'' which, rather than

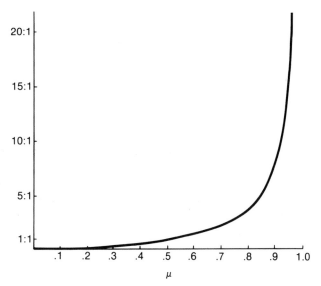

Figure 5.2   The relative contribution of domesticated plants to the diet r as a function of their relative abundance μ.

being a revolutionary event, was merely the change in state that occurs during a gradualistic evolution of agricultural systems.

To demonstrate that the procedure of expressing r_t in terms of μ_t is not arbitrary, it might be useful to see how μ changes over time. We have already seen in Equations 4 and 5 that

$$r_t = r_0 e^{\lambda t} \quad \text{and} \quad \mu_t = \frac{r_t}{1 + r_t}.$$

Hence,

$$\mu_t = \frac{r_0 e^{\lambda t}}{1 + r_0 e^{\lambda t}}.$$

But, following Equation 6,

$$r_t = \frac{\mu_t}{1 - \mu_t},$$

therefore,

$$r_0 = \frac{\mu_0}{1 - \mu_0}.$$

Substituting this last equation for r_0 in the earlier equation, we can see that

$$\mu_t = \frac{\dfrac{\mu_0}{1 - \mu_0} e^{\lambda t}}{1 + \dfrac{\mu_0}{1 - \mu_0} e^{\lambda t}} \;,$$

or   (7)

$$\mu_t = \frac{\mu_0}{1 - \mu_0} e^{\lambda t} \left(1 + \frac{\mu_0}{1 - \mu_0} e^{\lambda t}\right)^{-1}.$$

The general model is congruent with the evolutionary analysis. The abundance of domesticates, expressed over time, rises from near zero at t_0 and grows slowly until the takeoff point is reached (see Figure 5.3). During this period of rapid increase, the system grows explosively. Following the period of explosive growth, the system again slows down as it approaches its limit of $\mu_t = 1$ at t_∞. Another interesting implication of the model arises at this point: no matter how great the abundance of domesticates, at least some wild resources will always be used. This fits well with both the archaeological and the ethnographic data.

Introducing Variability

Most models for the origin of agriculture have been phrased in terms of the *rate* of the origin of agricultural systems. This is true, for example, for arguments holding that agriculture was an invention, or for those claiming that the implementation of preexisting knowledge about farming was a response to immediate conditions of population stress. All these models predict that substantial agricultural reliance will occur quickly; that is, that humans artificially increased the importance of domesticates in the diet. The earlier chapters of this volume have advanced and defended the idea that the rate of evolution of domesticates is independent of such intentionality; or, put in other terms, that any innovation or invention must be included in the parameter measuring the rate of domesticate evolution. Expressed in terms of the model developed thus far, k_d is not subject

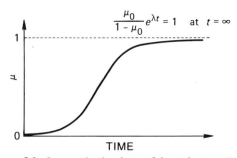

Figure 5.3   Increase in abundance of domesticates over time.

to artificial variation—the only manner in which humans can intentionally change the relative contribution of domesticates to the diet is by preferentially reducing consumption of wild resources.

This observation is central to understanding the evolution of agricultural systems and it bears repetition. The yield of domesticates that may be obtained at any given point in time is a function of all the genetic modifications that have occurred while these plants were adapting to a symbiosis with humans. This yield is also a function of any and all human activities affecting the plants in the agroecology. All the genetic and cultural factors are included, by definition, in k_d and they may not be arbitrarily reentered to provide an explanation for a change occurring in the agricultural system. At the level of the mathematical model itself, any change in r_t may equally well be modeled as either an increase in k_d or a decrease in k_w. If all possible increases in k_d are implicit in the value given this parameter, and if we wish to introduce variability into the model to compare the general model to other explanations for the rise of agriculture, we are forced to model this variability in terms of k_w.

Let us create a variant form of W that is capable of expressing the human ability intentionally to modify the diet so as artificially to increase the proportion of domesticates in the diet. An increase in the proportion of domesticates in the diet, as has already been noted, is the only manner in which we may functionally define agriculture. A variant form of W, W^*, is created here to express variability in potential dietary strategies:

$$W_t^* = W_t/\phi,$$

where ϕ is a variable expressing the amount of decrease in utilization of wild resources chosen in a particular variant dietary strategy. Holding D_t constant and recalling that $r_t = D_t/W_t$, we may express the effect of ϕ on the proportion of domesticates in the diet in this manner : The modified proportion of domesticates in the diet, μ_t^*, is

$$\mu_t^* = \frac{D_t^*}{W_t^* + D_t^*},$$

but because we have already determined that domesticates are immune from artificial increase, that is that $D_t^* = D_t$,

$$\mu_t^* = \frac{D_t}{W_t^* + D_t}.$$

The only difference between a variant strategy and the one expressed by the general model lies, as desired, in the utilization of wild resources. However, we may compare the variant diet to the general diet by recognizing that $W_t^* = W_t/\phi$. Hence:

$$\mu_t = \frac{D_t}{W_t/\phi + D_t} \cdot$$

Following the exact logic used in the derivation of Equations 5 and 6, it may be shown that

$$r_t = \frac{1}{\phi} \frac{\mu_t}{1 - \mu_t} \cdot \tag{8}$$

Expressed in words, the effect of ϕ on the diet is to increase the relative proportion of domesticates in any diet taking the W^* strategy. This is what would be expected to happen from intentionally decreasing the utilization of wild resources to favor domesticated ones.

Another way to express the general effect of ϕ is to note that it creates a distinction between *potential* and *actual* diets. Throughout the model as created thus far, I have been assumming that humans consume *all* the necessary edible resources present in the environment. This is clearly an assumption that needs further investigation. The variable ϕ may be therefore used to introduce a new level of realism into the model.

We define two new variables to express the relationships that may arise under conditions in which less than the total set of available resources is being consumed. First, consider the actual proportion of resources being consumed.

$$U = \frac{D + W^*}{D + W} \cdot \tag{9}$$

The utilization (U) of the environmental resources is the proportion of resources actually being consumed relative to those present in the environment and available for consumption. In this relationship we hold D constant because we (1) assume that domesticates are being utilized to the limit of their abundance, and (2) we know that if the utilization of any given domesticate is reduced too much, the coevolutionary pressures will be removed and that domesticate will pass out of the domesticated class of resources and revert to the wild, or even go extinct.

Second, consider the proportion of the total diet that is composed of domesticates. The relative contribution of domesticates in any diet that takes the W^*-strategy may be defined in these terms:

$$C = \frac{D}{D + W^*} \quad \text{or} \quad C = D/T^*,$$

where T^*, the total actual diet, equals $D + W^*$.

Now let us consider the effect of an intentional change in dietary composition upon the speed at which domesticates come to dominate the diet. Define the intentionally modified r as

$$r^* = D/W^* = \frac{D}{T^* - D} = \frac{1}{(T^*/D) - 1} \cdot$$

But, because T^*/D equals $1/C$,

$$r^* = \frac{1}{(1/C) - 1} \, .$$

Multiply by μ/μ

$$r^* = \frac{\mu}{(\mu/C) - \mu} \, .$$

However, μ/C may be expressed in the following manner:

$$\frac{\mu}{C} = \frac{D}{D + W} \frac{D + W^*}{D} = \frac{D + W^*}{D + W} \, .$$

Referring to Equation 9, the last relationship is identical to U. Hence (replacing subscripted t):

$$r^*_{t_i} = \frac{\mu_{t_i}}{U - \mu_{t_i}} \, . \tag{10}$$

We have already seen that U must be less than 1 and greater than 0. Thus we can see that r^* will always be greater than r for any given value of μ. The precise effect of this, however, is of major interest in understanding the rise of agricultural systems.

Because agricultural subsistence must be defined as a substantial dependence upon domesticates, r (or r^*) may be used, at any appropriate level, as the measure of agricultural subsistence. As equation 10 demonstrates, decreasing the amount of wild resources in the diet has the effect of bringing about agricultural subsistence at a lower value of μ. Recalling our earlier discussion showing that μ is proportional to time, we may understand that the W^*-strategy will indeed bring about a rise of agricultural systems at an earlier point of time. As desired, we have created a model for a fast, and an intentionalistic, origin for agriculture (see Figure 5.4).

However, by considering Figure 5.4, we may understand the fundamental contradiction inherent in any model for the origin of agricultural systems that relies upon population pressure as the driving variable. Population pressure, as cause for agricultural origins, relies upon what may only be described as an exogenous variable to control the development of agriculture—population pressure is something that arises independently of the agricultural system or domestication process. It may be contrasted to the model developed here that describes the origin and development of domesticatory and agricultural systems solely in terms of the natural and cultural processes inherent in the differentiation and intensification of the agroecology and the plants evolving within it. Clearly, population pressure arguments require that agricultural systems must develop at relatively low levels of domesticate abundance. The same is true for *any* model calling for a rapid pristine origin of agriculture as stimulated by an exogenous

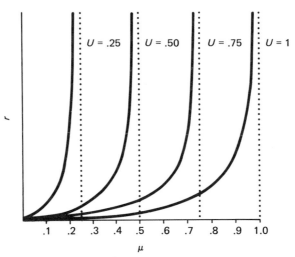

Figure 5.4   The effect of variable utilization of the total resource base on the timing of the rise of agricultural subsistence. As utilization of total resources decreases, agricultural subsistence arises at a lower abundance of domesticates (= earlier in real time).

force. However, the development of agricultural subsistence at low levels of domesticate abundance requires a reduction in feeding pressure upon wild resources. Yet, to reduce the utilization of wild resources in this manner is to contradict the fundamental principle of population pressure: put into bald terms, why would a culture experiencing population pressure abandon *any* resource?

An obvious retort to this argument presents itself: the populations experiencing population pressure did not abandon resources, but merely intensified the system while domesticates were at low levels of abundance. Unfortunately, this arguement is flawed. The general model presented here does not permit any intensification of the system above that already included in the general equations. This is because all intensifications have already been included in the argument when k_d was initially defined. The variable contains all intensifications (as well as all genetic and ecological events affecting the growth of domesticates as a class of resources) and therefore they may no longer be reentered into the process as causal variables.

Even though intensification cannot be reentered into the model, my analysis might be criticized by noting that population pressure caused intensification at an earlier point in time than would have been the case if it did not exist. This argument is also incorrect because of the manner in which we have corrected for time processes during the development of the model. By expressing the origin of agricultural systems in terms of the abundance of domesticates (itself proportional to time), we have expressed the general model in terms that substitute "domesticate time" for chronological time. Therefore, the only manner in which an event may occur earlier in real time is to have it occur at an earlier point in domesticate time. I return to this argument later in this chapter in the context of a more detailed and specific graphic model. As is pointed out then, population pressure is not totally discarded in this argument; instead it is explained as an *effect* of the development of agricultural systems. Population pressure is a corollary of agricultural evolution, not its cause.

Besides reducing immediate yield, the W^*-strategy has the effect of actually limiting the total growth possible for agricultural systems under different dietary traditions. We may easily see that the growth of the agricultural system must, ultimately, be measured in terms of μ—this parameter measures the abundance of domesticates in the environment and therefore the "size" of the agroecology. If we express μ_t in terms of r^* and U, we find (following the derivation of Equation 7):

$$\mu_t = \frac{\mu_0}{U - \mu_0} e^{\lambda t} \left(U + \frac{\mu_0}{U - \mu_0} e^{\lambda t} \right)^{-1}.$$

The size of U places a limit upon the total development of μ_t. This is illustrated in Figure 5.5.

Any theory for the origin of agriculture that seeks to transcend the limitations placed by evolutionary processes upon the development of the domestication symbiosis faces not only the problem of accounting for a reduction in the contribution of wild resources, but also must consider that this same limitation also acts to restrict the total potential development of the agroecology. An early establishment of agricultural subsistence by means of an artifical lowering of U will have the somewhat strange side effect of limiting future growth in the system.

Domestication and Demic Spread

The preceding disucssion of variability in U has had two purposes: First, I have tried to demonstrate that population pressure and similar arguments predicated upon a speeding-up of the evolution of agricultural systems face serious problems in accounting for the specific way in which the consumption of domesticates interacted with the consumption of wild resources; here, the general model serves to reduce such arguments to absurdity. Second, I am trying to point out that the development of agricultural systems may have important effects upon potential population growth in cultures possessing differing subsistence traditions; here, the concept of variation upon U may have some meaning and may generate some interesting hypotheses. We cannot assume that a culture will optimize its tactics to create the most efficient pristine development of agricultural systems. Instead, we should explore the effects of differing subsistence

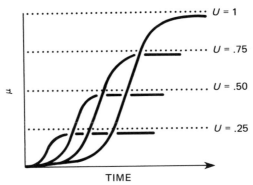

Figure 5.5 The effect of an early origin of agriculture on the ultimate development of agricultural systems. As U decreases below its limit of 1, the total abundance of domesticates at $t = \infty$ also decreases. The dotted lines indicate the value of $[\mu/(U - \mu)]\, e^{\lambda t}$ for selected values of U: at time $t = \infty$, μ will obtain these values.

strategies (expressed in U and hence r) upon the potential for growth and expansion of these systems.

Thus far the general model has had to treat the proportion of domesticates in the diet as totally independent of population size. This is neither logical nor true. At any given population size N, any particular abundance of domesticates μ is equally plausible from a purely logical viewpoint. Furthermore, as has been stressed in the earlier chapters of this volume, the absolute size and density of population is a major determinant in the nature and rate of change in domesticatory processes. We see that a definite cost must be paid by cultures choosing to restrict their consumption of wild plants. Although these cultures achieve agricultural subsistence at an early time, they are less fit than cultures possessing another tactic that includes maximization of total rate of return from *all* types of resources.

This observation is counterintuitive. Restricting consumption of wild resources would seem both logical and prudent. General and far-reaching constraints upon the consumption of wild resources in favor of domesticates would serve to buffer the subsistence system against occasional crises in agricultural production. In this sense, it would be an adaptive tactic for a culture to choose. However, the rate of growth of such rational subsistence systems would be far less than that of cultures possessing a less-adaptive strategy. Consumption of all the resources present in the environment to the limits of their perceived availability will permit a higher differential rate of increase for members of that culture. Artificial restrictions upon consumption of available resources will perforce decrease the differential rate of increase. Cultures adopting, for whatever reason, this latter tactic will have, over time, far fewer members and will come to be swamped by cultures with an "irrational" subsistence tradition.

This relationship may be illustrated by considering Figure 5.5 in a slightly different light. As t approaches infinity, the limit of μ approaches U. However, at the same time, as time progresses, r^* also rises to infinity (Figure 5.4). Hence, as time goes on, the diet becomes increasingly composed of domesticates and, within realistic time spans, we may speak of a time when domesticates comprise an extremely important part of the diet; indeed, they provide the bulk of the utilized resources—that "substantial proportion" of the diet that was earlier used to characterize agricultural subsistence. At this point, the implication should be clear: the quantity of food provided by agricultural subsistence serves, in essence, to define the carrying capacity of a region for humans. If, however, the absolute abundance of domesticates is limited, as it is when U is less than unity, the absolute carrying capacity will likewise be limited. Treating population as a function of available yield (as has been done throughout this work), it should be obvious that to restrict μ is to restrict population; and to restrict population is to limit the number of members in future generations. Whereas k_d expresses the fitness of domesticates, the interaction of μ and r^* will come to express, over relatively short periods of time, the fitness of cultures with alternative subsistence strategies.

This argument may be formalized (following Pileou 1977 and F. Smith 1963) and put into terms of individual fitness in the following manner: We know that the *rate* of population growth is dependent upon per capita yield; the greater this yield, the greater the potential fecundity (controlling for all other variables that may reasonably be assumed to be randomly connected to subsistence strategy). When a population is growing, food is being used both for maintenance of that population and for growth. However, when a population reaches the level defined by the carrying capacity, food is utilized solely for maintenance. Defining population as N and the equilibirum (carrying capacity) population as N_{max} we may state:

$$\frac{1}{N}\frac{dN}{dt} = (b - e)\left(\frac{N_{max} - N}{N_{max}}\right) = d\left(\frac{N_{max} - N}{N_{max}}\right), \qquad (11)$$

where b is the birth rate, e is the death rate, and d is their result (b-e). When the population is below N_{max}, it increases, and when it reaches this level, growth of population has been stabilized (no further change occurs).

However, the *actual* growth rate depends not upon the relationship of a given population value to the ultimate population that may eventually be obtained, but rather depends upon the rate with which available food is utilized. Below population saturation, food is extracted at a rate below that of maximal extraction. Hence another relation controls per capita population growth rates at any moment in time:

$$\frac{1}{N}\frac{dN}{dt} = d\left(\frac{F_{max} - F}{F}\right).$$ (12)

Here, F is the per capita rate of food utilization for *all* resources at a given time, and F_{max} is the highest possible rate of food utilization. Note that F and F_{max} are not equal to N and N_{max} but are variables that relate solely to the rate of extraction of resources from the environment. We now consider how changes in the rate of food utilization may affect growth rates, and therefore how different strategies affecting this parameter may influence the differential fitness of competing cultural traditions.

The rate of food utilization is related to the total population existing at any given time (that food being utilized for maintenance), and it is also dependent upon the rate at which the population is growing at that time. Thus F_t is dependent upon both N_t and $(dN/dt)_t$. At any given moment, the rate of utilization is unlikely to be the same for these two variables; that is, the rate of utilization determined by the standing population is not the same as the rate of utilization determined by the immediate capacity for population growth. Nevertheless, the total rate of utilization at any given time will be the sum of the contributions made by each of the separate processes. Hence,

$$F_t = c_1 N_t + c_2 \frac{dN}{dt} \qquad \text{(where} \quad c_1, c_2 > 0\text{)}.$$ (13)

The most important effect of taking a W^* strategy will be to limit the value of F_t relative to those cultural traditions that take advantage of all resources present in the environment.[1]

When the population is at equilibrium we know that it is no longer growing. Therefore, $dN/dt = 0$. We also know that the population has reached its maximum. Therefore, $N_{max} = N$. Finally, we know that the rate of food utilization has also been maximized. Hence $F_t = F_{max}$. Substituting these equilibrium values into Equation 13 we get

$$F_{max} = c_1 N_{max}.$$ (14)

---

[1] I will not consider the relative values of c_1 and c_2 assuming, for the purposes under consideration here, that these variables will be causally unconnected with differing subsistence strategies vis-à-vis domestication. (Indeed, the relationship of c_1 amd c_2 to N and dN/dt may possibly best be described as a function, and Equation 13 hence may have to be replaced by partial differentials.) Nevertheless, these variables are of major interest in modeling the natural selection of cultural traditions: cultural behavior affecting the ratio of energy invested in maintenance of a population to that invested in succeeding generations brings about natural selection between these traditions. Were this volume concerned with demography, *sensu stricto*, the implications would be spun out at length.

Having now determined an expression for F_{max} in terms of N_{max} (Equation 14) and having already defined F_t in terms of N and dN/dt (Equation 13), we may now express the relationship between per capita growth and the utilization of food in new terms. Substituting Equations 13 and 14 into Equation 12 (I will leave in all of the steps because the algebra is a bit tricky):

$$\frac{1}{N}\frac{dN}{dt} = d\left(\frac{c_1 N_{max} - c_1 N - c_2 \frac{dN}{dt}}{c_1 N_{max}}\right).$$

Simplify:

$$\frac{1}{N}\frac{dN}{dt} = d\left(\frac{N_{max} - N - \frac{c_2}{c_1}\frac{dN}{dt}}{N_{max}}\right),$$

let $c_1/c_2 = c$;

$$\frac{1}{N}\frac{dN}{dt} = d\left(\frac{N_{max} - N - \frac{1}{c}\frac{dN}{dt}}{N_{max}}\right),$$

$$\frac{1}{N}\frac{dN}{dt} = \frac{dN_{max} - dN - \frac{d}{c}\frac{dN}{dt}}{N_{max}}$$

$$\frac{1}{N}\frac{dN}{dt} + \frac{d}{c}\frac{dN}{dt}\frac{1}{N_{max}} = \frac{d(N_{max} - N)}{N_{max}}$$

$$\frac{dN}{dt}\left(\frac{1}{N} + \frac{d}{c}\frac{1}{N_{max}}\right) = \frac{d(N_{max} - N)}{N_{max}}$$

$$\frac{dN}{dt}\left(\frac{cN_{max} + dN}{NcN_{max}}\right) = \frac{d(N_{max} - N)}{N_{max}}$$

$$\frac{dN}{dt}\left(\frac{N_{max} + \frac{d}{c}N}{NN_{max}}\right) = \frac{d(N_{max} - N)}{N_{max}}$$

$$\left(\frac{dN}{dt}\right)\left(\frac{1}{N}\right)\left(\frac{N_{max} + \frac{d}{c}N}{N_{max}}\right) = \frac{d(N_{max} - N)}{N_{max}}$$

$$\frac{1}{N}\frac{dN}{dt} = \frac{d(N_{max} - N)}{N_{max}}\frac{N_{max}}{N_{max} + \frac{d}{c}N}$$

$$\frac{1}{N}\frac{dN}{dt} = d\left(\frac{N_{max} - N}{N_{max} + \frac{d}{c}N}\right).$$

(15)

In other words, the effect of rate of food intake on population growth is to depress the rate of growth in proportion to the quantity $(d/c)N$. The intrinsic rate of increase, d, may be assumed to be the same for all human populations. The variable c expresses F_t, the rate of food utilization at a moment in time. Therefore, reducing the value of F_{max} (the effect, at time t, of putting limits upon U) will have the effect of reducing the rate of population growth because it also decreases the value of c. (If this is not clear in this context, substitute higher and lower values for F_{max} in Equation 12).

Compare Equation 15 to Equation 11: the important difference is that Equation 15 includes the variable that controls the rate of food intake at any moment in time. As we have already seen, this variable is merely another way to express differences existing between a maximizing tradition and any other tradition that preferentially restricts the consumption of wild resources. Therefore, c has the power to express the type of variation that will exist between the maximizing strategy W and the W^* strategy where U is less than unity.

Figure 5.6 graphs population over time for two hypothetical cultural traditions, one of which has adopted the W^* strategy and the other of which maximizes its consumption of all resources at any moment in time. This figure may be read to represent the *fitness differential* existing at any point in time between these two alternative dietary strategies: subscripting N_t to indicate the two strategies, we may see that the population at any moment in time would be N_{tW} for the population with the maximizing tradition and N_{tW^*} for the population taking a tradition in which U is less than unity. However, $N_{tW} >> N_{tW^*}$ during the early period of time indicated on the figure. Although in the ideal world of mathematics both populations would ultimately reach the same population level, in the real world numerous factors will interfere with the development and spread of the cultural tradition maintaining a W^* strategy.

The difference in population growth rates expressed during early periods of time by the different cultural traditions favors the culture with the maximizing strategy. Several reasons exist for this. First, the cultural tradition having the higher growth rate also has a higher inherent rate of diffusion; any factor independent of subsistence behavior that encourages migration of individuals out of an area has a larger pool of individuals to work on. Therefore, given the same rate of diffusion (which as we see in Chapter 6 is unlikely), the maximizing tradition has a higher cultural "fecundity." This leads to the second reason why the maximizing tradition has a higher demic fitness. In the real world, land for colonization and settlement is a limiting factor. The cultural traditions discussed here may therefore be seen as competitors, over long periods of time, for the utilization of this resource. The cultural tradition having the greater rate of increase during early periods of time will have a competitive advantage simply because of its higher probability for an early colonization of unoccupied land. This competition need not be played out solely in terms of a passive competition

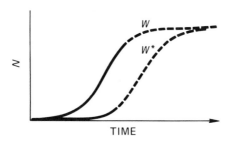

Figure 5.6 Fitness of different dietary strategies (W = maximizing strategy; W^* = W^* strategy). The solid lines represent the evolutionarily significant portion of the curve; the broken lines serve to show that we are, in fact, dealing with alternative strategies in the same framework that have the same limit.

for resources: the higher numbers achieved at an early point of time in the culture with a subsistence-maximizing strategy places that culture at a great advantage in any direct competition between the two traditions (such as warfare). Finally, any independent variable that decreases population is more likely to drive the smaller population to extinction merely because it is small for a longer period of time than is the maximizing tradition. At the very beginning of the sequence (shown at the left of Figure 5.6), both populations have an equal probability of extinction from such random variables; during the evolutionarily significant part of the curve, the probability of extinction from random variables decreases for the maximizing population but remains essentially constant for the population with the W^* strategy.

The preceding analysis of competition between two traditions has been in keeping with basic evolutionary theory. One matter, however, deserves mention lest misunderstanding occur. When speaking of relative fitness, we must assume that the two groups are in sufficient proximity that the effects of differential fitness may be expressed. Differential fitness may not be expressed if a group is isolated from the evolutionary changes that have been occurring in other populations. These groups are analogous to the "relic" or "primitive" species of evolutionary biology that, having been insulated from evolutionary events occurring in the rest of the lineage, maintain certain traits characteristic of the ancestors of the lineage as a whole. Hence the model is not to be read as claiming that a maximizing strategy *must* evolve; only that as it evolves it will displace other strategies.

Individual fitness *within* a particular cultural tradition gives us another avenue to approach the evolution of the maximizing strategy. Equation 15, as has already been noted, is expressed in terms of per capita reproduction. A given cultural tradition receives enhanced relative fitness solely because of the cumulative effects expressed over time of enhanced individual fitness. Therefore we may claim that an inherent reason exists for believing that a tradition maximizing consumption of resources will become established within any culture: those individuals maximizing their consumption of wild resources will have the greatest number of offspring and, in terms completely analogous to those discussed above for groups, will come to dominate the culture simply by means of enhanced fecundity. This observation is pregnant with implications for a better understanding of the evolution of subsistence traditions within cultures.

We have already seen that good reasons exist for believing that a maximizing strategy will be fit at both the individual and the demic (population) levels. Yet we also know that restrictions upon the consumption of resources are anything but uncommon in human societies (tabus, ritually unclean foods, etc). How are we to explain this?

Let us begin with the individual fitness problem. Equation 15 implies that minimizing c, that is, maximizing the diet, will also maximize fitness. Hence, if a food tabu is maintained within a lineage, we might hypothesize that the cost of including a particular resource within a diet is greater than the reproductive gain obtained from that resource. However, an implicit and seldom acknowledged assumption underlies this train of logic. We must assume that the important cultural inheritance system is strictly vertical (in the sense of Cavalli-Sforza and Feldman 1981); that is, the cultural trait is obtained from one's genetic parents and passed to one's offspring. Unfortunately, cultural inheritance is seldom as simple as genetic inheritance. Instead, cultural information is passed between genetically unrelated members of the same and differing generations. The "economic" interpretation for the maintenance of tabu must demonstrate that the specific mode of inheritance for the behavior under consideration is appropriate to the explanation given.

At the other extreme, we might analyze the origin and maintenance of tabu in terms that place emphasis upon the *cultural* fitness of the behavior. Here, we would begin by assuming that inheritance is multidirectional and that a particular trait (such as a food tabu) has a "fitness" that may be derived from the *symbolic* importance of that trait. Here our emphasis would be placed upon totally non-economic concerns, and phenomena such as prestige, symbolic meaningfulness, cultural identity, and status would come to the forefront. Within the confines of this analysis, the effect of a given behavior upon individual fitness would be irrelevant to its spread throughout the culture because "preselection" of traits would occur: individuals would adopt the behavior and maintain a tabu without recognizing that to do so is to reduce the number of offspring they might have. This analysis is not really contradictory to the one based upon individual fitness because it is proceeding upon the basis of another, equally valid, method for the maximization of fitness—here fitness is maximized within the symbolic system and, having been maximized, *all* members of the cultural tradition possess the trait. Within Darwinian analysis, if no difference exists within a trait in a population we cannot speak of fitness differentials; recall that variation underlies any possible selection. Hence it is not surprising that many tabus are accompanied by rather severe consequences if they are broken: the Orthodox Jew who intentionally eats pork disassociates himself from his cultural ingroup; in the process, uniformity in behavior is maintained within that group.

The symbolic and the economic, or the individual and the group, interpretations of cultural evolution are not really in conflict. Instead they are merely different levels at which cultural selection may occur. These levels are in no manner distinct; instead, interaction occurs between them. Therefore, even though both individual and group fitness would be maximized by a subsistence strategy that maximizes potential resource procurement, evolution may well produce an economically less-optimal result (indeed, as we have already seen, and as is stressed in the final chapter of this volume, the optimal strategy is merely optimal for the spread of the trait: dietary maximization, although fit, is, in fact, rather maladaptive over the long term). Because Darwinism makes no predictions concerning the potential adaptiveness of selected traits, we must treat problems such as food tabus at the empirical level. But in this we can gain some guidance from individual fitness as interpreted within the realities of a cultural-inheritance system. Although we may be certain that a tabu will not survive if it in fact lowers the fitness of individuals (decreases the number of their potential offspring), we must remain open to the fact that individual fitness within a cultural setting may be decreased by reasons totally unrelated to economic or caloric concerns—the individual who is killed by the group for eating a tabu resource has had his or her fitness lowered in a manner that is the evolutionary equivalent of the situation in which the individual had chosen to expand dietary preferences to include a plant that was a deadly poison.

Returning to our discussion of the evolution of a maximizing strategy, we may note that a final reason exists for believing that, over time, agricultural traditions that maximize their consumption of resources will be the most successful. Throughout this general model, I have taken the tactic of varying only one parameter at a time. Hence the changes in reproductive rate were discussed independently of the relationship between total utilization of the environment (U) and its effect on the development of the agroecology (expressed as μ). Yet a little reflection will show that the population trajectory shown in Figure 5.6 must be modified by the processes illustrated in Figures 5.4 and 5.5. Placing limits upon the development of μ places limits upon the total population that may be obtained. Likewise, the early adoption of agricultural subsistence is bought at the cost of a limited development of the agricultural system. I would now like to turn to a final matter that influences the evolution of the agricultural subsistence system and point out that a tendency exists in all agricultural systems for U to tend to unity.

The modeling of variability in subsistence patterns thus far, I hope, has produced the impression that taking a tactic of reduced dependence upon wild resources produces numerous problems both logical and demographic. Further, I hope it demonstrates that a fairly deterministic model for the timing and evolution of agriculture can be used to analyze the relative fitness of members of cultures having different subsistence traditions. The model as sketched out here is totally congruent with a Darwinian interpretation of cultural change: cultural traditions that maximize fitness during the evolution of agricultural systems will have an inherent tendency to spread differentially by means of the enhanced reproductive success of individuals within that tradition.

I now return to the issue of the coevolution of humans and domesticates and show that tendencies within the symbiosis also dictate a history for the evolution of agricultural systems under which U will tend, over time, to approach unity.

We have already seen (Equation 3) that the relative contribution of domesticates to the diet may be expressed as

$$ r_{t+1} = \left(\frac{k_d}{k_w} \right)^{t+1} \frac{D_0}{W_0} $$

Yet, because of the nature of coevolution between humans and plants, the rate of increase in domesticate (k_d) eventually comes to dominate the relationship k_d/k_w. This occurs for several reasons. First, the effective increase in the rate of extraction of wild resources levels off. As has been stressed in earlier chapters, and as is discussed at length in Chapter 6, one of the most important effects of human interactions with plants is an increase in the abundance of domesticates. Fitness is thereby increased for both members of the symbiotic relationship. The relationship between humans and wild plants, however, is purely predatory. Increased feeding brings with it no tendency for increased productivity in these resources. In fact, increased consumption may well *decrease* potential yields because of over-harvesting. I do not here consider the effect of decreases in the rate of increase in yield obtainable from wild resources; however, it should be recognized that such events are likely and dramatically enhance the process to be outlined here. Second, as human population levels and densities increase during the transitions from incidental to specialized and finally to agricultural modes of domestication, the rate of increase in domesticate productivity (itself a function of domesticate evolution) also increases. Again, as has been done throughout the development of the general model, this effect can be ignored by treating the phenomenon as part of an averaging procedure.

Let us consider the effect upon the change in the relative contribution of domesticates to the diet brought about by the decline in the rate of extraction of wild resources from the environment. At some point, which we may call t^*, increase in rate of extraction of wild resources will not change; k_w will, in effect, equal unity. At that point the relative contribution of domesticates to the diet will be controlled by

$$r_{t^*} = \left(\frac{k_d}{1} \right)^{t^*} \frac{D_0}{W_0} = k_d{}^{t^*} \frac{D_0}{W_0} \, . \tag{16}$$

Under these conditions, the relative increase in domesticates in the diet will equal the rate of increase in domesticates themselves. Hence, the rate of increase in domesticates will be the only factor influencing the equation

$$r_{t^*} = r_0 e^{\lambda t}.$$

Under these conditions, the harvested portion of W serves to define the abundance of W in the environment. The W^*-tactic, it may be recalled, reflects a reduction in the rate of increase in the contribution of wild resources to the human diet. Yet without a rate of increase, that is, without any growth in the rate of consumption of wild plants (and recalling that we are not concerned with declines in k_w below 1), no reduction is possible in that rate. Under these conditions W^* must be identical to W. If W^* is equal to W, U must be 1:

$$W = W^* \, , \qquad U = (D + W^*)/(D + W);$$

hence,

$$U = (D + W)/(D + W) = 1.$$

Substituting into Equation 10:

$$r_t^* = \frac{\mu_t}{U - \mu_t}, \qquad \text{at} \quad t = t^*,$$

$$r_t^* = r_{t^*} = \frac{\mu_{t^*}}{1 - \mu_{t^*}} . \tag{17}$$

We see subsequently that an analogous argument may be developed by considering the large quantitative increases that occur because of the coevolution of humans and domesticates. As agricultural systems develop, the yield obtained from domesticates overwhelms any possible yield obtainable from wild resources. Hence, the utilization of the total resources base approaches unity as a simple expression of the large additive yield made possible by agricultural modes of subsistence.

The Graphic Model

The previous model, although interesting, does not allow for certain operations of potential importance. Therefore I now take some of the insights developed in the general model and place them into a context in which we may discuss diet and dietary change in a more detailed manner.

The model advanced here is based upon a rather simple view of feeding behavior, but one that may be applied to humans with little danger of contradiction. We define *food* as that subset of total available resources culturally perceived as edible. Food may be divided into classes of resources. In general, we refer to any class of resources as R. Hence, the potential amount of plant resources available for inclusion in a diet is represented by P, and the potential amount of animals is A.

Each of these classes will be totally consumed only if *all* the potential food is included in a diet at a given time. Because this need not occur, we define the *actual* amount of a resource class that is consumed in this manner: r for any class; hence, p for plants and a for animals.

Classes of resources are obviously composed of a number of distinct foods. The *components* of a class are defined as sets

$$R = \{R_1, R_2, R_3, \ldots, R_k\}$$

for potential resource components, and

$$r = \{r_1, r_2, r_3, \ldots, r_l\}$$

for components included in the diet. Note that the sets need not be the same size (not all potential resources have to be included in the actual diet) but that they are ordered in the same manner. Thus if A_5 refers to rabbits, a_5 also refers to rabbits; however, the latter are actually in a diet.[2]

We assume that food is a relatively inflexible resource; that is, the amount of food consumed $(a + p)$ is not dependent upon the absolute abundance of resources in the environment, even though the specific foods that comprise the diet (the specific a_n's and p_n's) may be determined by the total amount of

[2]The components of a resource (our R_n, A_n, and P_n) do not have to represent species. In reality, most species of plants or animals contain several discrete resource components: for example, apples that may be picked from the ground will be more highly valued than (that is, likely to be consumed before) those growing at the top of the tree. However, because we are dealing with relationships between classes of resources, this phenomenon is rather unimportant.

[3]Of course, the total amount of available food—if we consider time periods congruent with a human generation—may well have effects upon the total number of people a given environment may support (the carrying capacity of that environment). When agricultural domesticates finally come to dominate the human diet, this increase in total yield becomes extremely significant. In the earliest stages of domestication, the total yield contributed by domesticates is small compared to that contributed by wild foods.

Recalling our earlier discussion on the role of the individual organism in evolutionary change, this discussion of feeding behavior permits us to see how the feeding activities of *individual* humans are congruent with the types of coevolutionary sequences advanced earlier in this volume. Thus it is heuristically useful and conceptually rigorous to separate long- and short-term demographic effects of yield on population growth from the short-term perceptions of an individual concerning food availability and valuation. The latter will determine the actual diet and type of feeding behavior adopted. The actual diet will be drawn from a universe (the potential diet) that is also, at least in part, a function of individual and cultural perceptions concerning what constitutes *food*.

available resources $(A + P)$.[3] (The precise amount of food consumed and the specific composition of the diet do not have to be known to be able to use this model.) An important corollary of this assumption of inflexibility in food consumption is that the percentage of that resource actually consumed also varies (although not in a simple way) with the total amount of that resource present in the environment. Thus, if a highly valued food is especially abundant one year, feeding on less-valued foods will decrease while feeding on more-valued ones remain unchanged.

Our second assumption is that not all resources are equally valued. In practice this means that certain foods are consumed to their level of availability in the environment whereas others are seldom eaten. It is very important not to confuse *availability* with *biomass* or some other "ecological" measure of abundance: deep-sea fish are not a readily available resource for humans who lack boats even if such fish occasionally enter shallow waters where they may be caught. Likewise, the valuation of resources reflects the technology, preferences, and habits of a culture. For the purpose of this model, we consider all these variables as integrated into our concept of *value*. The environment contains a large variety of potential food sources and the utilization of these resources is different for different species. Certain foods are never used because they are outside an animal's feeding pattern (wolves are not food for deer), such foods are not included in a class of resources (as R_n's). Other foods are resorted to only in times of emergency (given a choice, deer will eat grass in preference to leafless twigs); thus the components $A_1, A_2, A_3 \ldots A_n$ of the resource set A will be given differing values by the consuming animal. It is trivially obvious that man, like any other animal, tends to choose high-valued components of any resource over the lower-valued components.[4]

Our third assumption, the one that is most difficult to justify mathematically, is that value and abundance in diets are related in a particular manner. We assume that the most abundant resource in the potential set of dietary resources receives a valuation that is intermediate between the most and least valued potential resource components. That is, that valuation is a ranking system in which the most highly valued and the least-valued components are relatively scarce *in the diet*. Hence, the general way to represent a diet graphically would be a curve with a single maximum that represents the most abundant resource component of the potential diet as having a median value. One possible representation of such a curve is given in Figure 5.7.

[4]Although it may be trivially obvious that "high value" food will be preferentially eaten, analysis of the factors entering into the determination of value is an exceedingly difficult task. For an introduction to this subject (which has unfortunately been dominated by an "optimization" approach—Lewontin 1979a), see Covich 1976, Pearson 1976, Pulliam 1974, Schoener 1971, Orians and Pearson 1979, Emlen 1966b, and Pyke *et al.* 1977. One of the advantages of the model advanced here is that we need to determine valuation only in relative terms for it to function as a halfway decent (although highly abstract) approximation of reality.

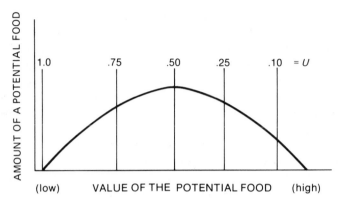

Figure 5.7   The relationship between food value and amount of food. Total area under the curve gives potential diet; lines show actual diets at five levels of U where U is the actual diet/potential diet.

   The relationship between the amount of a given resource and its value is depicted in such a manner that resource components having approximately the average value for the curve are most abundant in the diet. This is not so much an assumption as it is inherent in the way we define the concept of *value* for different sources of food within a class of resources: scarce and desirable resource components receive a higher value than do common and desirable ones. Similarly, scarce components of a resource that are less desirable than the average for that resource are given a lower valuation even if their caloric content or some other aspect of value is comparable to foods having an average valuation. Another way of viewing this is by noting that a resource component that is highly valued but difficult to obtain with existing habits and techniques contributes only a small amount of food to the actual diet even though it is common in the environment by some other measure. Conversely, the less-valued components of a resource that are functionally scarce or difficult to obtain enter the diet only after other components of the resource have been consumed. (This all follows from our assumption that animals try to include as many high-valued foods and as few low-valued ones as possible in their diets.) All in all, the contribution of a resource, or a component of a given resource, to the actual diet of humans reflect the valuation given it. At the same time, changes in diet affect valuation. Valuation is not caloric value or some other objective measure of value of a food to humans; rather *value* is human integrated perception of investment and return, availability, nutrition, and abundance. Thus for most animals, and especially for omnivores such as humans, the most-valued and least-valued components of the potential diet are relatively uncommon in relation to the components having an average value.

   Valuation is not, per se, related to the relative abundance of the competing *classes* of resources. Given any location on the graph for the two valuation curves, manipulation to account for all possible combinations of relative abundances is a rather simple matter.[5] Valuation is thus not synonymous with the absolute amounts of resources eaten. Under various regimes of resource valuation the same quantities of food eaten may reflect very different relative availabilities, resource utilization levles, and valuations. Valuation is just what it appears to be—the value placed upon different resources by the consuming animal, the order in which foods are consumed in terms of preferences within a real diet. Study of the valuation given resources by most animals is very difficult because we must deduce it from feeding behavior under natural and experimental conditions. The study of the valuation given resources by humans is somewhat easier because we may use cultural and individual attitudes toward resources to help establish our valuation curves. It is at least theoretically possible to construct valuation curves for different classes of resources by merely asking informants to rank foods and then by comparing the rankings with the food items utilized. (Again, food items are not necessarily species of food.)

Although the type of curve shown in Figure 5.7 is intuitively pleasing, it is necessary to consider several items or misunderstanding is likely to occur. First is the matter mentioned above: the curve does not seek to represent abundance of resource components in relation to ecological or objective criteria (we are speaking of abundance in potential diets). Hence, phenomena such as satiation must be brought into consideration. Because feeding proceeds from the most-valued to the least-valued resource components (from right to left on Figure 5.7) and because humans may only consume a limited amount of food, low-valued foods are under normal conditions brought into the diet in only limited amounts. Second, it must be recognized that Figure 5.7 does not attempt to express anything more than a heurestic: hence, the peak of the curve represents a maximum in abundance, and not an optimum of any sort. This is a graphic device and, like the general model, does not seek to make predictions concerning abundance in the environment and value. However, unlike the general model, it attempts to point out that highly valued resources are likely to be quite rare in diets; this is little more than another way of expressing the fact that desirable and rare resource components tend to have their perceived value enhanced simply as a function of their rarity.

Having constructed a curve of resources in terms of value, we may represent the realized diet by drawing a vertical line on Figure 5.7 such that the area to the right of the line and under the curve is equal to the percentage of available

---

[5]Of course, it may be argued that changes in abundance may actually cause changes in the shape and location of the value distribution curves—that a common resource may become more valuable simply because it becomes less available. There is some truth in this, and we return to it later. It is necessary to present this model in a highly simplified form before attempting realism.

resources that is consumed. This line is located in reference to the righthand endpoint of the curve; animals attempt to include as much of the available food that is highly valued as possible. For now, we shall not worry at all about the actual amount of food contributed by these highly valued resources because manipulation of proportional contribution of valued and less-valued classes of resources is quite simple. Instead we concentrate upon the location of this line of utilization (U). At $U = .1$ (10% of the potential resources have been consumed), only the most highly valued components are included in the diet,[6] whereas at $U = .75$, all but the least-valued foods are included. At $U = 1$, all potential food are eaten. Hence, at this level, and *only* at this level, $a = A$ and $p = P$. U is defined in this context in these terms.

$$U = \frac{\text{resources actually consumed}}{\text{potential resources}} = \frac{a + p}{A + P}.$$

Note that U in this context is the functional equivalent of U as derived in the general model.

Let us begin to explore the relationship between utilization of resources and classes of resources by considering an extreme situation in which the two classes of resources have a nonoverlapping valuation.

Figure 5.8 represents a case that never exists in reality, one in which two classes of resources have totally different mean and extreme values. Because it is reasonable to consider animal foods as generally more valuable to people than are plant foods, resource class P is located to the left of resource class A. In this case, the most highly valued componets of the less-valued class of resource are less valuable than the least highly valued components of the other class. Given the method of ranking we have used to establish value, this situation is also the extreme case for difference in value between two classes of resources: because value is established by ranking components of a class, ranking must be continuous (although, as is discussed subsequently, it is possible for components of *different* classes to have the same absolute valuation).

The contribution of the two classes of resources to the actual diet differs at various levels of resource utilization. At $U = .1$, only the most-valued animal resources are consumed, whereas at $U = .5$, all animal foods are totally consumed. Plant foods only begin to be consumed when $U > .5$.

In Figure 5.8, the potential availability of the two classes of food is shown as equal. We cannot assume, however, that this will always be the case. Thus it

---

[6]The relationship between potential and actual diets is not as independent as the treatment here seems to indicate. In reality, U must be relatively high except under the most extreme conditions, such as would exist, for example, following drastic and sudden drops in population in a region. Because we have defined our resource classes as "potential foods," it should be obvious that they must be perceived as such by people; unavailable, "inedible," and "tabu" resources are excluded.

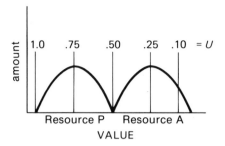

Figure 5.8 Two classes of food and resultant diets at differing levels of resource utilization.

is necessary to define a very important variable, μ, that expresses the *relative* abundance for the two classes:

$$\mu_a = \frac{\text{amount of available animal food}}{\text{total available food}} = \frac{A}{A + P}.$$

This variable is a subscripted form of the μ defined in the general model. Within the graphic model presented here, μ is a useful variable because it is equal to the relative area under either of the curves representing the classes of resources. Hence, by choosing appropriate levels for μ_a, we establish a complementary level of abundance for plants (μ_p). This is because relative abundance must be additive to unity ($\mu_a + \mu_p = 1$). Figure 5.9 represents several situations in which the value of μ_a (and thus μ_p) varies from 0 to 1.

Inspection of Figure 5.9 shows that as the relative abundance (μ) of the two classes of resource changes, the amount of either food that becomes part of an actual diet also changes: the contribution of a given class of food to actual diets is a function of the relative abundance of that food (μ) and of the level of utilization (U). This makes intuitive sense because as a valued food resource increases in abundance we would expect the diet to be increasingly composed of that food.[7] We may thus construct another variable, C, that gives the *relative* contribution of a particular class of food to the *total* actual diet:

---

[7]The problems of satiation for a particular source and thus of dietary composition cause no real problems at this point. We are merely trying to explain one aspect of feeding behavior and are establishing a framework in which to discuss it. The extremes of this model should not be expected to occur in real situations encountered by humans: they will not usually be consuming either 10% or 100% of available resources; neither will the ratio of plant to animal sources of food have relative abundances of .1 or .9. Also inherent in our concept of value is the recognition that certain plants (here, the low-value food) have a value comparable to certain animal sources of food. This becomes clear later when we deal with a more realistic description of availability and value curves.

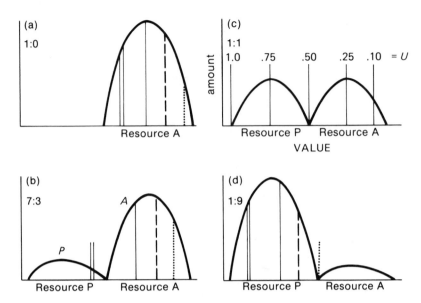

Figure 5.9   The effect of the composition of the diet from change in the abundance of food resources. The numbers on the graphs represent the relative proportions (μ) of animal (A) and plant (P) resources in the environment: (a) $\mu_a = 1$, $\mu_p = 0$; (b) $\mu_a = .7$, $\mu_p = .3$; (c) $\mu_a = .5$, $\mu_p = .5$; (d) $\mu_a = .1$, $\mu_p = .9$. The vertical lines represent the consumed proportion of total resources (U): $U = .75$ (double line) $U = .50$ (single line); $U = .25$ (broken line); $U = .10$ (dotted line).

$$C_a = \frac{\text{contribution of resource } a \text{ to diet}}{\text{total amount of food consumed}} = \frac{a}{a + p}. \tag{18}$$

The total amount of food consumed is a constant for any given level of resource utilization; both C_a and C_p are determined by the availability (μ) of the respective classes of food at any given level of utilization; and because C expresses the relative contribution of either food to the total diet, $C_a + C_p = 1$ no matter what level of utilization or relative abundance is posited.

Having investigated the relationship between abundance, feeding behavior, and resource utilization for two classes of resources with nonoverlapping valuations, we may now extend this same analysis to another, equally unlikely case. Here the two resources are given the same valuation. Although under these conditions it is manifestly absurd to speak of more or less "valued" resources, a relative valuation (ranking) is still maintained within each of the classes. The utility of recognizing this type of distribution is that it allows us to speak in a general manner of changes over time in the value of *one* class of resources. Thus if we have two classes of resources (here, domesticates and wild plants) that formerly were one class we may imagine a hypothetical equal valuation to be the starting point for their divergence. Conversely, if two resources of fomerly differing valuation were to converge in valuation, this type of distribution would be the limit of the convergence process. Assuming that we knew which resource was formerly the more valued, we might identify the two curves.

Figure 5.10 illustrates three selected cases (a–c) for this type of resource configuration. Because the two resources are varying only in their relative abundance, it is reasonable to expect that the contribution of each resource to the diet will be essentially the same as its abundance. Thus at all levels of resource utilization the contribution of each resource to the total diet is the same as its relative abundance in the environment.

Two interesting aspects of diet determination in relation to valuation become apparent in Figure 5.10. First, as the two resources become more equally valued, the desirability of the previously more highly valued resource falls off. Thus, for an example, if very little of the total available food is used (e.g., $U = .1$), total separation in valuation curves brings with it exclusive reliance on the more highly valued food over most conditions of relative abundance (compare Figure 5.9b–d). Conversely, if the two classes on resources are equally valued, then 10% of both resources will be consumed at a utilization level of $U = .1$. Thus, as classes of resources diverge in valuation, increasing pressure is placed upon the more valued class. Second, a utilization level of $U = 1.0$ (all resources consumed) for *any* configuration of valuation of classes of resources is mathematically equivalent to total equality in valuation for the classes of resources. Under either of these conditions (equality in valuation or $U = 1.0$), the contribution of either resource to the diet is a simple function of its abundance in the environment. Under these conditions we cannot speak of any real ''preference'' in feeding behavior. We cannot claim, for example, that people fed preferentially upon domesticates if people were consuming all available resources. Nor can we claim preferential feeding if domesticates as a class had the same valuation as did wild plant resources. However, the importance in an actual diet of two equally

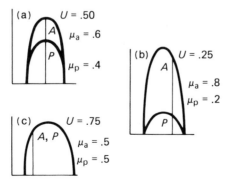

Figure 5.10   Resource patterns for totally overlapping resources. Animal resource indicated by A, plant resource by P; axes as in previous figures.

valued classes of resources may change radically over time because of the increasing abundance of one class. Changes in preference are not required to bring about these changes in importance. (This relates in a roundabout way to arguments advanced earlier against intentionalistic explanations for the rise of agriculture.) Thus, when the resources are equally valued, very little differential feeding pressure is placed upon resources at most levels of their abundance. The proportion of the available resources that is consumed is no more than a simple reflection of the general pressure placed upon all resources.

Our earlier discussion of two classes of resources having nonoverlapping value distributions makes a certain amount of intuitive sense—it is not hard to conceive of examples of such total separation of resources into high-value and low-value categories. The highly valued class provides the great bulk of the food and, at best, only the more valued members of the less-valued class are consumed with any frequency (except during periods of great stress, when U begins to approach 1.0 or when μ_a is very low). Hence, a clear and unambiguous preference exists for consumption of the more-valued class of resources.

In contrast, the reality of total overlap in valuation seems unclear. Yet there are two good reasons for considering this situation. First, total overlap in resources is the most general case by which we may relate changes in diet to changes in the relative abundance of resources. The second and at present more important reason for considering totally overlap in valuation is that it permits us to discuss changes in the valuation of components of a resource. The case of interest to us is the domesticated component of the plant class of resources. Thus the potential resource, P, is composed of two subclasses of resources P_w and P_d, and the actual diet p will be composed of the subclasses p_w and p_d. The above discussion may thus be extended to these two components of the plant resource; however, now the change in μ_{p_d} may be seen as the temporal dimension of domestication, for the major effect of domestication for people has been an increase in total recoverable yield (which, of course, need not be the same as increase in per capita yield). This increase in yield over time (which we know has occurred) is now modeled here as an increase in the relative abundance of subclass P_d, assuming that initially we had two subclasses of equal valuation. Because the dynamics of valuation change that occur because of domestication are quite complex, it is desirable to divide the process into two aspects: change in the yield of domesticates in relation to wild plants and change in the valuation curves for these two components of the class of plant resources.

We have already established that when resources are totally overlapping in value, the yield for each component is identical to the relative abundances of that resource:

$$C_{p_w} = \mu_{p_w} \quad \text{and} \quad C_{p_d} = \mu_{p_d} .$$

However,

$$C_{pd} = p_d/(p_d + p_w) \quad \text{and} \quad C_{pw} = p_w/(p_d + p_w),$$

thus,

$$\frac{C_{pd}}{C_{pw}} = \frac{p_d}{p_w}.$$

But we have already established that $\mu_{pd} + \mu_{pw} = 1$. Thus if we substitute μ_{pd} for C_{pd} and $(1 - \mu_{pd})$ for C_{pw} we end up with the equation

$$\frac{p_d}{p_w} = \frac{\mu_{pd}}{1 - \mu_{pd}}.$$

This equation is the graphic equivalent of Equation 6 of the general model, but here our notation is somewhat different. Because we are considering several classes of resources simultaneously, we use the notation p_d/p_w to indicate what was previously known as r (and considering variations in U, r^*).

As we have already seen in the general model, domesticate yields increase over time. It is unlikely that the increased yields coming from domesticates totally lack feedback into the cultural percpetion of the value of food sources. Figure 5.11 illustrates in a rough manner a hypothetical series of curves documenting increases in domesticates in relation to wild plants. I have been conservative in assuming that domesticates plants appear (at time t_0) in the *least* valued portion of the utilized plants. At this time, their numbers are extremely low and their contribution to the diet are also low. However, we must note that the food they are contributing to the diet is *additive*—in a sense, domesticates constitute (because of the symbiosis) a source of food that did not exist before. They may substitute for wild foods of even lower value that previously had to be consumed. (This earlier level of utilization is indicated by the broken line on Figure 5.11). Thus, as the human relationship with domesticates develops, its effect upon competing sources of wild food may be represented by the series of lines U_0, U_1, U_2, and U_3. The absolute amount of food consumed is not changing under these conditions; however, U is changing its intercept with the value axis. As we have noted, valuation takes into account availability, abundance, and nutrition. Thus we may let U equal an animal's perception of the ''normal'' range of foods required to maintain life. The animal's perception of the progression of U to the right over time will be of a regression of the P_w curve to the left. This also makes intuitive sense—as a new source of food comes to provide nutrition for people, previously desirable sources of food will lose their desirability and domesticates will begin to grow in importance.

A few words should be said about the shape of the curves given the domesticated resource in Figure 5.11. I have chosen this spikelike shape to point out that changing selective pressures tend to select different species of plants during the

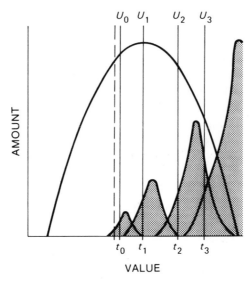

Figure 5.11   Changes in valuation of plant resources brought about by the rise of domesticates. Domesticated resources indicated by stippled curves, wild plant resources by open curves.

various phases of domestication. Although at any given moment U will be a line, it will not be completely stable over relatively short periods of time. Thus the greatest feeding pressure will be placed upon the more valued components of the domesticated flora. We have already discussed how the rise of the agroecology encourages the success of different species at different times. We can thus press Figure 5.11 into service to represent both the rise of the agroecology and the changing species composition of the domesticated flora that accompany this process: as domestication proceeds from incidental to specialized and even to agricultural, different species receive the most intense selective pressures because they receive higher valuation as a function of the feeding pressures placed upon them.

Getting Realistic

This discussion of changes in valuation for components of a resource leads us directly into a more reasonable view for human subsistence patterns. The eventual effect of the evolution of domesticates is the creation of a new curve of resources that overlaps at least the higher-valued components of the previously existing wild-plant resource curve. This same sort of overlapping pair of curves is the most realistic representation of human valuation curves for plants and animals taken as classes of resources. Thus we drop for the moment our consideration of domesticates and briefly look at the dynamics inherent in interacting valuation curves.

We consider the general effects of feeding upon resources with some but not a complete overlap in value, and assume that the lower-valued class of resources is plant foods. This assumption scarcely needs justification because this is the valuation expressed by the vast majority of human cultures, be they non-agricultural, agricultural, or industrialized. Plant foods are generally less valued than are animal foods, but at the same time certain types of plant foods may be more highly valued than certain types of animal foods. We further assume that domesticates evolved out of less-valued plant resources. I also seek to demonstrate that the additive value contributed by the domesticate suffices to explain change in diet during domestication and may help explain the change in human diet known as the *broad spectrum revolution* that frequently precedes the appearance of agriculture in the archaeological record. I also hope to show that the extremely low feeding pressure placed upon early domesticates tends to keep them from becoming significant parts of the diet for very long periods of time. Because the rate of evolution for domesticates is a function of the feeding pressure placed upon them, we can expect that a stable equilibrium between people and early domesticates could have existed for extremely long periods of time. I suggest that the development of the agroecology was eventually responsible for the destabilization of the situation, and thus that the rather explosive appearance of agricultural systems in the archaeological record is in complete harmony with the gradualistic, evolutionary processes underlying it.

Investigation of feeding behavior when the resource curves show some overlap is in accord with an intuitive understanding of feeding strategies. Figure 5.12 shows animal and plant resources under conditions of equal relative abundance ($\mu = .5$) but under different environmental conditions such that in Figure 5.12b both are half as abundant as in Figure 5.12a. If we keep the total amount of food actually consumed constant, the actual diet under conditions of relative scarcity includes more foods of low value. That is, more low-value plant foods are eaten under these conditions than would be under conditions of greater absolute abundance for both resources: as foods become scarce, people eat a greater variety of foods, and less-favored sources of food have to become part of the actual diet.

An important implication of Figure 5.12 for our model is that U may be used to define not only the level of utilization of resources, but, with appropriate modification, any desired abundance of resources. Thus a decrease in U is equivalent to an increase in the total available per capita resources and an increase in U is equivalent to a decrease in the available per capita resources (collective behavior defines U). We have already noted above that the relative abundance of resources (μ) is independent of the absolute amount of resources available in the environment. Thus by using the two variables μ and U, and by drawing resource curves with varying degrees of overlap, we may create a graphic representation for any possible diet and any possible relationship between potential and actual diets. The graphic model complements the general one

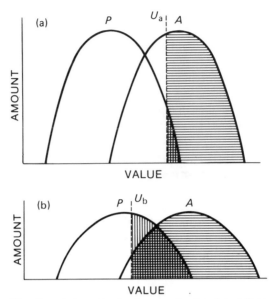

Figure 5.12   The effect of change in abundance on the diet. Actual diet indicated by shaded area, the animal contribution by horizontal hatching, and the plant contribution by vertical hatching.

in that it permits us to explore in greater detail the specific behavior of hypothetical diets; here concentrating upon how specific classes and components of diets may interact. In fact, we may calculate the precise contribution of various components of the diet because U represents both the resource component upon which feeding ceases and at the same time tells us how much of the area under the curves representing potential diets is significant in the actual diet.

Valuation and Domestication

If, as we have noted, one of the effects of plant domestication is the appearance of a regression of valuation on wild resources—that is, an apparent movement of the wild-plant curve left along the value axis—how will such a change in valuation occur? Figure 5.13 shows a plausible valuation curve for potential and actual human diets during the early phases of domestication. Animals are shown as contributing a substantial part of the original diet, but low-value animals are excluded and the animal curve has a higher average valuation. The original level of U is shown by the broken line at $x = 20$. Recalling our earlier discussion of classes of resources and their components, we may see that the last wild-plant resource that will be consumed may be referred to as P_{20} and the last animal resource consumed as A_{20}; the total diet is

$$[A_{75} + \cdots + A_{20}] + [P_{w50} + \cdots + P_{w20}] = a_0 + p_{w_0}.$$

Now, small amounts of domesticates are introduuced into the potential diet at times t_1 and t_2. This new component will be located very near the lowest limit of the valuation of consumed plant resources (p). The contribution of domesticates to the diet is approximately 13 units (vertical hatching) at t_1 and approximately 46 units (horizontal hatching) at t_2. We would expect the response of humans to this additive component of the diet to be a simple displacement of the plant resource curve to the left so that its limits would be first $x = (-1, 49)$ and then $x = (-4, 46)$. Such a relocation would remove at first 13 units and then 46 units of potential wild-plant food from the actual diet. Thus we would have a stable diet in which the amount of food remains constant. The theoretical diet would be $a_0 + (p_{w_0} - 13) + p_d$ at t_1 and $a_0 + (p_{w_0} 46) + p_d$ at t_2. Thus U would be stable at $x = 20$ and the devaluation of wild-plant resources previously posited would be accomplished. This model is congruent with our intuition that the valuation of foods is of major significance in diet determination and that changes in the abundance of domesticates (itself a result of the interactions of humans and plants) may be modeled to incorporate changes in the valuation of resources. Patterns of utilization have feedbacks upon the valuation of classes of resources.

Changing Values and Changing Diets

I have repeatedly stressed that the most important effect of the domestication of plants upon the human diet has been an increase in total recoverable yield. We can now see that the processes called upon to explain the increasing domes-

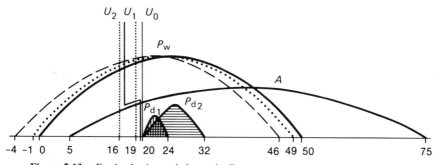

Figure 5.13.   Food valuation and change in diet.

Time	Theoretical diet	Actual diet	μ_d
t_0	$a_0 + p_{w0}$	same	0
t_1	$a_0 + (p_{w0} - 13) + p_d$	$a_0 + p_{w0} + p_d + A_{19} = a_0 + p_{w0} + 19$	.02
t_2	$a_0 + (p_{w0} - 46) + p_d$	$a_0 + p_{w0} + p_d + A_{19-16} = a_0 + p_{w0} + 63$	.04

tication of plants find their objective correlative in the feeding patterns of people. We have stressed that this change in diet is not necessarily the result of a choice by humans, but can be modeled simply as the mechanistic result of changes in resource abundances. We have seen how the valuation of resources may change over time: as domesticates begin to displace competing wild sources of food their ready availability and higher yield from the same investment of energy tends to bring about an increase in their valuation.

However, no simple or optimizing response by humans (or other animals for that matter) should be expected. Although we have depicted resources in these figures as if they had no temporal or spatial distribution except insofar as these paramaters affect valuation, the appearance of domesticates in the small amounts posited for incidental domestication would not be likely to change the day-to-day behavior of early domesticatory humans: additive changes in the subcomponents of a resource are unlikely to be perceived when they first occur. Put in other terms, the appearance of a few fruit trees around the settlement or an increase in edible seeds along a transhumant route would not be likely to divert humans from previously established patterns of foraging. Humans would not be likely to abandon traditionally important resource components merely because another component was beginning to increase in abundance. A certain amount of year-to-year variation in yield from any one food is to be expected, and it is unlikely that people would be able to distinguish such variation from those occurring because of domestication. Thus, during the early stages of domestication, people are not aware that the plants are undergoing incidental or early specialized domestication and that a different *type* of resource holding a great potential for increases in yield is appearing. Thus, instead of acting "appropriatly" and decreasing their consumption of wild resources people are most likely simply to accept the yield from domesticates while continuing their traditional foraging patterns. People do not change the resources they feed upon merely because coevolutionary interactions with plants are increasing the available quantities of certain select species. Put in the simplest terms, people eat the foods they are used to eating even though the availability of certain other foods is increasing. The effect of this will be that U maintains its relationship with the wild-plant resource curve despite its decreases in value. As modeled here, U—instead of at the theoretical level of $x = 20$—moves first to $x = 19$ and then $x = 16$ (shown by the dotted lines on Figure 5.12). People continue to consume wild-plant food originally located at P_{20} even though the relationship of plants and animals is changing. The conservative nature of human feeding patterns has major effects upon diet, yield, and demographics.

Resource Devaluation

What effect will the decrease in the point at which feeding stops (e.g., from $x = 20$ to $x = 19$) have upon the consumption of animal resources? I believe the most likely response of humans to the continued consumption of wild-plant resources that have become slightly less valued is not an increase in the average valuation of animal resources. Rather, the valuation of animal resources remains constant: the change in valuation of wild plants brought about by the rise of domesticates does not spread to animal resources. People are incapable of perceiving that a new class of resources is evolving; they only see a change in value within a general class of resources. (Here, of course, we are speaking of the early phases of domestication, not of highly developed agricultural systems.) Because the location of the valuation curve of animals has not shifted, previously uneaten animal sources of food enter the diet: in other words, previously undesirable sources of animal food become desirable. People find themselves eating plant foods with an absolute valuation of 19, or later 16, and they thus consume animal resources having these lower values: in essence, if P_{19} is eaten, A_{19} is also consumed.

The effect of this on the actual diet is rather interesting. At t_1, instead of having the theoretical diet of $a_0 + (p_w - 13) + p_d$, people adopt the actual diet of $a_0 + p_w + p_d + a_{19}$ that is equal to $a_0 + p_{w_0} + 19$ units—6 animal units above that contributed by domesticated plants. A similar process occurs at time t_2, but here the total increase is 27 units (see Figure 5.13). We may take this process further as the amount of domesticates in the environment increases. Thus we may see in Figure 5.14 a continuation of the processes that began in Figure 5.13. Although the amount of animal food consumed (a) is increasing, it eventually reaches its limit when all animal foods that are available are incorporated into the actual diet ($A = a$). Further, no change has occurred in the specific wild-plant resources (the specific P_{w_n}'s) that are consumed.

Is it reasonable to claim that foraging patterns for wild plants are conservative whereas those for animal resources are not? Why should humans not simply continue their existing foraging patterns and cease to consume animal resources when they reach A_{20} instead of beginning to include lower-valued animal resources? Several ways exist to approach this problem. First, the location of U is not a sharp line, but rather an average of periods of high and low relative availabilities for the two classes of resources; it represents the *probable* stopping point for feeding upon a class of resources. But because the resource curves represent available (''consumable'') resources, it is clear that all the resources must fall within the extreme limits of U as they are perceived by people. We have considered at length the various changes that may occur as the relative abundance of resources changes and have seen that rather small changes in abundance may be accompanied by rather great changes in utilization levels for the different resources. Animals, including humans, do not keep careful records of yield and intake, but rather consume and value the components of their diet as individual items are encountered in the environment. Resource A_{U-1} is not a resource *never* consumed, but only one not usually consumed; under conditions bringing

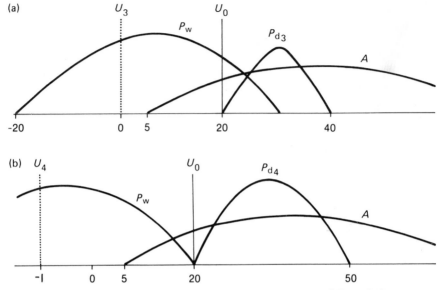

Figure 5.14   Food valuation and change in diet (continuation of Figure 5.13).

about a slight decline in total productivity for all plant resources (μ_p), animal resource A_{U-1} has a very high probability of being included in an actual diet. Now, if a slight devaluation has occurred such that a less-valued P_i terminates feeding, this will be interpreted as decreased productivity by the class of plant resources, and the only reasonable response will be compensatory feeding upon lower-valued animal sources of food. Because P_u and A_u are equally desirable, when P_{U-1} is seen as providing the termination point for feeding upon plant resources it is extremely probable that A_{U-1} not A_U will be the termination point for feeding upon animal resources. Thus a slight increase in the relative abundance of one component of a class of resources will be interpreted by the feeding animal as an apparent decline in the total productivity of the environment because it will appear that less-valued resources are being consumed.

Although at first this phenomenon seems to go against intuition, in fact it maintains the most conservative valuation of resources possible. Another way to approach this issue is to ask why U_2 does not assume the zigzag path indicated by the thick line on Figure 5.13. This configuration of utilization, however, is identical to a decrease in the valuation of all components of the animal class of resources (and also of P_d, a subset of the class P). We have seen how relative valuation within a class may change because of changes in relative abundances within that class, but it seems extremely unlikely that the absolute valuation of a competing resource class will change because of these changes in valuation occurring within another class. Thus we conserve the absolute *average* valuation of a class of resources when a rise in abundance of a component causes devaluation of the remainder of that class: if we are also to conserve the difference in average valuation between classes we cannot reduce the average value of a competing class of resources. Thus by keeping our valuation for two competing classes of resources as conservative as possible we also expect to find increased feeding on the less-valued components of the animal class of resources as a result of lowered valuation of wild plant resources.

Although domesticates are increasing the total potential yield available to people, the phenomena discussed above produces an actual diet that is not dominated by domesticated plants. Instead, the major portion of the increase in yield comes from nondomesticated plants and from animals. Furthermore, I have been extremely conservative in my depiction of the available less valued animal resources. The higher the animal resource curve at the intersection of U and A, the greater is the potential contribution of lower-valued animal foods to the diet as domestication proceeds. Greater familiarity with a resource and modifications in foraging techniques also increase the potential yield of these sources—as feeding pressure upon these resources is increased, innovation and greater efficiency in procuring and processing low-value animal resources (such as intertidal shellfish) may bring about increases in total yield that could not have been predicted from the apparent abundance of these resources when regular feeding upon them was first initiated. Also, as yield from domesticates reaches significant levels, great pressure will begin to be placed upon previously unutilized wild resources, with the result that a tendency for U to rise to 1 will become apparent under any type of foraging strategy (compare Equation 7).

The model for changes in feeding behavior advanced here is in harmony with a conservative interpretation of human foraging: the types of behavior that are being encouraged are those that already exist, namely, feeding upon wild animal and wild plant resources. The change in valuation of other resources induced by domesticates compensates for the rather small actual yields provided by domesticates. Thus a negative feedback system reducing human dependence upon domesticates is established during early domestication. If population growth were encouraged during the relatively long periods of time during which incidental and specialized domestication is occurring, most of the food supporting the increases in population will come from wild plant and animal resources, not from domesticates. The evolution of domesticates may encourage a "broad spectrum revolution" but it will *not* supply the resources upon which the revolution is to be based. Rather, people will continue and even expand their exploitation of traditional nondomesticated resources.[8] The effect of this over the short

[8]Hence the importance in certain regions of the world of montane zones, rather than the potentially agriculturally productive lowlands during this period?

run will be analogous to that of a brake upon the potential growth of the human–plant symbiosis and a great prolongation of incidental domestication and the early phases of specialized domestication. The time required to reach the point at which μ_{p_d} reaches levels measurable in terms of $p_d/(a + p_w)$ will be greatly prolonged.[9] The evolutionary aspects of domestication discussed earlier and the feeding aspects of domestication will go hand in hand, and the very slow growth of the human–plant interaction during early domestication is again predicted.

Agricultural Domestication and Diet

The common-sense definition of agricultural subsistence is a dependence upon domesticated plants for a substantial part of tbe diet. In terms of the model used here, agricultural subsistence implies a relatively high value for $p_d/(1 - p_d)$, where the quantity $(1 - p_d) = a + p_w$. A great deal of evolution must occur before plants will be capable of providing more than a very small amount of food.

Against the background of the general model and the discussion of feeding and changes in valuation just advanced, we now explore in some detail the changes that occur in diets as P_d approaches relatively high levels of abundance. Three major points are considered. In the first place, we note that a strong tendency exists for U to approach 1 (Equation 17). This is the result of two factors. First, in order for coevolutionary domestication to be successful, people must be utilizing this evolving class of resources at a high level (p_d must approximate P_d). Those plant species not utilized in this manner experience a reduction in selective pressure and end their relationship with humans. (Such plants represent crops that were domesticated at some point in history but later passed out of domestication; see Figure 5.11). Second, as domesticated plants begin to dominate the evolving agroecology, their absolute yield increases dramatically and the actual diet of humans comes to be dominated by them. The interaction of these two factors—high inherent utilization rates and increasing relative abundance—inevitably cause U to approach 1. Increasing domestication thus brings with it a fundamental shift in human feeding behavior. Rather than selecting their diet from the most highly valued resources, humans find their diet increasingly determined by the absolute abundance of resources (especially domesticates). This change in feeding behavior has far-reaching effects upon the stability of the

---

[9]We should not forget that the very best approximation of diet found in the archaeological record is not p_w but either $p_d/(a + p_w)$ or C_{p_d}: we do not find absolute abundances, but rather relative contributions, muffled by the fetters of archaeological preservation.

human diet over relatively short periods of time; for the moment it is sufficient to note that the broad-spectrum revolution initiated by the appearance of domesticates does not cease with the appearance of agriculture (although its effects come to be masked by the extremely high levels of yield derived from domesticated plants). We also see that although the valuation-regression hypothesis advanced above seems most logical, other modes of diet determination have little effect upon the rate of change in p_d, especially during the early stages of domestication. More important, the qualitative relationship between P_d and other classes of resources remains unchanged. Thus differences between possible patterns of utilization of nondomesticates have only a minor effect upon the rate of increase in the importance of domesticates, and the differences caused by these alternative feeding strategies is insignificant at low levels of μ_{p_d}. Finally, we note that the increasing abundance and relative value of domesticates provides the necessary positive feedbacks for the intensification of agricultural domestication. The most probable feeding pattern reinforces the significance we have given to the evolution of the agroecology. The interaction of feeding behavior and agricultural domestication brings about the rise of a ''new'' class of resources. If we were to define the moment when people became aware that they were involved in a relationship with domesticated plants, it would have to be after that point when the valuation of major domesticates and wild plants had separated sufficiently to permit recognition of two *distinct* classes of resources.

In the previous chapter we noted that agricultural domestication is coevolution within the developed agroecology and have seen that the characteristics of evolution within this setting permit an intensification of the symbiosis between people and plant. In the discussion just completed we have seen how feeding behavior interacts with valuation to create a negative feedback system repressing the development of the agricultural relationship and thus slowing the development of agricultural systems. We have also seen that a relatively high level for μ_{p_d} is prerequisite to the establishment of an obligate dependence upon agricultural plants. Finally, we have mentioned that the location of the takeoff point for agricultural yield in relation to the abundance of agricultural plants is described by the average value of U. We now explore the interaction of various feeding strategies in terms of increasing abundance of domesticates.[10]

Our discussion of the rise of agricultural modes of subsistence must consider several interrelated parameters, including absolute yield, the relative yield contributed by domesticates, human population levels, and the timing of the

[10]I want to remind the reader once again that increases in the yield of domesticates are not the result of teleological processes. The increases in yield shown here are meant to model increases that we know have occurred; they are based upon the increases in fitness produced by means of the coevolutionary relationship, and if that relationship for whatever reason were to become of sufficiently low intensity, the plant species in question or even the whole class of domesticates would stop evolving, and would either survive only in wild populations or become extinct.

agricultural takeoff point. We have already seen how regression in valuation inhibits the development of agricultural subsistence; thus another aspect of our study of agricultural origins must be investigation of the means by which human feeding behavior interacted with ecological and evolutionary factors to change feedback processes from negative to positive. The easiest way to approach these questions is by describing two basic types of feeding strategies and by comparing their effects upon the evolution of agricultural subsistence systems.

These two fundamental types of feeding strategies may be described as conservative and compensating in their relationships to domesticated plant resources. The *conservative* strategy in its simplest form (Strategy 1, Table 5.1) maintains previous consumption patterns for both animal resources (a) and wild plant resources (p_w) despite any change over time in the abundance of domesticates. If this strategy is modified to account for increased consumption of less-valued animal resources, we arrive at Strategy 2 (Table 5.1). Here, as domesticates increase in abundance over time, animal foods are added to the diet in amounts proportional to the given regression in valuation.

Compensating strategies are those in which conservative behavior or regression in valuation do not occur. Instead, under these strategies, additions to the plant diet from domesticated sources of food are compensated for by reduced feeding upon wild plant foods (Strategy 3) or by reducing feeding upon *both* plant and animal resources (Strategy 4). These strategies are also optimal ones for the appearance of agricultural subsistence because they maximize the rate at which agricultural subsistence is achieved (the value for μ_{p_d} at which the agricultural takeoff point occurs; see Figure 5.15).

When the relative abundance of domesticates (μ_{p_d}) is very low, alternative feeding strategies make little difference in the actual diet; domesticates are capable of making only a very small contribution to the diet when they are scarce, and there is little change in diet if domesticates are added to it or replace alternative forms of food. As domesticates increase in value and abundance, however, differences in the manner of feeding upon them will bring about different types of diets. A relatively great difference in potential contribution to the diet of various feeding strategies exists as domesticates begin to reach even modest levels of abundance. The contribution of domesticates to the diet differs substantially under various feeding strategies.

Feeding Strategies 1 and 2 (Figure 5.15) represent conservative feeding strategies in that no attempt is made to adjust the diet to compensate for the rise of domesticates (regression in valuation being merely the way in which the most conservative patterns of subsistence may be modeled). If compensation is made for increases in yield provided by increases in P_d, no net gain will be made in the total available yield, although the source of the yield is changing. Conversely, if the yield is merely additive, over long periods of time both net gains and changes in diet occur; but, looking at the process over short periods of time, the change in

TABLE 5.1

The Relationship between Several Feeding Strategies and the Value of $p_d/(a + p_w)$
in Terms of μ_{pd}

Type of diet	t_1 .02	t_2 .04	t_3 .13	t_4 .27	.38	.50	.67	.85	$= \mu_{p_d}{}^a$
Conservative (minimal reliance upon domesticates)									
1. a stable	.02	.97	.25	.52	.85	1.40	2.70	8.00	$p_d/(a + p_w)$
p_w stable	1.02	1.97	1.20	1.50	1.90	2.40	4.00	9.60	RY
	.71	.72	.75	.79	.81	.85	.92	.96	U
(regression on p_w and a)									
2. a compensates	.02	.06	.19	.49	.79	1.30	2.70	8.00	$p_d/(a + p_w)$
p_w stable	1.03	1.10	1.30	1.60	1.90	2.50	4.00	9.60	RY
	.72	.74	.78	.83	.85	.88	.92	.96	U
(regression on p_w)									
Compensating (reduced feeding upon wild foods and/or animals)									
3. a stable	.02	.08	.25	1.08	1.70	3.00	5.70	19.00	$p_d/(a + p_w)$
p_w declines	1.00	1.00	1.00	1.00	1.30	1.90	3.30	9.00	RY
optimal (theoretical)	.70	.62	.62	.52	.59	.66	.78	.91	U
(reduced feeding only on p_w)									
4. a compensates					.82	1.20	2.20	5.70	19.00	$p_d/(a + p_w)$
p_w compensates	b	b	b	1.20	1.60	2.00	3.30	9.00	RY
(slight)				.60	.69	.72	.78	.91	U
(reduced feeding on both p_w and a)									
Most conservative									
5. a compensates					.61	1.00	2.00	6.00	$p_d/(a + p_w)$
p_w compensates	b	b	b	b	2.30	2.80	4.20	9.80	RY
					1.00	1.00	1.00	1.00	U
(compensation to limit)									
Most compensatory									
6. a declines							9.00	24.00	$p_d/(a + p_w)$
p_w declines	b	b	b	b	b	b	3.10	8.80	RY
							.74	.89	U
(minimizes $a + p_w$)									

$^a RY$ = relative yield (see text); U = level of utilization.

^{b} Strategy does not exist until μ_{p_d} reaches a higher value.

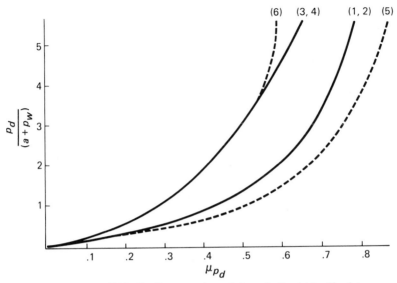

Figure 5.15   Feeding strategies and the agricultural takeoff point.

subsistence patterns is minimized. This concept does not contradict the fact of changes in scheduling or eventual changes in settlement pattern. Instead, it places emphasis upon the fact that, during the early phases of domestication, scheduling and settlement patterns are largely controlled by wild sources of food, not cultivated ones—domesticates being incapable of providing much food because their abundance is very low. For a subsistence strategy to undergo an increase in its rate of change, destabilization of the existing pattern must occur.

We have discussed regression of valuation and its resultant dietary patterns and pointed out that regression of valuation would be probable in human societies. Strategy 5 represents a modification of the processes occurring in Strategies 1 and 2. Under Strategy 5, the regression of valuation process is expanded to include the wild plant resources. Two developments might encourage this process. The first would be growth in human population densities. As populations begin to increase, and as the value of U increases, there is a decrease in the amount of potential food that goes unexploited. Feeding terminates in a region of the valuation curve that has only a small amount of any given resource component. Thus, small decreases in the productivity of any component of the curve having an average value (and thus a high potential yield) will cause not one, but several component resources to be fed upon as compensation for the losses encountered in the more favored food. The second development arises from the instability of yield of domesticated plants. As domesticates increase in abundance and their contribution to the diet increases, small percentage changes in yield have to be compensated for by feeding upon components of classes of resources other than domesticates. This is a simple reflection of the high inherent utilization rate for domesticates. There is reason to assume that the domesticated class of resources has a less stable yield than the wild class: as domesticates contribute increasing amounts of food to the diet they are subject to occasional episodes of dramatic decrease in yield that force people to rely upon the lowest-valued components of the wild plant class of resources. In other words, as domestication grows in importance, the variance on U increases. The result of this, for example, in Figure 5.14, would be the displacement of U_3 and U_4 to the left, further decreasing the relative contribution of domesticates to the diet and increasing the *total* recovered yield: previously undesirable wild-plant resources enter the diet.[11]

Feeding Strategies 3, 4, and 6 stand in contrast to Strategies 1, 2 and 5 and might be called *optimizing* strategies if people were to be aware of the potential benefits to be derived from agricultural production. Here, as agricultural productivity increases, an equivalent amount of food is removed from the wild plant class of resources. Strategy 3 specifically represents the compensating strategy just discussed; Strategy 4 is a modification in which some wild plant resources are retained; and Strategy 6 represents an extension of Strategy 3 so to include compensations from both animal and wild plant resources. The major differences between these three strategies and their conservative alternatives is the point at which agricultural subsistence arises and the speed with which it develops. As may be seen in Figure 5.15, total agricultural subsistence occurs at a much lower level of μ_{p_d} (that is, "earlier") under Strategies 3, 4, and 6 (μ_{p_d} = .5 vs. .8), and the relative contribution of domesticates to the diet is always maximized. These strategies are thus the ones that would bring about the earliest origin of agriculture and its fastest development, but they require a reduction in the consumption of wild plants relative to the conservative strategies. In essence, these compensating strategies are the ones that would be used were people consciously to adopt or invent agricultural subsistence because these strategies give the fastest returns and are based upon minimizing the contribution of wild plants to the diet. Feeding pressure upon wild plants is minimized, domesticates contribute as much food to the diet as is possible, and agricultural subsistence arises quickly.

One way for feeding pressure upon wild plants to be reduced during the early stages of domestication would be to posit that domesticates evolved from the most highly valued components of the wild plant class of resources. If domesticates were initially to appear far to the right of the line describing U, then

[11]An interesting implication exists here for understanding culinary traditions. A highly persistent, albeit unstable, agroecology, coexistent with high population levels, should produce a very strong tendency toward regression of valuation on all wild resources. This would yield a highly eclectic diet in which most all edible wild resources would be utilized. This seems applicable to the Chinese culinary tradition.

our previous arguments for the regression of valuation would become invalid: an unconscious preferential feeding upon domesticates would occur, along with a progression rather than regression in valuation of resources. There is, however, no indication that the plants on which agricultural systems were based were anything but relatively low-valued components of the wild plant class of resources. Agricultural plants require considerable processing, their yields have apparently increased relatively slowly, and they seem most closely related to "famine" or "adversity" foods. It is this observation of reliance upon previously undesirable resources that underlies the population-push hypothesis. Yet if we are to posit, on the basis of feeding behavior, that domesticates were used to solve problems arising from a relative lack of food, we immediately run into a major contradiction: we must claim that the protodomesticate suddenly became a highly valued source of food in this new situation while also maintaining that it was only brought into the diet because of an increase in pressure upon food resources. It is reasonable to posit that population pressure upon resources might encourage a greater utilization of previously ignored resources, but this cannot explain the initial increase in population. (If people weren't eating potentially domesticable sources of food, what were they eating?) It also cannot explain *how* low-valued "famine foods" suddenly changed in their desirability to people.

A second problem is the timing of the origin of agriculture. If agriculture were to have evolved by means of human interactions with plants, the location of potential domesticates among our most highly valued foods would have increased the intensity of the evolutionary pressures placed upon protodomesticates. Instead of the *negative* feedback processes described earlier as arising from regression of valuation (itself dependent upon a low relative valuation for the evolving domesticate), a *progression* of valuation would take place. As we have seen, regression of valuation places a "brake" upon the development of the human–plant symbiosis by encouraging the maintenance and even expansion of traditional foraging patterns. The existence of positive feedbacks by means of reductions in human dependence upon wild resources would have permitted intensified evolution of domesticated plants before human population levels rose to the point at which people induced those great changes in their environment that culminated in the development of the agroecology.

Progression of valuation also would tend to prevent demographic factors from influencing the rate of change in human subsistence patterns. Although producing the rapid origin of agricultural systems demanded by a population-push hypothesis, progression of valuation would cause another problem. Besides locating the source of domesticates in the "wrong" part of the valuation curve, it also contradicts the most important variable on which the population-push hypothesis is based. Progression of valuation implies a minimization of the quantity $(a + p_w)$; thus $p_d/(a + p_w)$ is maximized at all points in time. To claim that the increased yield of domesticates more than compensates for the loss of wild foods is to deny the fundamental premise of the whole demographic stress argument. If people were under food stress, why would they willingly abandon existing wild resources?

It may, of course, be argued that humans did not willingly abandon wild sources of food, but were forced to do so by the demands of scheduling and by changes in settlement patterns. Unfortunately, this implies a reduction in the importance of demographic stress as the cause of agricultural origins, displacing the focus from population size to foraging and settlement patterns. The implication would be that sufficient domesticated resources existed to make changes in settlement and foraging patterns the optimal choice in the situation. Foraging or settlement patterns, however, would not change until sufficient resources existed for support of the population. If change in *human foraging* patterns were to yield a net *decrease* in yield, people would not, *by definition,* be experiencing population pressures; if demographic stress existed, a change in foraging or settlement patterns that decreased yield would not be the optimal solution. In these terms, agriculture would have been an ''accident'' that arose from an error in which a change in settlement or scheduling yielded a system with lower initial yields despite a ''need'' for higher yields.

If we posit a relatively high value for protodomesticates, we are also unintentionally making domesticates the prime cause for increases in human population. A relatively high valuation implies domesticates were being consumed with great regularity. Furthermore, positing growing human populations implies that wild resources are being consumed toward the limit of their yield and that the major additive increases in yield must then come solely from the increasing yield of domesticates. Under these conditions increases in population would be best attributed to the additive yield of domesticates and changes in scheduling and foraging behavior would be permitted (not demanded) because wild plants were no longer the limiting resource.

If we claim that domesticates and agriculture had to be relied upon to compensate for the fact that wild yields were already near their limit for support of human populations (populations were near carrying capacity), we would also be implying that protodomesticates were located near the lower-valued end of the wild-plant resource curve. If protodomesticates were more highly valued, they would be consumed in their optimal form (domesticates) because the feeding pressure upon them would already be very high. Yet we already would have had to posit that domesticates must evolve among the most highly valued components of the plant resource curve. Put in other terms, if populations are to respond to the decreasing per capita yield of wild plants by increasing the yield derivable from domesticates, population pressure per se cannot exist: this option was always available to the population, and the ''most optimal'' solution would have been not to permit the situation to reach crisis proportions. This tactic, of course, would not be distinguishable on the basis of the evidence from a gradualistic tactic that concerned itself not with optimization, but only with maximizing each individual's food intake.

These arguments would seem confusing without an appreciation of the distinction between individual actions and goal-oriented adaptation. In apparently claiming that certain types of compensations are permitted whereas others are prohibited, I am really identifying the type of response that an individual human is likely to make when faced with specified options. It is logical to claim that an individual's response to an actual increase or decrease in the yield of a component of a resource will be compensatory feeding upon other components of the resource if this is possible, or on components of a differing class of resources if it is not. To claim, however, that people would reduce their consumption of wild plant resources *as a response* to the existence of protodomesticates implies a recognition of the *potential* value of this class of resources: if we are correct in our assumption that natural selection and genetics prohibit a more or less instantaneous evolution of the domesticated plant, we know that the appearance of domesticates was gradual and that during the early stages of domestication they were not distinguishable by people from wild plants. Thus people could not recognize protodomesticates as a class, much less recognize their potential, and people would have had no information on which to base compensatory responses.

We have already noted that the exponential growth characteristic of domesticate yields is compatible with any type of feeding strategy. The only difference introduced into the function by alternative feeding strategies is the relative abundance of domesticates (the specific μ_{p_d}) at which agricultural subsistence occurs. We can now see that the location of the agricultural takeoff point is determined by the valuation placed upon domesticates in relation to wild sources of food. Thus our variable U has the power to summarize the relative valuation given to domesticates. The closer U is to zero, the more quickly agricultural subsistence occurs. Conversely, as U approaches 1, the development of agricultural subsistence occurs more slowly (see Figure 5.4).

Yet what can we say about U and its potential value? Several important factors must be considered. The first is that the value of U for the domesticated class of resources must be realtively high. This is a simple outgrowth of the nature of domesticatory and agricultural interactions between humans and plants. If we grant that domestication and agriculture are a symbiosis between humans and certain plants, it is self-evident that a relatively high utilization of those plants by people must occur. Because the total U is partially dependent upon the value of U for domesticates alone, its value increases as domesticates come to dominate the diet. Thus it is illogical to claim that a high level of agricultural contribution to the diet can occur at low levels of abundance of domesticates. For example, for a potential takeoff point for agricultural subsistence to occur at $\mu_{p_d} = .1$, U must also equal approximately .1. This implies that the average utilization of resources is also .1. However, we have already established that such a low value for the utilization of domesticates cannot exist, and the only case in which such a low-average utilization function might exist would be one in which people intentionally ignored previously utilized sources of food and left most of the resources in the environment untouched. We might also note that if the vast majority of the resources in the environment were to be totally ignored, they would not be included in the list of foods from which the potential diet was constructed. Returning to our introductory remarks concerning food valuation, we must, therefore, recognize that only potential foods are included in the valuation curve, "objective" measures of their abundance are of no concern.[12]

Figure 5.15, although indicating the relative contribution of domesticates to the diet, gives little information concerning the absolute abundance of utilized resources. Yet we know that the evolution of agricultural systems has been accompanied by a rise in total yield. The easiest way to describe the relative increase in yield that occurs as domestication proceeds is by setting total diet at the time when domesticates did not exist at 1 and by expressing the increases in yield in relation to this. These figures have been computed for the hypothetical situation given here and they are shown as RY ("Relative Yield") on Table 5.1, and by the solid lines in Figure 5.16.

The quantity RY has major implications for an understanding of population pressure. We have seen that this theory would be incompatible with a shift from wild sources of food to domesticates, especially during the early phases of domestication when μ_{p_d} would be, by definition, low. Put back into terms of U, an early and steep rise in the contribution of domesticates to the diet is equivalent to a low value for U. A low value for U would contradict the concept of *population pressure* because it would imply that many wild resources were not being utilized.

As may be seen in Figure 5.16, minimal reliance upon domesticates provides the greatest realized diet, but, as we have seen, this tactic is correlated with the slowest rate of increase in the importance of domesticates. Thus the valuation-regression hypothes advanced earlier is not only a likely outgrowth of feeding behavior, but also describes how the total amount of resources could be maximized as agriculture develops. An intuitive approach for explaining the rise of agricultural subsistence does not provide the optimal strategy for increases in absolute yield; the conditions predicted for a population-pressure initiation for agricultural subsistence does not yield the results that are intended.

If we look a little more closely at the relationship between these two types of feeding strategies (conservative and compensating) another interesting fact

---

[12]This was touched upon earlier in our discussion of the broad-spectrum revolution. I have had to undervalue the potential contribution of lower-valued resources brought into the diet by valuation relocation because we have no way of approximating their potential abundance.

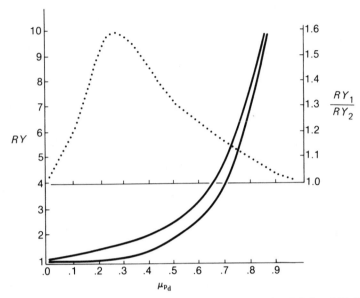

Figure 5.16 A comparison of benefits from two feeding strategies. Relative yields RY (solid lines) for feeding Strategies 2 and 3 (see Table 5.2 and text); advantage in yield, RY_1/RY_2, (dotted lines).

comes to light. We construct a final quantity RY_1/RY_2 that may be used to represent the advantage in yield that arises from adopting Strategy 2 rather than Strategy 3. This relationship is given by the dotted line in Figure 5.16. Perhaps the most interesting aspect of this curve is the location of the point of greatest advantage. Following a conservative and traditional pattern of behavior provides the greatest returns during the period of time when full agricultural subsistence has not yet arisen and when it would appear that the traditional forms of behavior are inhibiting, rather than encouraging, the growth of agricultural modes of subsistence. There is more to diet than the potential of a class of resources for further increases in yield, and it is fascinating that a conservative pattern of dietary construction provides the highest yields over time. Were one to adopt an intentionalist explanation for the origins of agriculture, one would have to conclude that people repressed the relative contribution of domesticates to the diet in order to maximize total yield. Of course, it is easier to claim that people merely took advantage of increasing yields derived from domesticates.

The peaking of the relative advantage curve at a time before domesticates dominate the diet is in complete harmony with my belief in a slow initial increase in the importance of domesticates in the diet and with the eventual rise and stabilization of the agroecology. A slight difference in yield may produce, over time, a great difference in human populations. A conservative strategy for diet determination maximizes not only actual yields, but also the potential rate of population growth because increases in yield over time are also maximized. Although during the early phases of domestication the increased human populations are not supported to any great extent by domesticated sources of food, these increased populations encourage the development of the agroecology.

I have described the development of the agroecology as the result of human manipulations of the environment that benefit the domesticated plant. The effectiveness of this manipulation is, to a great extent, a direct function of population size. As population increases, for whatever reason, the effect of people on their immediate environment also increases. The rate of evolution of the agroecology is thus a function of the rate of growth of human populations.

In this model, population pressure as a factor in the rise of agriculture is not denied but is given an entirely different, mechanistic, role. Rather than being *cause* for the development of agricultural systems, increase in population is *covariant* with the domestication of plants and the development of the agroecology. A dynamic process integrates foraging strategy, valuation, productivity, and coevolution. Increases in yield derived from the coevolution of people and plants initially cause regression of valuation, inhibiting development of the relationship with domesticates while encouraging population growth and the use of lower-valued resources. Over long periods of time this process encourages the slow growth of human populations. As population increases, the human effect upon the immediate environment also increases, bringing about the transition from incidental to specialized domestication. Increasing yields from domesticates now act in a positive feedback system that encourages the rapid growth of population and intensification of domestication. The eventual results are establishment of the agroecology, high domesticate yields, and agricultural subsistence. Population "pressures" exist, but they are only a partial expression of a fundamental dynamic.[13]

Given these observations, Figure 5.16 takes on a somewhat different meaning. The solid lines that previously identified the relative yield may now stand for the relative importance of wild plants to the diet under conditions of maximal and minimal utilization of wild plants for the hypothetical resource patterns (Figures 5.13 and 5.14). This is a reasonable interpretation of the lines because the utilization level for wild plants is the major factor contributing to their separation. The total additive contribution of animals is low for both strategies and quickly reaches its limit, whereas the contribution of p_w falls off relatively

---

[13]If we seek to understand why a given population did *not* become agricultural, we can see that conditions reducing the rate of growth in population are indistinguishable in their effects upon the rise of agriculture from conditions reducing dependence upon domesticates. Likewise, these two phenomena are indistinguishable from "chance" events that inhibit further development of the people–plant relationship (e.g., the lack of "suitable" plants in the environment on which to base an agricultural system). Such questions, however, are beyond the limits of this essay.

quickly under the compensating strategy. It may be seen, then, that the advantage in adopting a conservative strategy begins to fall off at about $\mu_{p_d} = .3$—the point at which the additive yield from domesticates begins to overwhelm conditions determining dietary contributions from other classes of resources. By referring to Figure 5.14b we may also see that $\mu_{p_d} = .3$ is also the same point at which the valuation curve for domesticates separates from that of the "parental" wild-plant resource curve. At this time, the difference between the total actual value of U (considering all classes of resources) for Strategy 2 (conservative) and Strategy 3 (compensating) is also maximized. The interaction of several paramaters—yield, relative yield, contribution of domesticates to the diet, and advantage to be gained from taking a particular strategy—all have a common nexus: in essence we have identified the approximate time when agricultural subsistence becomes possible (at least for the specific hypothetical situation given here—but note that changes in the specifics of any hypothetical case chosen only change the quantitative relationships; the qualitative relationship remains the same). The dotted line in Figure 5.16 illustrates the rise and eventual fall of a positive feedback system bringing about the rise of agriculture: thus we may hypothesize that agricultural domestication arose at a μ_{p_d} of approximately .3.

Timing and a Test

Just how much importance should be placed on this discussion of feeding strategies, relative valuation, and abundance of domesticates? Do the variables created to facilitate this discussion have any real meaning, or are they so abstract as to be totally unrelated to the processes that have occurred during the origin of agricultural systems? The answer lies somewhere between these two extreme positions. This model was created to show that agricultural origins can be approached as a natural phenomenon in a different way than that of earlier chapters of this volume. I have shown that by adopting the most conservative position for our description of human behavior (namely, that traditional behaviors are not usually abandoned) we arrive at a description of the origin of agricultural systems that contains few anomalies and that is completely congruent with the entirely different description of the development of agricultural systems advanced earlier. This model also allows us to relate the relative importance of the three major types of domestication to the development of agroecosystems and to the increase in importance of domesticates in the diet.

How literally should we take the resource valuation curves (Figures 5.13 and 5.14) used to describe the rise of agriculture? Not very literally, I would hope. These curves were invented using only a few self-evident principles—namely, that animal resources are more valued than plant resources, that wild plant resources are at least slightly more abundant than animal ones, and that domesticates arose from the less-valued components of the wild plant class of resources. It is important to recognize, however, that the precise shapes of the valuation curves (Figures 5.13 and 5.14), do not make great differences in the location of domestication events with respect to μ_{p_d} (summarized in 5.15) and make essentially no difference in the direction of the changes that are predicted (at least on the assumptions discussed above). If anything, the specific example explicated here overestimates the rate of evolution of domesticates, especially during incidental and specialized domestication, by underestimating the potential yield of wild resources. In a similar manner, this model slightly underestimates the general rate of change for causes in which the domestication of plants was accompanied by symbiotic interactions with animals. Although the subject is beyond the bounds of this essay, I might briefly note that if a domesticated component were to arise in the animal class of resources, I would expect that the rate of devaluation of resources would probably be much higher. Devaluation of wild plant resources would also proceed much more quickly, and a concomitant devaluation of wild-animal resources would be predicted. The rate in rise of U toward 1 would be much higher and dependence upon the domesticates would come relatively earlier. In that case we might expect agricultural subsistence to begin earlier than it would if only the domestication of plants were occurring.

Taking into account our discussion of feeding strategies and timing, I now attempt to place the three modes of domestication in a rough temporal framework. I attempt to sketch the progress of domestication and the rise of agriculture in making the x axis reflect both μ_{p_d} and time. To do this, several conditions have to be met; first, the time period representing incidental domestication (μ_{p_d} = 0 to approximately .1; see Figure 5.17, region I) must be determined. Whether this time span would be measured in centuries or dozens of millennia awaits meticulous studies of plant remains from preagricultural archaeological contexts as well as careful study of potentially symbiotic interactions between modern humans, both agricultural and nonagricultural, and "wild" plants. Were I to hazard a guess as to the time span of incidental domesticatory interactions, however, I would probably begin with the working assumption that incidental domestication has been occurring, at least in certain localities, throughout the better part of the history of our species. Clear proof of this apparently outrageous conjecture will of course be difficult to find. Demonstration of long-standing coevolved dispersal or protection relationships between humans and wild plants will require innovative techniques and approaches.

Second, we must note that the shifts in selective pressures as the agroecology develops cause a gradual shortening of the time periods between successive small intervals of μ_{p_d} (see Figure 5.17, region II). Much of the confusion that has existed in theoretical treatments of agricultural origins has been based on a conceptual confusion of the establishment of the agroecology with the domes-

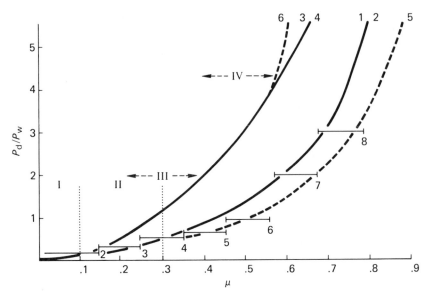

Figure 5.17 Feeding strategies (Lines 1–6) and the rise of agricultural subsistence. Bars 2–8 derived from MacNeish 1967 and 1970 (see text and Table 5.3). Phases of domestication indicated by I–IV: I, incidental domestication; II, specialized domestication; III, agricultural domestication; IV, recent agriculture. These phases do not have clear beginnings or ends but rather blend into each other.

tication of plants. Although domestication has taken new directions as it has become specialized, it is nevertheless an outgrowth of the evolutionary processes inherent in incidental domestication. Further research is required to demonstrate possible positive feedback systems that disturb the equilibrium established between people and incidentally dispersed and protected domesticates. Although the tendency toward increase in yield derivable from increased fitness is clearly an important component in the establishment of such positive feedbacks, the early stages in the establishment of the agroecology remain the most obscure aspect of the model advanced here.

As specialized domestication phases into agricultural domestication (Figure 5.17, region III), the intervals between successive points of μ_{p_d} will likely reach their minimum. Following this period, they stabilize or even reach a limit at which change in yield comes to a halt. The processes limiting the growth of agricultural systems is based upon a whole new series of tendencies that arise within the developing agroecology. The instabilities in potential yield induced by agricultural modes of production are responsible for both the spread of agriculture and the establishment of the negative feedback system that is to limit its ultimate productivity. Because the limits placed upon productivity are set by events occurring within the agroecology, it would seem clear that innovation and changes in technology may bring about changes in total potential yields. Thus peasant agricultural systems (Figure 5.17, region IV) experience dramatic increases in yield. All of this is rather obvious, but, in the context of my model, these changes in technology are intimately related to existing levels of productivity and, as an expression of this, to population. Also, increase in yield is invariably accompanied by increase in the variance of that yield.

The best way to approach the reliability of this model and the timing of domesticatory events is to compare the predicted events to data derived from archaeological investigations. Because the hypothetical example presented here has not been designed to account for domesticated animals, I turn to the New World for my data. The most accessible information on changes in New World subsistence patterns over time is found in the reports of the Tehuacan Archaeological–Botanical Project.

The horizontal bars on Figure 5.17 represent the values for $p_d/(a + p_w)$ computed by me from raw data presented in MacNeish 1967 and 1970 (see Table 5.2). Because my model deals only with *average* and *relative* contributions of various classes of resources, this type of computation is possible despite the numerous problems inherent in utilizing any archaeological source of data. Because I use an average statistic, seasonality and scheduling are controlled; and because it is relative, differential preservation of various types of foods is less troublesome (we need only assume that, on the whole, the various classes of resources have roughly equivalent amounts of preserved and nonpreserved resources). These computed bars fit the curve predicted from the general model and also the specific conservative feeding strategy (Tactics 1, 2 and 5) advanced above.

TABLE 5.2

The Rise of Agricultural Subsistence and the Relative Contribution of Domesticates to the Diet in the Tehuacan Valley of Southern Mexico[a]

Phase	Approximate years B.P.	$p_d/(a + p_w)$[b]
1. El Reigo	8800–7000	.05
2. Coxcatlan	7000–5400	.16
3. Abejas	5400–4300	.26
4. Purron	4300–3500	.53
5. Ajalpan	3500–2800	.60
6. Sta. Maria	2800–2100	.80
7. Palo Blanco	2100–1300	1.9
8. Venta Salada	1300–500	3.0

[a]Data derived from MacNeish 1967 and 1970.
[b]The contribution of domestic dogs and turkeys to the diet has been included in the animal class (a) of resources (see text).

However, to understand why the data fit this particular curve best, I should briefly mention how they were scaled to permit entry onto the figure. I first used the hypothetical resource-abundance curves presented above as a rough approximation for the resource availabilities during the earliest stages of domestication. Unfortuantely, reliable estimates for the relative contribution of plants and animals are not available for the earliest phases investigated by the project, but the values obtained from the earliest materials attributable to the El Riego phase is roughly comparable to that used in our hypothetical example. (C_a = .51, C_{p_w} = .49 calculated from MacNeish 1967: Tables 36 and 38; C_a = .48, C_{p_w} = .53 at time t_0 for our hypothetical example; see Figure 5.13.) The horizontal bars were then entered in this manner: first, it was noted that the final relative contribution to the diet of domesticates ($p_d/[a + p_w]$ = 3.0 in the Venta Salada phase [Phase Eight; see Table 5.2) could represent two possible values of μ_{p_d}, depending upon the nature of the feeding strategy. Using a conservative strategy (Tactic 2), μ_{p_d} would approximate .7, whereas, under maximal reliance upon domesticates strategy (Tactic 3), μ_{p_d} would approximately equal .5. However, the Abejas and Purron (Phases Three and Four) are the first ones for which clear evidence indicates that full-time agriculture was being practiced. Because the value calculated for MacNeish's Purron phase (Phase Four: $p_d/[a + p_w]$ = .53; see Table 5.2) intersects the predicted point for the period previously posited as representing the likely starting point for full agricultural domestication (μ_{p_d} = .3), the bar representing Phase Four was centered on μ_{p_d} = .3. Having determined two points, it was possible to convert the values of μ_{p_d} into a linear measure of time. Study of Figure 5.16 will show that if the Venta Salada phase (Phase Eight) is located on the other set of lines, the values of $p_d/(a + p_w)$ for the earlier phases will rise far above the predicted values irrespective of the scaling used to tie the values to the x axis. Not only do the data fit the curve, but they also fit the predicted feeding strategy.

Besides the relatively good fit of data from MacNeish's cultural phases to the predicted curve, one other matter deserves mention. As may be seen, none of the bar representing MacNeish's Phase One, nor the complete bar representing Phase Two will fit on the curve using the scaling methods utilized here. However, if we accept my logic that specialized and incidental domestication were processes requiring long periods of time (that is, that the number of years taken to get from μ_{p_d} = .0 to .1, and from μ_{p_d} = .1 to .2 were very long compared to the time taken to go between equal intervals of μ_{p_d} during the final phase of agricultural domestication), then the fit of observed with the predicted data would also include these earlier phases of domestication. The faster rate of evolution during agricultural domestication (region III) causes the bars representing Phases One and Two of MacNeish's sequence to seem far too long to fit properly into the period of time representing very late incidental and most of specialized domestication (late region I and most of region II on Figure 5.16).

We may make the proper correction by mentally "stretching" the graph to the left of region III so that the time period between the origin (μ_{p_d} = 0) and region III (μ_{p_d} = .3) becomes extremely long. MacNeish's data representing his Phases One and Two now comfortably fit onto our distorted graph. By taking my discussion of the variable rate of domestication into account, the observed data can be made to fit the predicted with a high degree of accuracy. Further detailed studies permitting quantification of $p_d/(a + p_w)$ may thus eventually yield precise quantification for changes in the rate of change in domestication. Such information will be of interest not only to students of human cultural change, but to all natural scientists interested in the dynamics underlying the evolution of animal–plant symbioses.

Throughout this volume I have stressed the necessity of separating *domestication* (human interaction with plants) from *agricultural origins* (the origin and evolution of the agroecology). In this chapter we have seen why the separation is conceptually useful. Domestication has a long, gradualistic history whereas the origin of agricultural systems was a relatively sudden phenomenon that was to have radical effects upon human social and cultural systems. However, it is now the time to hedge and point out as clearly as possible that these are interrelated phenomena. It is unfortunate that clear discussion of a process often requires the creation of classificatory system that may later come back to haunt the person who created the system. Nothing could be worse for our understanding of people than if, having found this model for domestication and the origin of agricultural systems a useful approximation of the history of human interaction with plants, students of the topic were to relegate domestication to the status of a nonissue. Attention cannot be placed solely upon the takeoff point in the origin of agricultural systems, even though it correlates with interesting social events. To appreciate fully the nature of agricultural systems, we must clearly define the processes by which they arose and that limit their functioning. Thus it is of extreme importance to remember that domestication (in its larger sense) does not cease with the development of the agroecology, but merely functions within a new, highly complex, environment.

CHAPTER 6

Instability, Cultural Fecundity, and Dispersals

> Look at the most vigorous species; by as much as it swarms in numbers,
> by so much will it tend to increase still further.
>
> CHARLES DARWIN, *THE ORIGIN OF SPECIES*

A model for the spread of agricultural systems may be proposed that directly relates the evolutionary tendencies present in the man–plant interaction to the dispersal of domestication and agriculture as modes of human subsistence. This model is phenomenological and behavioral. It is concerned only with certain types of biological interactions and their possible effects on the success (that is, increase in numbers over time) of various human populations. It is also mechanistic in that it attempts to view cultural change as the result of differential reproductive success under differing conditions. It does not approach the question of why changes in subsistence pattern have occurred, but focuses on *how* certain interactions of culture and environment (in the largest sense) might contribute to the dissemination of new cultural behaviors.

In this model I make no assumptions about the ultimate source of cultural variation and instead am concerned with the appearance and fixation of variant behaviors. There is no need to posit that cultural change involves any type of genetic change, or even that there is a genetic component to the appearance of new interactions of humans and their environment. I merely assume that cultural behavior is variable yet essentially conservative; that is, that it *normally* exhibits little error in transmission from one generation to another; I discuss a mechanism for the appearance of new behaviors as "errors". Given the assumption of behavioral conservatism, we are free to explore those factors that might encourage the fixation of behavior and those that spread change in the performance of cultural actions. Behavior, like any other phenotypic trait of an organism, is amenable to selection. Thus behaviors may influence the differential reproductive success of a lineage over time. If the presence of a new behavior increases the probability that a lineage will prosper (in numerical terms), the change in behavior has increased the fitness of that lineage. I hope to demonstrate that changes in subsistence patterns may have important effects on fitness.

Domestication, as we have seen, is an outgrowth of certain types of subsistence patterns; it is a natural evolutionary process resulting from predator–prey interactions. Domestication is thus both process and resultant state. It is a process that changes the morphology, physiology, and distribution of organisms, and it is a *mutualism*—a relationship that benefits genetically unrelated organisms. Domestication is, however, neither inevitable nor orthogenetic. It is only one of a large number of differing relationships that may evolve between a predator and its prey, and there are many ways in which its development can be subverted.

Like any model, this one is based on abstraction. One of its simplifications is that it includes only those aspects of the environment that are direct functions of subsistence. I treat populations here as a function of available yield and believe this is justifiable for these purposes, though clearly it is an inadequate summary of the forces acting on a population. Thus one way in which the intensification of domestication might be subverted is through radical change in climate, pathogenic attacks on either plant or people, or even warfare. Besides the influence of factors arising outside the model, there is also no guarantee that fitness will always be maximized by similar adaptations.

The purpose of the model is to elucidate the development of agricultural lineages. Thus little attention can be given to lineages that were once agricultural or domesticatory but abandoned the behavior later. To model such events is, in essence, to model general cultural change, a task far beyond our present capabilities. My aim is to describe a likely means by which successful agricultural societies arose and spread. In a sense, I am trying to follow one particular thread of cultural change backward in time; I do not attempt to describe the fabric. If the limits of the model are exceeded, the society is neither domesticatory nor agricultural and it falls outside my area of concern. Thus, although I give little attention to the loss of behaviors, this should not mislead the reader into believing that the limits of the model have never been exceeded.

The focus on the interaction of humans and plants opens a new avenue to the problem of domestication, one that treats the process as a natural outgrowth of subsistence behavior. Agriculture is seen as one type of interaction between humans and the environment that may have important effects on the environment and thus on the way people survive. I explore the demographic results of this interaction in detail here. First, however, it is necessary to define agriculture in greater depth.

The origin of agriculture involves the appearance and spread of certain behaviors relating people, plant, and environment. We have just considered the dynamics of this process in terms of changes in human feeding behavior. It should now be stressed again that agriculture is only a *level* or *type* of behavior; it cannot be defined as a set of specified actions or practices.

Most earlier definitions have attempted to limit agriculture to a particular set of techniques (the tilling of the ground, the planting of seeds, or the artificial selection of plants for improvement); this kind of definition may have utility in certain cases but may easily lead to error. For example, the notion of *cultural stages* is based on the assumption that the mere acquisition of technology is sufficient to explain the transition to an agricultural way of life. Again, by focusing attention on a given tool or technique we are giving a false impression of the importance of any particular technique within the overall framework of agricultural subsistence. Frequently this leads to a mistaken belief that cultures sharing a particular technique of production must therefore have a common origin. In essence, we define a trait as conservative and proceed to construct taxonomies of relationship on the basis of that trait. Also, because it is very easy to assume that correlations exist between techniques that may, in fact, be unrelated, we then tend to assume that agriculture arose only once, or at most a few times, in human history. Yet we lack independent confirmation for any such statement.

It is imperative that any definition of agricultural systems make as few assumptions as possible about its characteristics. To do this it seems best to define the process in broad terms. If one characteristic held in common by all recognizably agricultural systems could be abstracted, it might be this: *agriculture* is an integrated set of animal behaviors that affect the environment inhabited by domesticated plants throughout the whole life cycle of those plants. Admittedly, this is an extremely broad definition, and it may have implications troubling to some people. For one, it gives no consideration to the motivation of the behaviors. Arguments against this viewpoint have been presented earlier and are not restated here. Further, it includes certain nonhuman forms of plant–animal mutualism, such as that exhibited by fungus-cultivating ants. I have already given my reasons for believing that agriculture cannot be meaningfully restricted to the mutualistic relationships that have developed between people and plants.

Finally, the definition permits no "indicator" activity by which we might be able to separate agricultural from nonagricultural societies. I believe this is an advantage, rather than a drawback. If agriculture was not invented, but evolved out of existing patterns of behavior, it seems clear that it will never be possible to identify the moment of transition using specific indicator behaviors (however, see Chapter 5). At an early stage in the development of agricultural techniques it is obviously impossible to distinguish agricultural behavior from a highly developed form of specialized domestication. Agriculture is seen here as the result not of any one behavior but of a set of behaviors. A society may be more or less agricultural, and it may have differentially strong effects on the environment inhabited by its domesticated plants (thus my classification of incidental, specialized, and agricultural modes of domestication). The quantitative aspects of

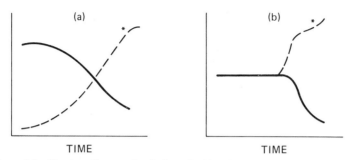

Figure 6.1   The time of onset of agriculture. Incidental and specialized domestication (——);
agricultural domestication and behaviors (---); the point of agricultural subsistence (*)

agricultural behavior may arise from two sources: the acquisition of new traits and the intensification of existing ones.

It is advantageous to recognize that agriculture, like domestication, is a dynamic phenomenon, one that has no stable characteristics but in any given situation can be characterized in many different ways. We should also recognize that distinctions are created to permit analysis. My definition of agriculture places the emphasis where it will be most useful—on the interaction of society, domesticated plant, and their environment.

If we need to define for any particular society the moment of transition from a domesticatory to an agricultural way of life, I suggest that we visualize it in this way: in the history of any pristine agricultural society, there was a period in which environmental manipulation (specialized domestication) became of greater importance to the subsequent development of that culture than ordinary co-evolutionary interactions with plants. Thus we may note that the origin of agriculture preceded agricultural subsistence (see Chapter 5). This is shown in graphic form in Figure 6.1a. For a culture that did not evolve an agricultural way of life but adopted one as a result of outside influence (conquest, imitation, "choice"), the same statement may be made with the reservation that the agricultural function arises de novo at an already high level (Figure 6.1b).

It might be noted that the importance of these behaviors may be read in several different ways. We may consider their relative significance for the development of the society. We may also view their contribution to the overall subsistence of the society at any given moment. This latter interpretation is of critical importance in understanding the evolution and subsequent spread of agricultural behavior. It should be noted that this interpretation, although suggesting *when* the change to agricultural behavior occurred, is not concerned with *how* it might occur. If the latter question is not addressed, the former is, at best, trivial.

I assume here, first of all, that cultural behavior is not genetically transmitted, although the potential for the behavior obviously has a genetic basis. In the case of domestication and agricultural origins, this basis is one that has arisen independently in numerous species of animals. Further, I assume that cultural behavior is transmitted from gneration to generation by means of imitation and learning. This transmission is conservative; that is, there is little intrinsic variation from one generation to antoher. Most scholars would agree that cultural behavior tends to be stable over short periods of time—indeed, the major problem presented by the study of cultural change is to understand why any change occurs at all. Finally, I assume that the relative success of certain cultural traits, such as agriculture, is at least in part a reflection of a fitness (numerical success) that the traits are capable of inducing in societies. Taking a comparative approach, I contrast the effects of agricultural behavior with the effects of non-agricultural behavior on the demography of societies that practice these differing modes of subsistence. For simplicity's sake, I consider *consciousness* or *intentionality* irrelevant to an understanding of the demographic and environmental processes that condition the origin and spread of agriculture, and I consider only that contribution to population growth made by domesticatory and, later, agricultural behaviors.

Domestication and Demography

Domestication has been present, to a greater or lesser extent, in all cultures and at all times. It has gained its major significance, however, from its interrelationship with the origin of agriculture, and this in turn is tied to the development of modern human societies.

We have established that coevolution is common in nature and that it has had major effects on animal survival by drastically altering the relationship between an animal and the plants on which it feeds. Domestication changes the competitive relationship that usually exists between a predator and its prey into one that could fairly be described as *cooperative*. By this I mean not that cooperation in any intentional sense occurs, but that advantage accrues to both members by virtue of the relationship. Recasting this argument in demographic terms, increase in numbers of one party to the relationship will be accompanied by a potential increase in the numbers of the other. The mathematical analysis of symbiotic relationships has been inhibited by this beneficial effect of species upon each other (May 1974; Pielou 1977): because the two members of the relationship have a positive effect on each other's fitness, "the mutualists embark upon an orgy of benevolence, and their numbers expand to infinity. Needless to say, this is not a meaningful statement about nature, but rather a flaw in the . . . equations" (D. Wilson 1980:145). Although recent work is beginning

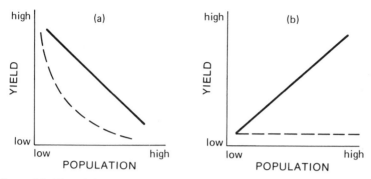

Figure 6.2   The relationship between predator population and yield. Total available food
(——); per-capita yield (---).

to confront this critical issue head on, any attempt to express the demographic
consequences of the developing agricultural relationship in strict mathematical
terms would be premature. Instead, we use a graphic approach in our modeling.

Figure 6.2a illustrates the relationship that exists between most predators
and their prey. As the total number of predators increases, both total and relative
amounts of food available to the predator decrease. The total amount of food
decreases because the predators, acting at higher densities, extract an ever-
increasing percentage of the prey, thus interfering with the ability of the prey to
replenish its numbers. Relative yield then decreases at an even faster rate as a
result of the absolute increase in predator population. Of course, in a simplification such as this, it is necessary to assume that the area inhabited and the total
variety of prey consumed remain constant. Nevertheless, it should be clear that
any new foods brought into the subsistence pattern, or any increases in the
feeding range, can be graphically compensated for by shifting the y axis of the
graph to the left. Thus, besides representing the relationship between a predator
and a particular prey, this graph may be used to represent roughly the type of
relationship that exists between any predator and its total potential range of prey
species. It may also be used to approximate the relationship existing even when
the predator is expanding its ranges: infinite expansion of range is impossible and
limits are eventually placed on the available food. It can be easily seen, therefore, that a point must exist at which reproduction is limited by an absolute lack
of nutrients per individual. The point at which most resources consumed are used
in maintenance and not for successful reproduction is widely known as the
carrying capacity of the environment. Obviously, beyond this point further
increases in population are impossible.

Figure 6.2b illustrates a mutualistic relationship of predator and prey. In this
case, as the number of predators increases, the total amount of available food
also increases. Success by one member of the relationship is tied to a high

probability of success for the other member. This is in keeping with the generalization that coevolution increases the fitness of both parties. Highly mutualistic relationships evolved over time when ancestral populations diverged or were replaced by new ones because of heightened fitness. The actions of the predator in such relationships tend to increase the numbers ofavailable prey. In coevolved relationships, the cost to the prey of providing subsistence to the predator is balanced by actions of the predator that provide increased opportunities for the survival or dispersal of the prey.

Another way of viewing highly evolved mutualistic systems is by noting that domestication has the effect of increasing the carrying capacity of the environment in relation to the predator. Domestication permits larger numbers of prey to occupy the area by maximizing the probability that the prey species will be able to colonize any suitable open space in the environment. Coevolutionary dispersals, by being integrated with the predator's behavior pattern, are of higher quality. Highly evolved protective relationships increase the probability that the prey will survive to reproductive age and may also allow for a greater total life span for the prey population. All this means a greater total potential population for the prey species. The predator's numbers may, in turn, increase to take advantage of this greater productivity. As a consequence of mutual increases in fitness, the evolution of a mutualism may (assuming, for the moment, no extrinsic limitations) involve increase in population for both predator and prey. An obvious corollary to this is that, over time, the yield of the prey organisms tends to increase. We can express this as selection for yield: the prey organisms best adapted to the behavior of the predator will prosper and in time will dominate the population. But insofar as the relationship is mutualistic, adaptation to the prey's behavior implies ability to contribute to the prey's survival. The prey that maximizes the numbers of predators will be most fit.

Of course, limits are placed on the continued growth of both the predator and the prey population; the situation represented by Figure 6.2b cannot continue indefinitely. For example, there may be a limited number of open areas for the prey species to colonize. This automatically restricts the total range of that species and concomitantly the range of the obligate predator species. Again, a new threat to the prey species—against which the protective behaviors of the predator are ineffective—may enter the environment. In this case, both members of the coevolved relationship will suffer. Competition may also limit the growth of the relationship. The feeding upon the prey species by agents that cannot successfully disperse it will reduce the amount of food available to the coevolved agent. This is an extremely common problem, and successful coevolved dispersal relationships minimize this threat. Eventually, then, the compensatory behavior of the predator will become ineffective and the total yield will level off or even assume a negative slope. This situation is illustrated in Figure 6.3a, where point A is the point at which the compensatory behavior becomes ineffective or

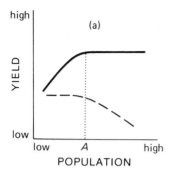
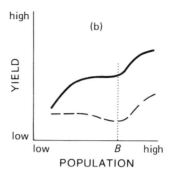

Figure 6.3 Changing limitations on mutualistic interactions: (a) a typical interaction; (b) the effect of an increase in mutualism. *Solid line,* total available resources; *broken line,* per-capita yield.

all available regions favorable to the prey became filled. Any factor that interferes with the continued growth and survival of either member of the mutualism will have effects on the growth and survival of the other member.

This stabilization of yield, and thus of predator numbers, is also subject to change. Most of the factors that place limits on the success of the prey species, and thus indirectly on that of the predator, are rooted in the environment; they are external to the relationship itself, although, of course, intimately connected with the form it takes at any given moment. Few if any species are reproducing at their intrinsic rate of increase; all organisms have a reproductive capacity far in excess of that needed for replacement of the parents. They prey species thus responds to any "relaxation" of the environment: if conditions more favorable to its growth and reproduction arise, the prey population increases and there is an increase in the potential number of predators. We describe this by saying that the carrying capacity of the environment for both members of the relationship has increased. Figure 6.3b illustrates a change in the overall environment that satisfies such conditions, with point *B* indicating the moment of effective change.

The carrying capacity may be increased either by a change in the morphology of the prey, or by a change in the behavior of the predator. It is most important to recognize that point *B* of Figure 6.3b may indicate either a change in the morphology of a plant or a modification of the behavior of its coevolved agent. The relationship between humans and the plants on which they feed is one of predator and prey. In such relationships, coevolutionary schemes may emerge, and from them mutualisms may develop. Yet, because we are dealing with a relationship, or interaction, evolutionary change in a plant may be as important to the further development of the relationship as any change in the behavior of the predator. Thus point *B* may stand for an event other than a behavioral change in a culture. It may indicate the effects on the carrying capacity for humans of a mutation in a plant's morphology; for example, the appearance of indehiscence in a cereal crop that, by exlcuding previously effective dispersal agents, could create a greater recoverable yield for humans. The change in the plant's morphology, by permitting higher human densities, could enhance the coevolutionary relationship existing between humans and plants: larger numbers of humans will create more opportunities for successful dispersal, which permits larger populations of both plants and humans.

The changes in the behavior of humans that can increase the carrying capacity of the environment for plants have been the major focus of all studies of agricultural origins. However, very minor modifications in human behavior, not directly acting on the environment, may also have major effects on the interaction of humans and plants. We have already seen how changes in feeding behavior may affect the symbiosis. Scheduling may also be of major importance; this is an area with great potential for further research.

In practice, it is extremely difficult to separate changes in the behavior of people from changes in the morphology of the plant. Feedback is continually occurring between the members of the mutualism: a mutation in a plant may allow for the successful expression of variant forms of behavior, and a change in behavior may allow a mutation to spread throughout the plant's gene pool. The analysis that follows nevertheless aims at the identification of the changes that were eventually to yield and disperse developed agricultural systems.

Agricultural behavior is here defined as behavior that, by acting directly on the environment, increases the carrying capacity within a specific local environment, the agroecology, for the domesticated plant. Agricultural behavior permits an intensification of the plant–human mutualism by expanding the size of the effective environment available for colonization by the domesticated plant. Although agricultural behavior does not bring an end to the domestication of plants, its demographic consequences arise from the interaction of the mutualism with the local environment rather than from interactions that occur solely within the mutualistic relationship.

In Figure 6.3b we see that the major effect of domestication is the potential for population growth. For the purposes of this model, I assume that this potential is realized (see also Chapter 5). The ultimate effect of mutualistically induced increases in domesticatory populations is that, over time, such populations will dominate any given geographical area by virtue of mere numbers. It is this domination that has caused most workers to characterize domestication as an adaptation by people, but it should be clear by now why I have stressed its *nonadaptiveness*. It seems less than accurate to describe an interaction of predator and prey such as that outlined here as an adaptation *of* or *by* the predator to the demands of continued survival—we can just as easily describe the process as an adaptation of or by the *prey*. In emphasizing the actions of humans, we neglect the fundamental contribution made by the plant. Reducing the process to an exploitation of the environment by humans ignores a limitation placed on further

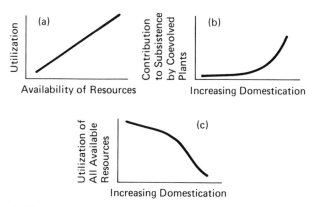

Figure 6.4 The interaction of changing resource utilization patterns with the rise of domesticates.

population growth for the human component of the relationship. As domestication increases in intensity, the continued survival of human populations at their new and higher levels is dependent on an ever-decreasing number of environmental factors. Concurrently, the development of an obligate relationship between humans and plants is accompanied by an increasing dependency on the plants with which they have established the relationship.

The major effect of the mutualism between humans and plants is a greater average productivity of the local environment. This increase is not based on equal increases in the productivity of every component of the environment; a very small subset of all potential sources of food is providing the total increase. As domestication proceeds, so does this reliance by people on an ever-smaller subset of resources (see also Chapter 5). Figure 6.4a illustrates the obvious relationship: as a resource increases in availability, the utilization of it will increase. The effect on total diet of increasing utilization of a particular set of resources is illustrated in Figure 6.4b. As domestication (and thus the abundance of a particular type of resource) increases, the percentage of the total diet derived from coevolved plants will also increase. As a direct result, the contribution of nondomesticates to the diet will sharply decline (Figures 6.4c and 5.7). This is the ultimte simplification of the diet mentioned in the last chapter, and it may have major effects on the stability of the coevolving system.

We may approach the potential instability brought into subsistence patterns by increasing reliance upon domesticated plants in terms of the relationship between humans and any one, particular, coevolved plant. The overall effect of increasing domestication is an increase in the yield that permits larger numbers of humans to survive in any given area. However, this is only the *average* increase over time. The basis for the increase is intrinsic to the coevolving relationship,

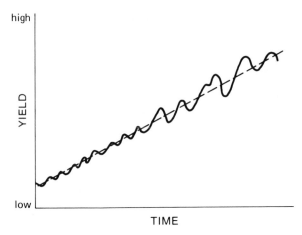

Figure 6.5   The relationship between the average and actual yields. Actual yields (——);
average yields (---).

but that relationship exists in an environment that is subject to periodic normal
fluctuation. The yield of any plant in any environment will vary about its mean
from year to year because of environmental conditions. These conditions are
external to the relationship and are not affected by the behaviors that characterize
incidental and specialized domestication (although they may be affected by agricultural behaviors). Thus the component of total yield contributed by any given
plant will vary in response to extrinsic environmental factors. This relationship is
diagrammed in Figure 6.5. Although the *average* yield of domesticates (P_d of
Chapter 5) is increasing over time because of the elaboration of coevolutionary
domestication, the *absolute* yield at any given time is determined by environmental conditions.

In summary, we can say that domestication is an interaction with the following evolutionary tendencies:

1. Mutualistic relationships allow increases in the abundance of both predator and prey. Coevolution may increase the populations of both members
   of the relationship over time.
2. The environment limits the continued growth of the populations. This
   limitation may be overcome by change either in the morphology of the
   plant or in the behavior of the animal.
3. Increase in the available yield of any coevolved taxon causes an increase
   in the contribution of that taxon to subsistence. Increasing elaborations
   of coevolutionary sequences increase the importance to people of the
   plants involved in the relationship.
4. Although domestication causes an increase in the *average* yield of a

taxon, environmental fluctuations will affect the relative yield of any given taxon from one year to another.

Under incidental domestication, the fluctuation in domesticate's yield should have little effect on the total yield obtained by people from the environment. Of primary importance, the contribution of any given incidental domesticate to the total subsistence will be relatively small, and the total contribution of domesticates will also be relatively small compared with their contribution once agricultural systems are established. Also (and this was not considered in Chapter 5 in the feeding model), incidental domesticates are likely to be rather uniformly distributed throughout the environment, no particular niche having established itself as primary. The incidental domesticates are likely to grow in most of the habitats through which people pass. Thus, although environmental fluctuation will have adverse effects on the yield of any particular incidental domesticate, the overall yield from incidental domesticates will remain relatively constant; a bad year for one plant is likely to be a good year for another with a different physiology or for one growing in a different habitat. Compensation in yield will permit production to remain relatively stable from year to year, much as it does in any nondomesticated ecology. Under specialized domestication this relationship will change dramatically.

Agricultural Origins

Inherent in the assumption that available food is a major determinant of population size is the implication that the nonfood limitations on population growth were similar for early agricultural and nonagricultural societies. There seems no reason to assume otherwise. In the past, when the belief that agriculture arose only once or a very few times was acceptable, it was reasonable to assume that some factor external to the human relationship with plants was the prime mover. Given only one or two centers for the development of agricultural techniques, it seemed likely that the societies inhabiting those centers had developed a religious or political system that could have been the direct cause of agriculture. But, as the numbers of independent centers of agricultural origin increase, the probability decreases that all those centers shared some cultural factor other than agricultural behavior. Thus it seems justifiable to focus on some aspect inherent to agriculture (*sensu lato*) as the factor responsible for the elaboration of developed agricultural systems.

There is, of course, an important and continuous interaction between agriculture and other aspects of a society. Numerous other variables are potential influences on the structure of any population and on the cultural traits that population manifests, but the model developed here is not concerned with general cultural change. Factors such as disease, religious beliefs, warfare, and political and kinship systems, to mention only some that are obvious, may have been of crucial importance for any society's development, but there is no evidence that they were important in the development of the fundamentals of agricultural production. Another reason to concentrate on the role of the components of agricultural behavior in the development of agricultural societies may be appreciated by considering existing agricultural systems. Agricultural societies do not appear to show any greater similarity among themselves than can be explained as a function of agriculture itself.

A final reason to look only to agricultural behavior as the underlying process in the elaboration of agricultural systems is to be found in the understanding that agriculture is a direct outgrowth of processes already existing under domestication. The transition from domesticatory to agricultural behavior can occur by an intensification of already existing processes; we need not posit any outside force to account for it. The transition is a change in emphasis, a change that is rooted in an existing economy and it is smooth rather than abrupt. If the entry of an outside force were responsible for the origin of agricultural systems, the record of pristine agricultural societies would show an abrupt discontinuity such as is sometimes found in the historical record when agriculture is introduced into an existing nonagricultural society. Pristine agricultural societies have not shown any such discontinuity. The development of cultivated plants seems to have been a continuous process. Likewise, the elaboration of agricultural processes seems to have been fairly continuous.

It is in keeping with the model developed here to view agriculture as a quantitative change in behavior over time. I distinguish domestication from agriculture in order to elucidate the interactions involved in each, but, though useful, the division is artificial. If we were able to go back in time and study a society in transition from specialized domestication to agriculture, we would face the same problems in classification that we would have if we were viewing it as it first embarked on domestication. A society cannot be classified as domesticatory or agricultural until the transition is completed. The utility of the classification is that it helps us to discuss the processes involved.

Agriculture as practiced today involves various techniques that affect the environment in which a cultivated plant grows: plowing, weeding, harvesting, storage, and planting. All modify the environment encountered by the plant and its propagules. It is clear the the origin of agriculture is in part the origin of these techniques of environmental manipulation, and it is reasonable to assume that the possession of these techniques in incipient form distinguished developing agricultural societies from contemporaneous (incidental) domesticatory and non-domesticatory societies. If we are to understand the transition from domesticatory to agricultural behavior, we must examine the origin of these techniques and consider their effects on the society practicing them.

The primary effect of agriculture on a society is an intensification of the dependence of that society on domesticated plants. In investigating the effects of domestication on a society, we did not need to assume that all or even a significant proportion of its subsistence came from domesticated plants. Highly developed agricultural systems, however, are based on a fairly limited number of cultivated plants, which provide most of the society's food. This is to be expected. The development of greater exclusivity in diet is a logical outgrowth of maximization of fitness in coevolutionary relationships. Also, any behavior that counteracts the limitations on the continued elaboration of mutualistic interactions will bring with it an extremely powerful selective advantage. Any given environment has fairly well-defined niches. Thus the potential number of places in which any particular plant might grow is determined by the parameters of that environment. For example, the total amount of naturally disturbed ground is both limited and a function of forces such as wind, fire, water, and gravity. The absolute amount of disturbed ground determines the maximum number of colonizing plants, such as the weedy heliophytes, typical of the ancestors of many agricultural plants, that may grow in any given area. The distribution of areas of disturbance will also be a function of the topography and climate of the region, and this distribution affects the probability of successful dispersal by plants requiring a disturbed habitat for colonization.

Any behavior that reduces environmental limitations on the potential number of domesticated plants that survive will bring with it a substantial increase in fitness for both members of the mutualism. Returning to Figure 6.3, we can see that a change in behavior (indicated now as Point B in Figure 6.3b) will do much to increase the carrying capacity of the environment for the plant, and this increases the carrying capacity of the environment for people.

The total yield of domesticated plants places a limit on the potential for human population growth. Thus, any behavior that increases the size of the niche available for plant colonization increases the effective yield. Similarly, any behavior that decreases the heterogeneity of the environment also increases the effective yield. Decreased environmental heterogeneity also increases the probability that an area will be colonized by a specialized domesticate rather than by another plant with the same edaphic requirements but not involved in a mutualism with humans. Protective behaviors intensify this relationship. As the domesticated plant is increasingly successful, the limitations placed on human population growth are raised to a higher level. Populations practicing agriculture come to be more successful relative to both domesticatory and nondomesticatory populations. These populations not only will be generally larger but will also be dispersing at far greater rates.

Any human behavior causing environmental changes that increase the probable reproductive success of the domesticated plant, and thus its yield, has important effects on the further evolution of the domesticated plant. Within any locality, the following tendencies guide the evolution of the cultivated crop: (1) reduction in diversity, both genetic and phenotypic, (2) an increase in productivity, and (3) autecological convergence. At the same time, the following tendencies become manifest within the total set of crops comprising the agroecology: (1) a reduction in the number of species on which people rely for their subsistence, (2) an increase in total crop yield, and (3) autecological convergence among all the crops that are important in the relationship. These interrelated tendencies from the basis for the elaboration and dispersal of agricultural systems.

Changes in the direction of plant evolution within the early, human-modified ecology are based on one rather simple factor: such human behaviors create an environment for plant growth and reproduction that is structurally homogeneous and relatively stable compared with the overall environment (see Chapter 4). Agricultural systems may be described as simplified ecologies: typically, the agricultural plant is a *weedy heliophyte,* a colonizer of disturbed habitats, and the agricultural field or garden is an environment in which the earliest stages of ecological successions are maintained. The predominance of this colonizer system for agriculture is based on the fact that there are numerous ways in which a disturbed habitat can be created by human behavior. Fires, disturbance of the soil, the clearing of forests or the ringing of trees, latrine practices, and the creation of dump heaps all are different routes by which a disturbed, and thus open, habitat may be created. Early domesticated plants preadapted to this niche are thus favored in their further evolution into agricultural plants.

Various agricultural behaviors serve to reduce the intensity of natural selection for characteristics of plants that are necessary for survival in the wild. Irrigation reduces the necessity for plants to maintain mechanisms for survival during periodically recurring droughts. Clearing of the land and weeding decrease the importance of competitive mechanisms. Planting encourages specific and uniform germination and seedling physiologies. The techniques of agriculture ultimately reduce the variation to be found in a species of domesticated palnt. Both the wild and the early domesticated plant must maintain pleiotropic responses in the face of an unpredictable environment; with a reduction in overall environmental unpredictability within agricultural systems, this variability is no longer being maintained by natural selection. Physiological and autecological convergence occurs in agricultural plants within any given locality. As we shall see, this decrease in variability ultimately increases vulnerability.

Besides freeing the cultivated crop from many of the demands for survival in the wild, however, agricultural behavior will have very important effects on the amount of variation to be found in the population of an agricultural plant in any given locale. Techniques of environmental manipulation, like other cultural traits, are transmitted conservatively, and the agricultural environment inhabited by the cultivated plant varies little over fairly long periods of time. Within the agricultural environment, selection occurs for those phenotypes best adapted to these relatively constant environments. The evolutionary effect of this is a loss in the variation that was present in either the wild or the early domesticated taxon. This causes, in essence, a physiological convergence within the cultivated plant species.

At the same time, selection for higher yields continues and even increases. Plants yielding the greatest number of propagules—those best adapted to the agricultural environment—are most likely to survive and spread. The relaxation of selective pressures for traits such as competitiveness with plants *not* adapted to the agroecology will also bring about a new means by which plants may increase their yield. Energy that was previously diverted by the plants to such tasks as seed protection, the manufacture of long internodes, perenniating structures, and the like can now be utilized for further increases in yield. Released from the requirement for possessing such structures, the plant that diverts energy into the production of propagules is the most fit within the agroecology.

Processes analogous to those occurring at the intraspecific level also occur at the interspecific level. Those plants best adapted to survival and reproduction within the agroecology come to dominate the system. Given the homogeneity and stability introduced into the environment by agricultural behavior, it is reasonable to expect a reduction in species diversity. We will find a tendency toward increasing reliance upon fewer and fewer species to provide the basis for agricultural production. This reduction in species numbers is, however, neither permanent nor absolute. Certain species may secondarily adapt to continued survival within the agroecology by changing their strategy. An example would be those crops assumed to have been domesticated for their propagules but later used for their vegetative organs (for example, lettuce, flax, and mustards). Although they could not survive as seed crops in competition with other, more prolific grains, they were not lost from the cultivated flora. Similarly, new plants may enter the agroecology as weeds or colonizers and become secondary domesticates. Thus we may note that secondary domestication is tied, in part, to the selection for yield that is occurring within the evolving agroecology (see Chapter 4).

The same selection posited for greater yield within a given taxon acts to decrease the diversity of species present in the cultivated flora. We find a tendency toward increasing reliance on fewer and fewer taxa to provide the basis for subsistence and a concomitant tendency toward increase in the productivity of the surviving taxa. The obvious culmination of such a process is monoculture:

agricultural behavior produces an intensification of a tendency already present in domesticatory relationships but now given far greater emphasis and direction because of the effects of environmental manipulation.

Another factor also reinforces this tendency towards uniformity and simplification of developing agricultural systems. As a food source becomes more common, feeding on it will increase. At first this will be merely a function of availability, but as time passes techniques of production and consumption will tend to improve. Availability and efficiency interact in a positive feedback manner, and further specialization in diet is likely.

Specialization in diet is also encouraged by the localization of agricultural production. Reduction of usage decreases the probability that coevolutionary relationships will be established with new species of plants. As more time is spent in the agricultural environment, people will come into contact with fewer species of alternative food sources. Changes in time allocation or scheduling reduce the importance of enterprises competing with agriculture. Thus the establishment of agriculture will intensify and direct tendencies already existing under domestication.

Agriculture creates a new type of climax formation. The agricultural flora tends toward stability as long as human behavior is interfering with other successional processes. Yet this mutualistic climax formation is unusual in that it is largely comprised of species with a colonizing evolutionary heritage. As we have noted earlier in this investigation, plants involved in domestication relationships with people are unusual in that they are fundamentally opportunistically dispersed plants that have established an obligate relationship with a generalized predator. The development of agricultural systems brings an equally fascinating ecological system into being: a climax formation composed of highly specialized colonizer species. This combination of genetic heritage and evolutionary setting goes far toward helping us understand both the productivity and the small number of species typical of agricultural systems.

Although the agroecology is both stable and simple compared with the overall ecology, it nevertheless will show change over time. Probably the most important long-term change that occurs in agroecologies is the creation of new niches within them. As time passes and plants respond to the options created by the existence of agricultural areas, subdivision of the existing space occurs. This subdivision permits the entry of new, not necessarily domesticated, plants into the agroecology. Weeds that evolve to utilize agricultural fields during fallow periods are one example. These plants begin to be subject to many of the same selective forces as the domesticates. Also, because we now have geographically stable ecologies, new types of domesticates may become established in them, among them plants whose utility is in their vegetative organs. Edible plants that can establish themselves in the agroecology need not to develop special coevolved means of distribution in order to survive in this environment; they need only scatter their seeds to persist from year to year (see Chapter 4). However, because they are growing in the same environment with other early agricultural plants, these secondary domesticates are still subject to the same selection for high propagule yield. They also exhibit the same tendency toward edaphic and autecological convergence as any other agricultural crop. The secondary domesticate evolve under the same selective pressures as the primary domesticate: it is evolving in the same environment.

The development of the agroecology will, because of the interaction between it and the general ecology, have major effects on the divergence of the cultivated plant from its progenitor species. Although diversity within the agroecology is likely to be, at least initially, less than that outside it, the result is nevertheless an increase in the heterogeneity of the entire region. This permits more opportunities for disruptive selection (Thoday 1958a,b) and intensifies the divergence of the *agricultural* plant from the early *domesticated* plant. Human interaction with the originally domesticated taxon will decrease as the interaction with the agricultural plant develops. Reduction of the amount of interaction between people and many of their earliest incidentally domesticated crops will, in the presence of highly developed agricultural systems, leave the early incidental domesticate "stranded." The human agent with which it had developed a coevolved dispersal or protection system no longer functions as dispersal agent or protector, and its extinction is almost inevitable. It is not surprising that it is so difficult to identify the "progenitor species" for many agricultural plants. Two major opportunities for extinction have occurred in the history of primary domesticates: (1) the wild, uncoevolved portion of the ancestral gene pool may have become extinct in the period during which coevolutionary domestication occurred, or (2) the portion of the gene pool that evolved under the conditions of early domestication may have become extinct during the period of intensified evolution that led to the development of the agricultural plant.

Dispersals of Agricultural Systems

We have already noted that increasing human reliance on cultivated plants brings with it a decrease over time in the absolute number of taxa producing the major portion of an agricultural society's subsistence. We have also noted that many domesticated taxa will vary in productivity from year to year. We have described how convergence occurs among agricultural crops. Finally, we have seen that, by removing certain environmental limitations on the carrying capacity of the environment for domesticated plants, agricultural behavior permits a tremendous increase in potential yield and thus in potential human population. I would now like to explore the interaction of a contracting subsistence base,

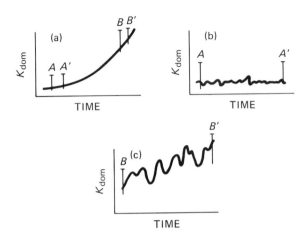

Figure 6.6   The evolution of the contribution to total yield derived from domesticates (see text).

increase in carrying capacity, and variation in productivity, an interaction whose demographic effects led to the spread of agriculture as a mode of human subsistence.

The introduction of domesticated plants brings an increase in the carrying capacity (Figure 6.6a). This increase in the carrying capacity of the environment K contributed by domesticated plants will hereafter be referred to as K_{dom}. Although K_{dom} tends to increase greatly over time, for reasons outlined above, the change in K_{dom} is not always positive from one moment to another. The short-term behavior of K_{dom} does not demand a year-to-year increase in absolute yield.

The components of the environment show seasonal and longer-term variation. Climate, for example, is not uniform from year to year, and over relatively short periods of time differing average climatic conditions will prevail. Figure 6.6b indicates the hypothetical variation in K_{dom} that might exist between times A and A' of Figure 6.6a for a society practicing specialized domestication. There is some variation in productivity, but variability is restrained by several factors. The early domesticated plant is growing in a variety of microenvironments, and a climatic change need not affect all the microenvironments in the same way. For example, domesticates growing in sites that are wetter than optimal will prosper during a period of reduced precipitation.

Domesticatory societies also tend to have a larger number of species contributing to K_{dom} than do agricultural societies. Thus any given change in the environment is not likely to affect all species of domesticates equally. Both of these intrinsic factors tend to reduce the effects of changes in the environment on the productivity of the agroecology. Perhaps the most important factor limiting the effect of changes in K_{dom}, however, is external to the symbiosis. The domesticatory society is relying upon domesticates for only part of its food supply. Thus it is possible to compensate for declines in K_{dom} by increased reliance on other components of the total carrying capacity. The failure of one domesticated plant, or of one domesticated species of plants, is compensated for by increased reliance on other plants, both wild and domesticated.

Figure 6.6c indicates the variation in K_{dom} that might be found between Times B and B' in an early agricultural society. The greater variance in K_{dom} for the agricultural system is an outgrowth of the evolution of the agroecology. The evolution of the cultivated plant in the agricultural environment increases not only productivity but also the susceptibility of that productivity to environmentally induced crop failure. The autecological convergence brought about by agricultural selection brings with it an increased uniformity in the response of agricultural plants to environmental parameters. Thus what is a bad year for any given member of a cultivated species is likely to be bad for all members of that species. Because the convergence is also occurring between, as well as among, plant species, bad years for any given agricultural staple are likely to be bad for other staples also. Increase in productivity has been bought at the price of uniformity in response to the environment.

Contributing to and intensifying the effects of autecological convergence is the greater localization of the agroecology. Localization, although it increases yield by allowing a more complete harvest, intensifies the effects of microclimatic differences on total yield. Localization, although increasing recoverable yield by allowing a more complete harvest of the plants, also serves to intensify the potential instability of the system. Localization intensifies agricultural instability in several ways. Prime among them are the effects of microenvironmental and microclimatic variation. A garden area that optimizes drainage during average years will be too wet during periods of high precipitation and too dry during periods of drought. A hailstorm just before the harvest has vastly different effects depending on whether it falls on the cultivated fields or in the woods. Because hail is frequently a highly localized phenomenon, the effect of hailstorms on the total available yield is far greater for agriculturalists than for either a domesticatory or a nondomesticatory society. Thus localization of resources increases the possibility that all the resource may be lost to a catastrophe.

Agricultural subsistence is also accompanied by a growing specialization in diet. Thus a decrease in the yield from a staple crop has major effects on the total *perceived* food supply, as well as on the absolute food supply. Food preferences and techniques of preparation will have placed certain food sources in positions of dietary prominence. Information concerning the edibility and processing of alternative food supplies may be lost. Thus decreases in yield from cultivated plants may create the appearance of food shortage even though the total available food supply in the region may not have fallen to the point at which it is actually limiting the survival of the population. Animals respond to the perceived food supply, not to some objective measure of total available calories, and perceived food shortage may be as effective as actual food shortage in causing the emigration of humans from the locality.

The increased susceptibility of the agroecology to the extremes of normal climatic variation may be viewed as an effect of environmental manipulation itself. Many of the simpler forms of environmental manipulation control relatively constant, predictable parameters of the environment. They increase K_{dom} by mitigating the effects of fundamental restrictions on the carrying capacity of the environment for the plant. Increasing control over any limiting aspect of the environment brings with it an increase in vulnerability of the newly increased K_{dom} to those aspects of the environment left unaffected. For example, localization of production increases K_{dom}, and this allows for a larger number of human beings. At the same time, however, it increases the susceptibility of the system to negative microclimatic and microenvironmental effects. A change in precipitation that was previously unimportant will now have major effects on the total yield from cultivated plants. The relationship between climate and total production is becoming more finely tuned. Thus the introduction of techniques such as drainage or irrigation brings with it a large increase in K_{dom}. The removal of one factor as a limit on the total production, however, will leave other limitations unaffected. Thus agriculture aims at ever-increasing yield, and yet the removal of any given limit on yield allows other, uncontrolled limits to become evident.

Agriculture allows previously nonlimiting factors in the growth and productivity of plants to express themselves. For example, a plant suffering from drought will not have a major limitation placed on its productivity by minor insect infestation. Correction of the drought condition will allow for the expression of the limitation on productivity contributed by the insect predation. Thus we may note that techniques of environmental manipulation allow the environment to affect yield adversely in new ways. At any given time, techniques of environmental control increase the negative effects of conditions that cannot be controlled.

Agriculture also creates entirely new opportunities for limitations on productivity and thus for instability in yield. The increasing genetic uniformity of a crop, reduction in species diversity, edaphic and ecological changes created by agricultural practices, and the concentration of resources in a limited area all contribute to new potential instabilities in productivity.

Increasing genetic uniformity increases the susceptibility of the crop to attacks by pathogens. Polymorphisms in the production of secondary metabolites acting as biochemical defenses discourage the evolution of specialized pathogens. A reduction in this defense strategy encourages the evolution of pathogens that may seriously damage a plant species. This potential source of damage is especially clear when we consider crops in which secondary metabolites distasteful to people also serve to protect the plant from attack by other animals. Loss of these substances thus makes the plant more palatable to other herbivores.

Reduction in species diversity within the agricultural ecology may in itself have important effects on the susceptibility of the crop plant to pathogen attack. Escape from predation may be aided in many plant communities by the association of many different plants. Thus the plants "hide" from potential predators by being hard to find in the mosaic of diverse plant species (Feeny 1973; Tahvanainen and Root 1972). Competition between agricultural crops that results in tendencies toward both increased yield and reduction in species diversity thus may work to counteract defenses acquired during the evolution of the plant in the nonagricultural environment.

Agricultural practices may also have extremely important effects on the land—for example, erosion and changes in soil structure and drainage patterns. These are well documented and are not given further consideration here. Less well studied are the effects of agricultural systems on the feeding patterns of animals other than people. Especially in its later phases, agriculture may have significant effects on the local ecology simply by replacing areas of wild vegetation with cultivated fields. This destruction of wild habitats creates food shortages for animals feeding on the plants of these habitats. They may turn, out of necessity, to feeding on agricultural crops, even though these plants may not provide favored sources of food. Finally, the increasing concentration of resources encourages predation. The same concentration of resources that facilitates harvest of the crop by humans also facilitates its harvest by nonhuman predators. Although humans may delay consumption of the plant to optimize harvest, most of these predators will not and therefore may attack the field before the crop ripens. They may also be capable of utilizing a crop at a period during its life cycle when it cannot be consumed by people; thus predation may occur during early seedling or vegetative stages. Loss of the total crop may occur before any yield has been recovered.

The fundamental cause of agricultural instability is agriculture itself. All the new adverse effects the environment may have on agricultural productivity are induced by agricultural practices. Yet, at the same time, agriculture is responsible for greatly increased average yield and thus permits greatly elevated human population levels. Over the long term it would seem justified to say that, despite the greater instability of agricultural production, this increase in population levels is evidence of "progress."

However, the increase in population that accompanies increases in K_{dom} over long periods of time is a "success" for the system *only at the moment of change*. The agriculturally enhanced population levels now require continually elevated levels of production for their maintenance and, over relatively short periods of time, the enhanced population level becomes nothing more than the normal population level. The increases in productivity brought about by agriculture are absorbed by a growing population. Increasingly sophisticated techniques of environmental manipulation are required for the maintenance of the same rate of growth and, because of instabilities, improvements in techniques may be required for the maintenance of the same level of population. Human populations are growing in proportion to the effectiveness of agricultural procedures in raising the carrying capacity of the environment. Yet, the more effective agricultural procedures are in reducing environmental limitations on the productivity of the agricultural ecology, the more likely they are to create new opportunities for failures of the system. Increased productivity, as we have noted, is bought at the cost of increased vulnerability of production. As agriculture creates higher and higher population levels, the effect on the society of the failures of the system will become increasingly tragic. Successful agricultural systems require increasingly successful techniques of environmental manipulation merely to maintain the status quo.

We have noted that the two major effects of agriculture are an increase in productivity and a concomitant increase in the instability of that productivity. The interaction of these two factors has been responsible for the tremendous spread of agricultural techniques. To understand this somewhat paradoxical situation, it is necessary to understand how any animal population responds to changes in the carrying capacity of an environment. As is illustrated by Figure 6.7, no population of animals is capable of instantaneous change in population numbers in response to change in carrying capacity. Instead, the potential population must "track" the changes in carrying capacity. It will be somewhat out of phase with actual changes in carrying capacity because of lags in reproduction or behavioral response to perceived changes in the environment. Changes over time in K_{dom} are responded to by delayed changes in the potential population: this is shown by curves 1 and 2.

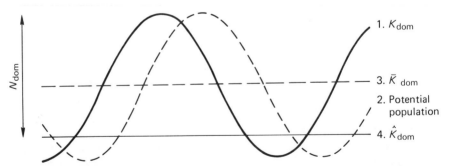

Figure 6.7 Productivity and the tracking behavior of a population (see text). Line 4 is the minimal effective carrying capacity ($\hat{K}_{\text{dom}}$); line 3 the average carrying capacity ($\overline{K}_{\text{dom}}$); curve 1 the contribution to carrying capacity from domesticates (K_{dom}); curve 2 the potential human population taking into consideration lag in response; the arrow represents the increase in potential human population permitted by domestication (N_{dom}).

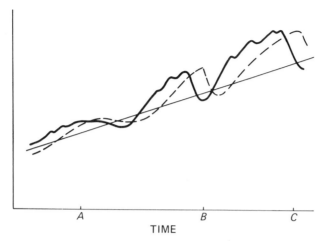

Figure 6.8 The relationship between productivity and population (see text). Contribution to carrying capacity from domesticates K_{dom} (——); minimum effective carrying capacity $\hat{K}_{dom}$ (——); actual population (---). That portion of the population above the line representing $\hat{K}_{dom}$ is "excess" population.

The actual average population over long periods of time, however, will much more closely resemble the population level give by curve 4. This is because the average population level cannot be determined by the average productivity (shown by curve 3 as $\overline{K}_{dom}$). Instead, actual average population levels will be determined by a recurrent lower level of productivity (shown by curve 4 as $\hat{K}_{dom}$). Part of the potential human population produced by recurring increases in K_{dom} cannot be successfully maintained during those periods when K_{dom} drops to its lowest levels. We may view this component of the population as agriculturally induced "excess production" of people (N_{dom}). N_{dom} is that component of the population determined by increases in K_{dom} above its recurring lower levels of yield ($\hat{K}_{dom}$). The amplitude of N_{dom} is clearly determined by the interaction of a large number of factors. These would include generational time, amplitude of K_{dom}, frequency in variation of K_{dom}, the availability of alternative food sources, and the means by which the tracking of K_{dom} is accomplished. Nevertheless, for present purposes treating N_{dom} as determined by the interaction of K_{dom} and $\hat{K}_{dom}$ has considerable heuristic value.

Figure 6.8 is a reinterpretation of the demographic effects of agricultural instability taking into account the tracking behavior of human populations. The hypothetical productivity and resultant demographic changes represent conditions such as might have existed in nearly pristine agricultural society. As in all the earlier illustrations, there is an average increase in K_{dom}, but in this case the potential populations and the effective minimal K_{dom} (= $\hat{K}_{dom}$) are also indicated. At three points (A, B, and C) the actual population is above $\hat{K}_{dom}$. We must consider the fate of this temporarily excess production of human beings. In essence, we are seeking to identify that factor that acts to decrease the population in a particular locality when environmental crises cause a decline in K_{dom}.

The most frequent response of any animal population to a decrease in carrying capacity is emigration. Emigration (rather than starvation or decline in per capita consumption) is especially likely if only one component of the carrying capacity is suddenly reduced. Part of the animal population will leave the original area in search of a place where the limiting resource is more abundant. Agricultural humans experiencing a decline in food supply could be expected to respond in the same way. A decrease in K_{dom} encourages part of the population to leave and establish itself in a new area. Slow declines in relative productivity, such as that occurring at time A, might be due to ecological or edaphic degradation caused by agricultural practices. Although large changes in population would not necessarily occur, emigrant groups might leave in search of better environments in which to farm. Sudden and drastic declines in relative productivity, such as are shown at times B and C, also cause emigration. In these cases, we assume that emigration will be encouraged by another factor in addition to the search for a better agricultural environment. If a drastic drop in production (a famine or sudden great scarcity of cultivated foods) should occur, it is likely that the population will return to the exploitation of nondomesticated food sources. We have already noted that the famine need not be an objective lack of potential calories in order to be perceived as a food shortage by the population. However, because an agricultural economy can support higher population levels than can a gathering economy, and because at least certain nonagricultural domesticates have been lost, it is logical that part of the population will be forced into new areas in search of wild resources. Thus declines in K_{dom} will be especially noticeable to an agricultural population because of higher population levels in any given area, losses of wild food sources and nonagricultural domesticates, and changes in the perception of what constitutes food scarcity.

The only other likely and effective response of a society to a decrease in K_{dom} is increased demand for environmental control. If this occurs, however, it may simply be incorporated into the illustration by increasing the amplitude of the K_{dom} curve. Because increases in productivity in this model already incorporate environmental control, it cannot be treated as a new variable. Also, as I have said, environmental control is intimately tied to instability in yield: drops in K_{dom} are inevitable effects of an agricultural system existing in a variable and evolving environment. At best, environmental manipulation delays the need for a solution to the problem of the excess population induced by the periods of "successful" agricultural production.

Besides the conservatism inherent in any society, another factor facilitates the emigration of agricultural populations without loss of agricultural technology. The environmental controls inherent in agricultural behavior permit easy colonization of new regions. Emigrant groups from most nonagricultural societies require a definite set of preexisting conditions if they are to maintain their mode of subsistence. Of course, agricultural societies also require certain ecological conditions, but the relatively slow spread of agricultural societies proposed here allows time for adaptation by crops to gradients in the environment. Secondary domestication permits colonization of the fields and gardens by new, better-adapted plants, and eventually a whole new ecological zone may open up to the agriculturalist. Also, although the activities of most nonagricultural peoples do not interfere with subsequent utilization of the land by agriculturalists, the converse is far from true. Thus it is reasonable to assume that most of the emigrating subpopulations maintain their agricultural ways of life. Some will probably abandon it, but any population that abandons agriculture and adopts a new means of subsistence can no longer be called an agricultural population and is not of interest here.

The spread of agricultural behavior may best be understood by briefly considering two hypothetical societies, one agricultural and one nonagricultural. We can assume that the effective carrying capacity is comparable for the two societies—a condition that could be satisfied if, for example, the nonagricultural society were based on a diversified maritime subsistence. The relationship between carrying capacity and population could be equivalent in both cases. Any factors acting to limit populations in the two settings (for example: fertility, age structure, disease, and warfare) could be assumed to be roughly comparable. The fundamental distinction between the two societies would be in the fluctuation in carrying capacity introduced by domestication.

In a situation in which both societies have achieved relatively high population levels, the maritime society, like the agricultural one, would experience periodic fluctuations in carrying capacity. However, its subsistence resources would be wild, and the effect of climatic variation on total carrying capacity would be rather unimportant. Sufficient variability in resource choice would exist that temporary shortage of one food might be compensated for by alternative sources. Thus an equilibrium would be established between carrying capacity and population. Excess production of humans, if it existed, would be both stable and a direct function of the total population.

Under agricultural conditions the situation would be radically different; agricultural production would intensify the effects of environmental variation. Environmental crises would be both more frequent and more severe, and the relative intensity of the environmental crises would increase with increases in the carrying capacity. As population increased the incidence of environmental crises would increase. Excess population, in part a function of K_{dom}, would increase over time.

The dynamic interaction between an increasing K_{dom} and N_{dom} places the agricultural society at an advantage in terms of potential for growth and dispersal. By inducing environmental instability, agriculture creates the conditions favorable for its own spread; the environmental control inherent in agriculture not only spawns new populations and sends them off into new environments, but also allows these populations to continue in the same subsistence pattern.

The close connection between environmental manipulation and instabilities in production was of major importance for the evolution of domesticatory interactions into genuine agricultural systems. Environmental instability was a potent factor in the early evolution of agricultural systems. The earliest agricultural techniques were well-integrated parts of preexisting subsistence strategies. Activities performed for other reasons—the clearing of brush for shelter construction, the destruction of trees for defensive or economic reasons, the burning of grasslands as an aid to hunting, the selective preservation of an immediately useful plant tree, the creation of dump heaps near human habitations—could not help but have effects on the local environment. However, although the significance of these forms of behavior has long been recognized, precisely how they brought about the transition to an agricultural way of life has been less than clear. Especially confusing is the recognition that many of the techniques may be known to people without their necessarily utilizing them in a subsistence *system*. The model thus far developed for the origin and spread of agricultural systems may be extended, with certain qualifications, to an understanding of how domesticatory interactions evolved into agricultural systems.

Again, the interaction of productivity, crisis in production, population growth, and dispersal can account for the evolution of domestication into agriculture. We may first consider a null case: a given technique increases the carrying capacity of the environment for the early domesticated plant. There is an increase in human population in that area. However, if *only* the carrying capacity of the region for humans were to increase—that is, if now the same area of land could support a larger number of humans—population would remain constant following the increase. However, the "cultural fecundity" resultant from this change in technique would be extremely low. If emigrant groups were to be sent out by this population, and if emigration were a fixed proportion of reproductive capacity, then emigration would only increase in direct proportion to the increase in population brought about by the increase in the carrying capacity. If emigration were the result of extrinsic factors, emigration rates would remain unchanged. If, however, another form of behavior were to increase both carrying capacity and the instability of yield, the same sequence of events as given for the dispersal of agricultural population would be expected. We earlier spoke of the *success* attendant upon techniques of environmental manipulation. We may now modify the concept of success to include the probability that certain types of activity will provide the cause for their own dispersal. We may even speak, allegorically, of the "reproductive potential" of a culture practicing a given form of behavior.

This potential is an optimization of both productivity and instability. Theoretically, an improvement in the reproductive potential of a culture could come about without increases in absolute productivity if an increase in instability alone were to occur. It seems unlikely that this would have happened in the development of agriculture because of the pressures toward the natural selection of continually higher-yielding domesticated plants, but, it should be recognized as a possibility.

We may restate this idea by noting that those systems that maximize N_{dom} will spread most successfully. N_{dom} is the direct expression of the interaction of K_{dom} and $\hat{K}_{dom}$. Thus, from the very first stages in which agriculture appeared, a tendency toward the proliferation of unstable, albeit productive, systems was evident.[1]

Perhaps the greatest anomaly for those accepting the revolutionary view of agricultural origins is the observation, from the archaeological record, that well-domesticated plants are found in even the most primitive agricultural contexts. This is not in keeping with the view that early agricultural techniques provided the selective forces that transformed the wild plant into the domesticate. Another confusing issue for those believing in a few centers of agricultural origin is the great number of species of plants that seem to have independently entered agricultural systems over vast areas. Both of these problems are solved if we accept the coevolutionary interpretation of the evolution of the domesticated plants. As we have noted, domesticatory systems, especially to the extent that they are developing into agricultural systems, disperse into new areas by means of the same instability processes characteristic of the dispersal of agricultural systems. The dispersal is accompanied by opportunities for the evolution and acquisition of new domesticates. New domesticates may evolve a domesticatory symbiosis with people who move into new areas. Also, because humanly coevolved plants are not necessarily culturally bound, the movement of already coevolved domesticates between coexisting domesticatory societies is usually easy. This permits a society to acquire new domesticates and perhaps to move them into new regions. Of course, not all domesticatory societies have to develop into agricultural societies, but those that do will bring with them the plants with which they already have coevolved relationships. Early agricultural societies should be characterized by well-developed domesticated plants from several different sources.

Implicit thus far has been the assumption that domesticatory societies evolve more slowly than do agricultural ones. This assumption deserves comment. "Pure" domesticatory relationships, as I have already noted, lack the

---

[1]Of course, this is not to claim that $\bar{K}_{dom}$, the average productivity of a system, is totally unimportant. It is probably highly significant, at least today, where periodic shortages in one area are at least potentially amenable to solution by the import of food from other agricultural regions. However, the incorporation of $\bar{K}_{dom}$ into a model for agricultural origins is inappropriate.

important instability present in agricultural systems. Thus, the spread of domesticatory societies is dependent on factors extrinsic to the domesticatory relationship. The spread of domesticatory societies would be much more like that of nondomesticatory societies; in comparison, agricultural systems are literally driven into new localities by periodically recurrent environmental crises. It is evident that my feeding model and this dispersal model are totally congruent.

Another important distinction between domesticatory and agricultural systems is related to a final aspect of environmental control. Although agriculture directly affects the carrying capacity of the environment for the domesticated plant, nonagricultural societies must rely on the inherent carrying capacity of the environment for their early coevolved plants. The population levels attainable by the incidental-domesticatory society are, initially, little different from those of nondomesticatory societies.

Finally, an incidental-domesticatory society is totally dependent upon natural, albeit human-aided, colonization by the domesticated plant of new regions. This must of necessity be a slow process. The number of places available for colonization by any plant in an existing environment are both stable and limited. As I have already noted, one of the distinctive features of agriculture is the creation of a specialized environmental patch available to colonization by domesticates. This preexisting limitation on the dispersal of domesticates was probably the greatest limitation placed on the rapid spread of incidental-domesticatory societies. Indeed, we might wonder if incidental domestication was not one of the less "successful" human subsistence strategies.

One final aspect of the relationship between N_{dom} and K_{dom} deserves mention. Environmental crises and their demographic consequences may have implications for the understanding of innovations in agricultural techniques. When subsets of an agricultural population are dispersed because of an environmental crisis, an opportunity arises for the origin and establishment of new techniques and traditions. This increase in innovation, it must be stressed, need not be the only means for innovation within a society. However, in a manner perfectly analogous to that used in the description of the *added* human carrying capacity of the environment contributed by domesticated plants, agriculturally induced innovation may be added to any and all sources of innovation already present in that society. The innovation characteristic of many early agricultural societies may have been related to this interaction between dispersals and change.

A society involved in a symbiosis with plants may be described as a set of individuals that possesses a given set of domesticatory and/or agricultural information—beliefs, techniques, and traditions. Clearly, this set of information is not distributed uniformly among all members of the society: each individual does not possess all the information present in the society. The information set is also somewhat self-contradictory: differences in interpretation and emphasis will exist among members of the society in relation to any given trait.

If we were to extract a random subset of individuals from the society, we would also be extracting a sample of information from the total information set. If this random subset of individuals were the emigrant group sent off in times of agricultural crisis, then, when the sample of information it possessed was put into practice, changes (analogous to the "founder effect") in the techniques of production might occur. Techniques or traditions that previously placed limits on the productivity of the agricultural system might not be present in the new set of information. Limiting techniques or traditions might be lost. Variant techniques increasing the productivity or instability of agricultural production might appear. For example, minor changes in scheduling based on preferences atypical of the parent population might have major effects on K_{dom}, instability, or both. A simple sampling error might thus have major effects upon the subsistence system.

The loss of old behavior and the appearance of new behavior can both be considered "errors" in the performance of conventionalized tasks. Social interactions, learning, and social pressure all limit the performance of any behavior. However, the intensity of these forces is, in part, a function of the size of the group. The likelihood that any error will be corrected varies directly with the size of the group. The error of one individual cannot be corrected unless another individual is present. That a single individual is likely to make an error is an obvious corollary to our assumption of cultural heterogenity and uniformity: there is little chance that any individual will possess all the information present in a culture. The probability that any given bit of information will be present in the subset of the population given off at the time of agricultural dispersals is also a function of the size of that subset. As the size of the dispersed subset decreases, the likelihood of an error's being both made and maintained also increases. Thus we may begin to appreciate that the absolute size of N_{dom} may also be optimized if we are to maximize the rate of innovation. We might also note that the *rate* of environmental crisis will also, in theory, affect the amount of innovation: the more frequent the crises and their resultant emigrations, the more numerous will be the opportunities presented for agricultural innovation. Thus we may see that not only can N_{dom} be optimized for maximal innovation, but also innovation can be encouraged by other factors acting over time.

Agriculture brings with it a tremendous potential for change. The model given here may be described as a way of understanding the evolution of the rate of evolutionary change. Any daughter population that incorporates changes in the existing domesticatory or agricultural tradition that increase its "reproductive potential" will become more quickly dispersed than the parental population. The development of agriculture tends to feed on itself, encouraging its own further development. Besides all the evolutionary tendencies in the agricultural relationship toward instability, increased productivity, and more frequent dispersals, a final factor enhances the rate of change in agricultural societies: innovation in the techniques and traditions of agricultural subsistence is encouraged by the dynamics of agricultural dispersal.

There can be little doubt that agriculture is an exceptionally dynamic evolutionary relationship. That it should be correlated with sweeping changes in the life and subsistence of human societies is not in the least surprising.

Implications

Modern plant-breeding projects and agricultural-development schemes are a response to the shortages of food and resources that plague so much of the world. All agricultural systems are prone to occasional crises in production. Besides the immediate effects of these crises on human health and survival, shortages of food frequently have major (frequently justifiable) effects upon the stability of political and economic systems. The importance of the interaction of population growth, dwindling natural resources, and recurrent shortages of food places the study of agricultural systems in the mainstream of human concern. Proper understanding of the dynamics of the origin and function of agricultural systems may help us to mitigate, or perhaps even to counteract, many of the ecological and evolutionary tendencies that have thus far dominated the development of agricultural systems.

The prospect of food shortages encourages the breeding of improved crops and the development of more efficient cropping systems. Success in these programs encourages reliance on fewer and fewer species of plants and on an ever-decreasing number of varieties of a given species. Agricultural production also tends to become more localized, to take advantage of ever-more-specialized environmental conditions, and to allow for the exploitation of economies of scale. Yet monocultures are notorious for their vulnerability to catastrophic failure from disease, pests, and climatic extremes.

Today, people rely on about 20 species of plants for most of their food (National Academy of Sciences 1975). Yet hundreds of species of plants have been domesticated and thousands utilized. Remarkably, no major crop has been domesticated from the wild since the early days of agriculture. Even our best attempts at improvement of existing crops have been less than totally successful. Recently we have begun to understand that plant breeding and the successful introduction of improved cultivars may inadvertently be accelerating the loss of much of the variation in the crops grown by people. Much-belated attention has come to be directed toward the conservation of gene pools for cultivated plants. We are also beginning to understand that the breeding of improved varieties of crop plants is, paradoxically, often accompanied by increased susceptibility of the crop to previously unknown or unimportant pests and diseases (Lupton 1977; Pimentel 1977).

Only parts of the agroecology are completely under human control, but as we begin to understand its functioning and its evolution our control over it should, and is likely to, increase. The direction agricultural development will take is, at least to some extent, open to an enlightened view of the directions it should take. Our understanding of how agricultural systems function today must be based on an understanding of both their hisotry and their present structure. Although understanding of the modern structure is of major significance, it is, in and of itself, insufficient. Part of our understanding must also come from a careful investigation of the factors that have controlled and limited the absolute yield obtainable from domesticated plants. Study of the functioning of agricultural systems cannot help but be enhanced by a better understanding of its origins and development.

If this theory of agricultural origins and the domestication of plants proves useful, its greatest contribution may be the recognition that we are not facing a qualitatively different set of problems than did our distant forebears. The history of agriculture is a history of instability in production and of agriculturally induced crises. Identification of the factors responsible for the widespread and tragic instability of agricultural production may allow for actions to reduce that instability. Yet, we must remember that it was the less stable, *not* the best-adapted agricultural systems, that flourished.

From the perspective of this theory, many of our best efforts at increasing agricultural productivity should be expected to increase the vulnerability of the system; the best efforts of our ancestors to improve the productivity of their systems had a similar effect. Whereas in the past we have survived agricultural crises and even have benefited from them, we are no longer in a strictly comparable situation. Our situation is not new, but its consequences are. The response to agricultural crises can no longer be emigration. We are now faced with the possibility—and increasingly the reality—of political disruption, economic crisis, and starvation. By appreciating the evolutionary history underlying existing agricultural systems, we may be better prepared to develop our agricultural systems in new ways—that is, to minimize or perhaps even to counteract evolved instability. This will not be an easy task because, today, famine for the many has come to mean profits for the few. Although I call for a nonintentionalistic interpretation of the evolution of agricultural systems, this is not to be read as support for the status quo; indeed, the reverse is true. The spread of agriculture resembles the spread of a pathology such as rabies in that the symptoms facilitate the dissemination of the disease; I would hope that an awareness of the processes by which agriculture developed may act as a spur to us to gather the information that may permit us to become as successfully intentional as we have so glibly claimed to be.

References

Abernathy, V.
    1979   *Population pressure and cultural adjustment.* New York: Human Science Press.
Alexander, Richard D.
    1971   The search for an evolutionary philosophy of man. *Proceedings of the Royal Society of Victoria, Melbourne* **84:**99–120.
    1974   The evolution of social behavior. *Annual Review of Ecology and Systematics* **5:**325–383.
    1975   The search for a theory of behavior. *Behavioral Science* **20:**77–100.
    1977   Evolution, human behavior and determinism. *Proceedings of the Biennial Meeting of the Philosophy of Science Association* (1976) **2:**3–21.
    1979a   *Darwinism and human affairs.* Seattle: University of Washington Press.
    1979b   Evolution and culture. In *Evolutionary biology and human social behavior: An anthropological perspective,* edited by N. A. Chagnon and W. G. Irons. North Scituate, Mass.: Duxbury. Pp. 59–78.
Alland, A., Jr.
    1970   *Adaptation in cultural evolution: An approach to medical anthropology.* New York: Columbia University Press.
    1972   Cultural evolution: The Darwinian model. *Social Biology* **19:**227–239.
    1973   *Evolution and human behavior* (second edition). Garden City: Anchor, Doubleday.
    1975   Adaptation. *Annual Review of Anthropology* **4:**59–73.
Alland, A., Jr., and B. J. McCay
    1973   The concept of adaptation in biological and cultural evolution. In *Handbook of social and cultural anthropology,* edited by J. Honigmann. Chicago: Rand McNally. Pp. 143–178.
Allee, W. C., A. E. Emerson, O. Park, T. Park, and K. P. Schmidt.
    1949   *Principles of animal ecology.* Philadelphia: W. B. Saunders.
Allen, G.
    1897   The origin of cultivation. *The Fortnightly Review* **61:**578–592.
Altieri, M. A., A. Van Schoonhoven, and J. D. Doll
    1978   A review of insect prevalence in maize (*Zea mays* L.) and bean (*Phaseolus vulgaris* L.) in polycultural systems. *Field Crop Research* **1:**33–49.
Ames, O.
    1939   *Economic annuals and human cultures.* Cambridge: Botanical Museum of Harvard University.
Anderson, E.
    1956   Man as a maker of new plants and plant communities. In *Man's role in changing the face of the earth* (volume 2), edited by W. L. Thomas. Chicago: University of Chicago Press. Pp. 763–777.

1960   The evolution of domestication. In *Evolution after Darwin*, edited by Sol Tax. Volume 2: *The evolution of man, culture, and society*. Chicago: University of Chicago Press. Pp. 67–84.

1969   *Plants, man, and life* (revised edition). Berkeley: University of California Press.

Antonovics, J., and A. D. Bradshaw

1970   Evolution in closely adjacent plant populations. VII: Clinical patterns at a mine boundary. *Heredity* **25**:349–362.

Athen, J. Stephen

1977   Theory building and the study of evolutionary process in complex societies. In *For theory building in archaeology*, edited by Lewis R. Binford. New York: Academic Press. Pp. 353–384.

Bachofen, J. J.

1861   *Das Mutterrecht*. Stuttgart: Krais und Hoffman.

Bachofen, J. J.

1967   *Myth, religion and mother right: selected writings of J. J. Bachofen*, translated by Ralph Manheim. Princeton: Princeton University Press.

Baker, H. G.

1965   Characteristics and modes of origin of weeds. In *The genetics of colonizing species*, edited by H. G. Baker and G. L. Stebbins. New York: Academic Press. Pp. 147–168.

1970   *Plants and civilization* (second edition). Belmont, Mass.: Wadsworth.

Baker, J. M.

1963   Ambrosia beetles and their fungi, with particular reference to *Platypus cylindrus* Fab. *Symposia of the Society of Genetic Microbiology* **13**:232–265.

Baldwin, J. M.

1896   A new factor in evolution. *American Naturalist* **30**:441–451, 536–553.

Barash, David

1976   *The whisperings within*. New York: Harper and Row.

Beadle, G. C.

1977   The origins of *Zea mays*. In *Origins of agriculture*, edited by Charles A. Reed. The Hague: Mouton. Pp. 615–636.

Beatty, J.

1980   Optimal-design models and the strategy of model building in evolutionary biology. *Philosophy of Biology* **47**:532–561.

Berlinski, D.

1976   *On systems analysis: An essay concerning the limitations of of some mathematical methods in the social, political, and social sciences*. Cambridge: MIT Press.

Bernal, J. D.

1971   *Science in history* (third edition, four volumes). Cambridge: MIT Press.

Binford, L. R.

1968   Post-Pleistocene adaptations. In *New Perspectives in archaeology*, edited by S. R. Binford and L. R. Binford. Chicago: Aldine. Pp. 313–341.

Birdsell, J. B.

1972   *Human evolution*. New York: Rand McNally.

Blute, Marion

1979   Sociocultural evolutionism: An untried theory. *Behavioral Science* **24**:46–59.

Boaz, N. T.

1977   Paleoecology of early Hominidae in Africa. *Kroeber Anthropological Society Papers* **50**:37–62.

Bonner, J. T.

1950   The role of toxic substances in the chemical interaction of plants. *Botanical Review* **16**:51–65.

1980   *The evolution of culture in animals.* Princeton: Princeton University Press.

Boserup, Ester

1965   *The conditions of agricultural growth.* Chicago: Aldine.

Bowler, P. J.

1975   The changing meaning of "evolution." *Journal of the History of Ideas* **36**:95–114.

Boyd, R., and P. J. Richerson

1980   Sociobiology, culture and economic theory. *Journal of Economic Behavior and Organization* **1**:97–121.

Braidwood, R. J.

1951   *Prehistoric men* (second edition). Chicago: Chicago Natural History Museum.

Braidwood, R. J., and B. Howe

1960   *Prehsitoric investigation in Iraqi Kurdistan.* University of Chicago Oriental Institute *Studies in Ancient Oriental Civilization,* No. 31. Chicago: University of Chicago Press.

Braidwood, R. J., and C. A. Reed

1957   The achievement and early consequences of food production: a consideration of the archaeological and natural-historical evidence. *Cold Springs Harbor Symposia on Quantitative Biology* **22:19–31.**

Braidwood, R. J., and G. R. Willey

1962   Conclusions and afterthoughts. In *Courses towards urban life,* edited by R. J. Braidwood and G. R. Willey. Viking Fund Publications in Anthropology No. 32. Chicago: Aldine. Pp. 330–359.

Brandon, R. N.

1981a   A structural description of evolutionary theory. *PSA Journal 1980* **2**:427–439.

1981b   Biological teleology: Questions and explanations. *Studies in the History and Philosophy of Science* **12**:91–105.

Bray, Warwick

1973   The biological basis of culture. In *The explanation of cultural change: Models in prehistory,* edited by C. Renfrew. Pittsburgh: University of Pittsburgh Press. Pp. 73–92.

1976   From predation to production: The nature of agricultural evolution in Mexico and Peru. In *Problems in economic and social archaeology,* edited by G. deG. Seveking, T. H. Longworth, and K. E. Wilson. London: Duckworth. Pp. 73–95.

1977   From foraging to farming in early Mexico. In *Hunters, gatherers, and first farmers beyond Europe,* edited by S. V. S. Megaw. Leicester: Leicester University Press. Pp. 225–249.

Brenan, J. P. M.

1959   *Flora of tropical East Africa* (no. 90, part 1). London, Kew: Royal Botanic Gardens.

Broadhurst, P. L., D. W. Fulker, and J. Wilcox

1974   Behavioral genetics. *Annual Review of Psychology* **25**:389–415.

Bronson, B.

1972   Farm labor and the evolution of food production. In *Population growth: Anthropological implications,* edited by B. Spooner. Cambridge: MIT Press. Pp. 190–218.

1977   The earliest farming: Demography as cause and consequence. In *Origins of agriculture,* edited by C. A. Reed. The Hague: Mouton. Pp. 23–48.

Brothwell, Donald R.

1975   Salvaging the term "domestication" for certain types of man-animal relationships: The possible value of an eight point scoring system. *Journal of Archaeological Science* **2**:397–400.

Brush, S. B., H. J. Carney, and Zósimo Huamán

1981   Dynamics of Andean potato agriculture. *Economic Botany* **35**:70–88.

Burian, R. M.
1981   Human sociobiology and genetic determinism. *The Philosophical Forum* **13**:43–66.
in   Adaptation. In *Dimensions of Darwinism: Themes and counterthemes in twentieth*
press   *century evolutionary theories,* edited by M. Grene. Cambridge: Cambridge University
         Press.

Bye, R. A., Jr.
1979   Incipient domestication of mustards in Mexico. *Kiva* **44**:237–256.
1981   Quelites—ethnoecology of edible greens—past, present and future. *Journal of Eth-nobiology* **1**:109–123.

Campbell, D. T.
1956a   Adaptive behavior from random response. *Behavioral Science* **1**:105–110.
1956b   Perception as substitute trial and error. *Psychological Review* **63**:330–342.
1960   Blind variation and selective retention in creative thought as in other knowledge. *Psychological Review* **67**:380–400.
1965   Variation and selective retention in sociocultural evolution. In *Social change in under-developed areas: A reinterpretation of evolutionary theory,* edited by R. W. Mack, G. I. Blanksten, and H. R. Barringer. Cambridge: Schenkman. Pp. 19–48.
1970   Natural selection as an epistemological model. In *A handbook of method in cultural anthropology,* edited by R. Naroll and R. Cohen. New York: Natural History Press. Pp. 51–85.
1972   On the genetics of altruism and the counter-hedonic components of human culture. *Journal of Social Issues* **28**:21–37.
1974a   'Downward causation' in hierarchically organized biological systems. In *Studies in the philosophy of biology,* edited by F. J. Ayala and T. Dobzhansky. New York: Mac-millan. Pp. 179–186.
1974b   Evolutionary epistemology. In *The Philosophy of Karl Popper,* (volume 1), edited by P. Schilpp. Lasalle, Ill.: Open Court. Pp. 413–463.
1975   On the conflicts between biological and social evolution and between psychology and moral tradition. *American Psychology* **30**:1103–1126.
1976   Comment on Richards' ''Natural selection model for conceptual evolution.'' *Philosophy of Science* **44**:502–507.

Caplan, A. L.
1979   Darwinism and deductivist models of theory structure. *Studies in History and Philosophy of Science* **10**:341–353.
1981   Say it just ain't so: Adaptational stories and sociobiological explanations of social behavior. *The Philosophical Forum* **13**:144–160.

Carniero, R. L.
1967   Introduction. In *The evolution of society: Selections from Herbert Spencer's Principles of Sociology,* edited by R. L. Carneiro. Chicago: University of Chicago Press. Pp. ix–lvii.
1972   The devolution of evolution. *Social Biology* **19**:248–258.
1973a   Classical evolution. In *Main currents in cultural anthropology,* edited by R. Naroll. New York: Appleton-Century Crofts. Pp. 57–121.
1973b   Structure, function and equilibrium in the evolutionism of Herbert Spencer. *Journal of Anthropological Research* **29**:77–95.
1974   Comment. In ''The evolutionary theories of Charles Darwin and Herbert Spencer,'' by Derek Freeman. *Current Anthropology* **15**:222–223.

Carter, G. F.
1977   A hypothesis suggesting a single origin of agriculture. In *Origins of agriculture,* edited by C. A. Reed. The Hague: Mouton. Pp. 77–88.

Cavalli-Sforza, L. L.
    1971   Similarities and dissimilarities of socio-cultural and biological evolution. In *Mathematics in the archaeological and historical sciences*. Edinburgh: Edinburgh University Press. Pp. 535–541.
Cavalli-Sforza, L. L., and M. W. Feldman
    1973a   Cultural versus biological inheritance. Phenotypic transmission from parents to children. *American Journal of Human Genetics* **25**:618–637.
    1973b   Models for cultural inheritance. I: Group mean and within group variation. *Journal of Theoretical Population Biology* **4**:42–55.
    1976   Evolution of continuous variation. Direct approach through joint distribution of genotypes and phenotypes. *Proceedings of the National Academy of Sciences (USA)* **73**:1689–1692.
    1979   Towards a theory of cultural evolution. *Interdisciplinary Science Review* **3**:99–107.
    1981   *Cultural transmission and evolution: A quantitative approach*. Princeton: Princeton University Press.
Cavalli-Sforza, L. L., M. W. Feldman, K. H. Chen, and S. M. Dornbusch
    1982   Theory and observation in cultural transmission. *Science* **218**:19–27.
Cherret, J. M.
    1968   The foraging behavior of *Atta cephalotes* L. *Journal of Animal Ecology* **37**:387–403.
Childe, V. Gordon
    1925   *The dawn of European civilization*. New York: Knopf.
    1941   *Man makes himself* (revised edition). London: Thinker's Library.
    1951   *Social evolution*. London: Watts.
Choudhury, B.
    1976   *Solanum melongena*. In *Evolution of crop plants*, edited by N. W. Simmonds. London: Longmans. Pp. 278–279.
Clark, J. G. D.
    1952   *Prehsitoric Europe: The economic basis*. New York: The Philosophical Library.
Cloak, F. T., Jr.
    1975   Is a cultural ethnology possible? *Human Ecology* **3**:161–182.
    1976   The evolutionary success of altruism and urban social order. *Zygon* **11**:219–240.
    1977   The adaptive significance of cultural behavior: Comments and reply. *Human Ecology* **5**:49–52.
Clutton-Brock, T. H. (Editor)
    1977   *Primate ecology: Studies of feeding and raning behavior in lemurs, monkeys and apes*. London: Academic Press.
Cody, M. L.
    1974   Optimization in ecology. *Science* **183**:1156–1164.
Cody, M. L., and J. M. Diamond
    1975   *Ecology and evolution of communities*. Cambridge: Harvard University Press.
Cohen, M. N.
    1975   Population pressure and the origins of agriculture. In *Population, ecology and social evolution*, redited by S. Polgar. Chicago: Aldine. Pp. 79–121.
    1977a   *The food crisis in prehistory: Overpopulation and the origins of agriculture*. New Haven: Yale University Press.
    1977b   Population pressure and the origins of agriculture: An archaeological example from the coast of Peru. In *Origins of agriculture*, edited by C. A. Reed. The Hague: Mouton. Pp. 135–178.
Cohen, M. N., R. S. Malpass, and H. G. Klein (Editors)
    1980   *Biosocial mechanisms of population regulation*. New Haven: Yale University Press.
Colinvaux, P.

1978   *Why big fierce animals are rare.* Princeton: Princeton University Press.

Connell, J. H.
1978   Diversity in tropical rain forests and coral reefs. *Science* **199:**1302–1310.

Connell, J. H., and E. Orians
1964   The ecological regulation of species diversity. *American Naturalist* **98:**399–414.

Covich, A. P.
1976   Analyzing shapes of foraging areas: Some ecological and economic theories. *Annual Review of Ecology and Systemics* **7:**235–257.

Crompton, A. W., and K. Hiiamäe
1969   How mammalian molar teeth work. *Discovery* **5:**23–34.

Darlington, C. D.
1969   *The evolution of man and society.* New York: Simon and Schuster.
1973   *Chromosome botany and the origins of cultivated plants* (third revised edition). London: Allen and Unwin.

Darwin, Charles
1859   *The origin of species.* London: John Murray.
1860   *The origin of species* (second edition). London: John Murray.
1868   *The variations of animals and plants under domestication* (two volumes). London: John Murray.
1882   *The variation of animals and plants under domestication* (two volumes) (second edition). London: John Murray.

David, N.
1976   History of crops and people in north Cameroon to A.D. 1900. In *Origins of African plant domestication,* edited by J. R. Harlan, J. M. J. deWet, and Ann B. L. Stampler. The Hague: Mouton. Pp. 223–268.

Davis, Tilton, IV, and R. A. Bye, Jr.
1982   Ethnobotany and progressive domestication in *Jaltomata* (Solanaceae) in Mexico and Central America. *Economic Botany* **36:**225–241.

Dawkins, R.
1976   *The selfish gene.* New York: Oxford University Press.

de Candolle, Alphonse L. P. P.
1886   *Origin of cultivated plants.* New York: Hafner Publishing Company.

Demetrius, L.
1975   Reproductive strategies and natural selection. *American Naturalist* **109:**243–249.
1977   Adaptness and fitness. *American Naturalist* **111:**1163–1168.

de Wet, J. M. J.
1975   Evolutionary dynamics of cereal domestication. *Bulletin of the Torrance Botany Club* **102**(6):307–312.

Dobzhansky, T.
1968   Adaptedness and fitness. In *Population biology and evolution,* edited by Richard Lewontin. Syracuse: Syracuse University Press. Pp. 109–122.

Doggett, H.
1965   Disruptive selection in crop development. *Nature* **206:**279–280.

Dole, Gertrude E.
1973   Foundation of contemporary evolutionism. In *Main currents in cultural anthropology,* edited by Raoul Naroll and Frada Naroll. Englewood Cliffs: Prentice-Hall. Pp. 247–280.

Downhower, J.
1979   Introduction. In *Analysis of ecological systems,* edited by D. J. Horn, G. R. Stairs, and R. G. Mitchell. Columbus: Ohio State University Press. Pp. vii–ix.

Dunbar, M.
    1960   The evolution of stability in marine environments: Natural selection at the level of the
            ecosystem. *American Naturalist* **94**:129–136.
Dunnell, Robert C.
    1980   Evolutionary theory and archaeology. In *Advances in archaeological method and
            theory* (volume 3), edited by M. B. Schiffer. New York: Academic Press. Pp. 38–99.
    1978a   Archaeological potential of anthropological and scientific models of function. In *Ar-
            chaeological essays in honor of Irving B. Rouse,* edited by R. C. Dunnell and E. S.
            Hall, Jr. The Hague: Mouton. Pp. 41–73.
    1978b   Style and function: A fundamental dichotomy. *American Antiquity* **43**:192–202.
Dunnell, R. C., and R. J. Wenke
    1980   Cultural and scientific evolution: Some comments on "The decline and rise of Meso-
            potamian civilization." *American Antiquity* **45**:605–609.
Durham, W. H.
    1976a   The adaptive significance of cultural behavior. *Human Ecology* **4**:89–121.
    1976b   Resource competition and human aggression. Part 1: A review of primitive war.
            Quarterly Review of Biology **51**:385–415.
    1977   The adaptive significance of cultural behavior: Comments and reply. *Human Ecology*
            5:59–66.
    1978   Towards a coevolutionary theory of human biology and behavior. In *The sociobiology
            debate,* edited by A. L. Caplan. New York: Harper and Row. Pp. 428–448.
Elton, C. S.
    1958   *The ecology of invasions by animals and plants.* London: Methuen.
Emerson, A. E.
    1960   The evolution of adaptation in population systems. In *Evolution after Darwin,* edited
            by Sol Tax. Chicago: University of Chicago Press. Pp. 307–348.
Emlen, J. M.
    1966a   Natural selection and human behavior. *Journal of Theoretical Biology* **12**:410–418.
    1966b   The role of time and energy in food preference. *American Naturalist* **100**:611–617.
    1973   *Ecology: An evolutionary approach.* Reading, Mass.: Addison-Wesley.
Falk, Arthur E.
    1981   Purpose, feedback and evolution. *Philosophy of Science* **48**:198–217.
Feeny, P.
    1973   Biochemical coevolution between plants and their insect herbivores. In Coevolution of
            animals and plants, edited by L. E. Gilbert and P. H. Raven. Austin and London:
            University of Texas Press. Pp. 3–19.
Feldman, M. W., and L. L. Cavalli-Sforza
    1975   Models for cultural inheritance. A general linear model. *Annals of Human Biology*
            2:215–226.
    1976   Cultural and biological evolutionary processes: Selection for a trait under complex
            transmission. *Journal of Theoretical Population Biology* **9**:238–259.
Flannery, K. V.
    1965   The ecology of early food production in Mesopotamia. *Science* **147**:1247–1256.
    1968   Archeological systems theory and early Mesoamerica. In *Anthropological archaeol-
            ogy in the Americas,* edited by B. J. Meggers. Washington, D.C.: Anthropological
            Society of Washington. Pp. 67–87.
    1969   Origins and ecological effects of early domestication in Iran and the Near East. In *The
            domestication and exploitation of plants and animals,* edited by P. J. Ucko and G. W.
            Dimbleby. London: Duckworth. Pp. 73–100.
    1973   The origins of agriculture. *Annual Review of Anthropology* **2**:271–310.

Francke-Grosman, H.
    1967   Ectosymbiosis in wood-inhabiting insects. In *Symbiosis* (volume 2), edited by S. M.
              Henry. New York: Academic Press. Pp. 141–205.
Frazer, J. G.
    1912   *The golden bough* (volume 1: part V). New York: Macmillan.
Freeman, D.
    1974   The evolutionary theories of Charles Darwin and Herbert Spencer. *Current Anthropol-*
              ogy **15:**211–237.
Friedman, D.
    1979   *Human sociobiology.* New York: The Free Press.
Fryer, J. D., and R. J. Chancellor
    1970   Herbicides and our changing weeds. In *The flora of a changing Britain,* edited by F.
              R. Perrig. Hampton: Classey. Pp. 105–118.
Gadgil, M., and W. H. Bossert
    1970   Life history consequences of natural selection. *American Naturalist* **104:**1–24.
Galdikas, Biruté M. F., and Geza Teleki
    1981   Variations in subsistence activities of female and male pngids: new perspectives on the
              origins of hominid labor division. *Current Anthropology* **22:**241–256, 316–320.
Gale, F. (editor)
    1970   *Women's role in Aboriginal society.* Australian Aboriginal Studies, number 36; Social
              Anthropology Series 6. Canaberra: Australian Institute of Aboriginal Studies.
Galinat, Walton C.
    1965   The evolution of corn and culture in North America. *Economic Botany* **19:**350–357.
    1974   The domestication and genetic erosion of maize. *Economic Botany* **28:**31–37.
    1975   The evolutionary emergence of maize. *Bulletin of the Torray Botanical Club*
              102:313–324.
Gartlan, J. S., and C. K. Brain
    1968   Ecology and social variability in *Cercopithecus aethiops* and *C. Mitis.* In *Primates:*
              Studies in adaptation and variability, edited by P. D. Jay. New York: Holt, Rinehart
              and Winston. Pp. 253–292.
Gaulin, S. J. C., and M. Konner
    1977   On the natural diet of primates including humans. In *Nutrition and the brain* (volume
              1), edited by R. J. Wortman and J. J. Wurtman. New York: Raven Press. Pp. 1–
              86.
Geertz, C.
    1963   *Agricultural involution: The process of ecological change in Indonesia.* Berkeley:
              University of California Press.
Gerard, R. W., C. Kluckhohn, and A. Rapoport.
    1956   Biological and cultural evolution. Some analogies and explorations. *Behavioral Sci-*
              ence **1:**6–34.
Ghiselin, M.
    1969   *The triumph of the Darwinian method.* Berkeley: University of California Press.
    1974   A radical solution ot the species problem. *Systematic Zoology* **23:**536–544.
Glander, K. E.
    1975   Habitat description and resource utilization of howler monkeys. In *Socioecology and*
              psychology of primates, edited by R. H. Tuttle. The Hague: Mouton. Pp. 37–
              57.
Glassow, M. A.
    1980   *Prehsitoric agricultural development in the northern Southwest.* Socorro, New Mex-
              ico: Nallena Press.

Gliessman, S. R., R. Garcia E., and M. Amador A.
   1981   The ecological basis for the application of traditional agricultural technology in the
            management of tropical agro-ecosystems. *Agro-Ecosystems* **7:**173–185.
Good, R.
   1964   *The geography of flowering plants* (third edition). London: Longman.
Goodall, Jane (VanLawick)
   1971   *In the shadow of man*. London: Collins.
Gould, S. J.
   1977a   *Ever since Darwin: Reflections in natural history*. New York: W. W. Norton.
   1977b   *Ontogeny and phylogeny*. Cambridge: Harvard University Press.
   1980   *The panda's thumb: more reflections in natural history*. New York: W. W. Norton.
Gould, S. J., and R. C. Lewontin
   1979   The spandrels of San Marco and the Panglossian paradigm: A critique of the adapta-
            tionist programme. *Proceedings of the Royal Society (London), Series B*, 205–
            581–598.
Gould, S. J., and E. S. Vrba
   1982   Exaptation—a missing term in the science of form. *Paleobiology* **8:**4–15.
Green, S. W.
   1980a   Broadening least-cost models for expanding agricultural systems. In *Modeling pre-
            historic subsistence change,* edited by T. Earle and A. Christensen. New York: Aca-
            demic Press. Pp. 209–214.
   1980b   Toward a general model of agricultural systems. In *Advances in method and theory in
            archaeology* (volume 3). New York: Academic Press. Pp. 311–355.
Grigg, D. B.
   1976   Population pressure and agricultural change. *Progressive Geography* **8:**135–176.
Gruber, H. E.
   1980   *Darwin on man* (second edition). Chicago: University of Chicago Press.
Grümmer, G., and H. Beyer
   1960   The influence exerted by species of *Camelina* on flax by means of toxic substances. In
            The biology of weeds, edited by J. L. Harper. Oxford: Blackwell Scientific Publica-
            tions. Pp. 153–157.
Gunda, Bela
   1981   Comment to Rindos 1980. *Current Anthropology* **22**(1):81–82.
Hahn, E.
   1909   *Die Entstehung der Pflugkultur*. Heidelberg: C. Winter.
Hairston, N. G., D. W. Tinkle, and N. M. Wilbur
   1970   Natural selection and the parameters of population growth. *Journal of Wildlife Man-
            agement* **34:**681–690.
Haldane, J. B. S.
   1932   *The causes of evolution*. London: Longmans.
Hamilton, W. D.
   1964   The genetical evolution of social behavior: I, II. *Journal of Theoretical Biology*
            7:1–16, 17–52.
   1975   Innate social aptitudes in man: An approach from evolutionary genetics. In *Biosocial
            anthropology,* edited by R. Fox. New York: Wiley. Pp. 133–135.
Handel, S. N.
   1978   The competitive relationship of three woodland sedges and its bearing on the evolution
            of ant-dispersal of *Carex pedunculata*. *Evolution* **32:**151–163.
Hardesty, D. L.
   1977   *Ecological anthropology*. New York: Wiley.

Hardin, G.
    1960   The competitive exclusion principle. *Science* **131**:1291–1297.
Harding, Robert S. O.
    1975   Meat-eating and hunting in baboons. In *Socioecology and psychology of primates*,
               edited by R. H. Tuttle. The Hague: Mouton. Pp. 245–257.
Harding, R. S. O., and G. Teleki (Editors)
    1980   *Omnivorous primates: Gathering and hunting in human evolution*. New York: Colum-
               bia University Press.
Harlan, J. R.
    1965   The possible role of weed races in the evolution of cultivated plants. *Euphytica*
               14:173–176.
    1967   A wild wheat harvest in Turkey. *Archaeology* **20**:197–201.
    1970   The evolution of cultivated plants. In *Genetic resources in plants: Their exploration
               and conservation*, edited by O. H. Frankel and F. Bennett. Oxford: Blackwell Scien-
               tific Publications. Pp. 19–32.
    1971   Agricultural origins: centers and non-centers. *Science* **174**:468–474.
    1976   Plant and animal distribution in relation to domestication. *Philosophical Transactions
               of the Royal Society of London and Britain* **275**:13–25.
    1977   Origins of cereal agriculture in the Old World. In *Origins of agriculture*, edited by C.
               A. Reed. The Hague: Mouton. Pp. 357–384.
Harlan, J. R., and J. M. J. de Wet
    1965   Some thoughts about weeds. *Economic Botany* **19**:16–24.
    1973   On the quality of evidence for origin and dispersal of cultivated plants. *Current
               Anthropology* **14**:51–62.
Harlan, J. R., J. M. J. de Wet, and A. B. L. Stempler
    1976   Plant domestication and indigenous African agriculture. In *Origins of African plant
               domestication*, edited by J. R. Harlan, J. M. J. de Wet, and A. B. L. Stempler. The
               Hague: Mouton. Pp. 3–19.
Harlan, J. R., and D. Zohary
    1966   Distribution of wild wheats and barley. *Science* **153**:1074–1080.
Harner, M.
    1970   Population pressure and the social evolution of agriculturalists. *Southwestern Journal
               of Anthropology* **26**:67–86.
Harper, J. L.
    1956   The evolution of weeds in relation to resistence to herbicides. *Proceedings of the
               Third British Weed Control Conference* **1**:179–186.
    1965   Establishment, aggreshion, and cohabitation in weedy species. In *The genetics of
               colonizing species*, edited by H. G. Baker and G. L. Stebbins. New York: Academic
               Press. Pp. 243–264.
    1967   A Darwinian approach to plant ecology. *Journal of Ecology* **55**:247–270.
Harper, J. L., I. H. McNaughton, and G. R. Sagan
    1961   The evolution and ecology of closely related species living in the same area. *Evolution*
               15:1205–1227.
Harris, David R.
    1967   New light on plant domestications and the origins of agriculture: A review. *Geograph-
               ical Review* **57**:90–105.
    1969   Agricultural systems, ecosystems and the origins of agriculture. In *The domestication
               and exploitation of plants and animals*, edited by P. J. Ucko and G. W. Dimbleby.
               London: Duckworth. Pp. 3–14.
    1972   The origin of agriculture in the tropics. *American Scientist* **60**(2):180–193.

1973   The prehistory of tropical agriculture: An ethnoecological model. In *The explanation of cultural change: Models in prehistory*, edited by C. Renfrew. Pittsburgh: University of Pittsburgh Press. Pp. 311–356.

1976   Traditional systems of plant food production and the origins of agriculture in West Africa. In *Origins of African plant domestication*, edited by J. R. Harlan, J. M. J. de Wet, and A. B. L. Stempler. The Hague: Mouton. Pp. 311–356.

1977a   The origins of agriculture: Alternate pathways toward agriculture. In *Origins of agriculture*, edited by C. A. Reed. The Hague: Mouton. Pp. 173–249.

1977b   Settling down: An evolutionary model for the transformation of mobile bands into sedentary communities. In *The evolution of social systems*, edited by J. Friedman and M. L. Rowlands. London: Duckworth. Pp. 401–417.

Harris, M.
1968   *The rise of anthropological theory*. New York: Harper and Row.

1974   Comment. In "The evolutionary theories of Charles Darwin and Herbert Spencer," by Derek Freeman. *Current Anthropology* **15**:225–226.

1979   *Cultural materialism: The struggle for a science of culture*. New York: Random House.

Hartung, J.
1976   On natural selection and the inheritance of wealth. *Current Anthropology* **17**:607–622.

Hartzell, A.
1967   Insect ectosymbiosis. In *Symbiosis* (volume 2), edited by S. M. Henry. New York: Academic Press. Pp. 107–140.

Hassan, Ferki A.
1978   Demographic archaeology. In *Advances in archaeological method and theory* (volume 1), edited by M. Schiffer. New York: Academic Press. Pp. 49–105.

1981   *Demographic archaeology*. New York: Academic Press.

Hawkes, J. G.
1969   The ecological background of plant domestication. In *The domestication and exploitation of plants and animals*, edited by P. J. Ucko and G. W. Dimbleby. London: Duckworth. Pp. 17–29.

Heiser, C. B., Jr.
1969   Systematics and the origin of cultivated plants. *Taxon* **18**:36–45.

1973   *Seed to civilization: The story of man's food*. San Francisco: W. H. Freeman.

Herbert, Sandra
1971   Darwin, Malthus and selection. *Journal of the History of Biology* **4**:209–217.

Herklots, G. A. C.
1972   *Vegetables in South-East Asia*. George Allen & Unwin.

Herskovitz, M. J.
1940   *The economic life of primitive people*. New York: Alfred A. Knopf.

Higgs, E. S.
1976   Archaeology and domestication. In *Origins of African plant domestication*, edited by J. R. Harlan, J. M. J. de Wet, and A. B. L. Stempler. The Hague: Mouton. Pp. 29–39.

Higgs, E. S., and M. R. Jarman
1969   The origins of agriculture: a reconsideration. *Antiquity* **63**:31–41.

1972   The origins of animal and plant husbandry. In *Papers in economic prehistory*, edited by E. S. Higgs. London: Cambridge University Press. Pp. 3–13.

Himmelfarb, G.
1962   *Darwin and the Darwinian revolution*. New York: W. W. Norton.

Hinde, R. A.
1975   The concept of function. In *Function and evolution in behavior,* edited by G. Baerends, C. Beer, and A. Manning. Oxford: Clarendon. Pp. 3–15.

Hladik, C. M.
1975   Ecology, diet, and social patterning in Old and New World primates. In *Socioecology and psychology of primates,* edited by R. H. Tuttle. The Hague: Mouton. Pp. 3–35.
1977   Chimpanzees of Gabon and chimpanzees of Gombe: Some comparative data on the diet. In *Primate ecology: Studies of feeding and ranging behaviour in lemurs, monkeys and apes,* edited by T. H. Clutton-Brock. London: Academic Press. Pp. 481–501.
1981   Diet and the evolution of feeding strategies among forest primates. In *Omnivorous primates: Gathering and hunting in human evolution,* edited by R. S. O. Harding and G. Teleki. New York: Columbia University Press. Pp. 303–343.

Hladik, C. M., and A. Hladik
1967   Observations sur le rôle des primates dans la dissemination des vegetaux de la fôret Gabonaise. *Biologica Gabonica* **3**:43–58.

Hocking, B.
1975   Ant–plant mutualism: Evolution and energy. In *Coevolution of animals and plants,* edited by L. E. Gilbert and P. H. Raven. Austin and London: University of Texas Press. Pp. 78–90.

Hofstader, R.
1955   *Social Darwinism in American thought.* Philadelphia: University of Pennsylvania Press.

Hollings, C. S.
1973   Resilience and stability of ecological systems. *Annual Review of Ecology and Systematics* **4**:1–24.

Horn, David J., and R. V. Dowell
1979   Parasitoid ecology and biological control in ephemeral crops. In *Analysis of ecological systems,* edited by D. J. Horn, G. R. Stairs, and R. D. Mitchell. Columbus: Ohio State University Press. Pp. 281–308.

Hull, D. L.
1976   Are species really individuals? *Systemaic Zoology* **25**:174–191.

Hutchinson, G. E.
1959   Homage to Santa Rosalia, or why are there so many kinds of animals? *American Naturalist* **93**:145–159.

Hymowitz, T.
1972   The trans-domestication concept as applied to Guar. *Economic Botany* **26**:49–60.

Igbozurike, M. J.
1971   Ecological balance in tropical agriculture. *Geographical Review* **61**:519–529.

Iltis, H. H., and J. F. Doebley
1980   Taxonomy of *Zea* (Gramineae). II: Subspecific categories in the *Zea mays* complex and a generic synopsis. *American Journal of Botany* **67**:994–1004.

Immelmann, K.
1975   The evolutionary significance of early experience. In *Function and evolution in behavior,* edited by G. Baerends, C. Beer, and A. Manning. Oxford: Clarendon. Pp. 243–253.

Irons, W. G.
1979a   Investment and primary social dyads. In *Evolutionary biology and human social behavior: An anthropological perspective,* edited by N. A. Chagnon and W. G. Irons. North Scituate, Mass.: Duxbury. Pp. 181–213.

1979b   Natural selection, adaptation and human social behavior. In *Evolutionary biology and human social behavior: An anthropological perspective,* edited by N. A. Chagnon and W. G. Irons. North Scituate, Mass.: Duxbury Press. Pp. 4–39.

Isaac, E.
1970   *The geography of domestication.* Englewood Cliffs, N.J.: Prentice Hall.

Iversen, J.
1949   The influence of prehsitoric man on vegetation. *Danmarks Geol. Undersoegelse,* series 4, volume 3, no. 6.

Jackson, G., and J. S. Gartlan
1965   The flora and fauna of Lolui Is., Lake Victoria. *Journal of Ecology* **53:**573–597.

Janzen, D. H.
1967   Fire, vegetation structure, and the ant–acacia interaction in Central America. *Evolution* **20:**249–257.

1969   Seed-eaters *vs.* seed size, number, toxicity and dispersal. *Evolution* **23:**1–25.

1970   Herbivores and the number of tree species in tropical forests. *American Naturalist* **104:**501–528.

1971   Seed predation by animals. *Annual Review of Ecology and Systematics* **2:**465–492.

1973   Tropical agroecosystems. *Science* **182:**1212–1219.

1980   When is it coevolution? *Evolution* **34:**611–612.

Jett, S. C.
1979   Peach cultivation and use among the Canyon de Chelly Navajo. *Economic Botany* **33:**298–310.

Jolly, Clifford J.
1970   The seed-eaters: A new model of hominid differentiation based on a baboon analogy. *Man* (NS) **5:**5–26.

Jones, Rhys
1975   The neolithic, paleolithic and the hunting gardeners: Man and land in the Antipodes. In *Quaternary studies,* edited by R. P. Suggate and M. M. Cresswell. Wellington: The Royal Society of New Zealand. Pp. 21–34.

Kaplan, L.
1981   What is the origin of the common bean? *Economic Botany* **35:**240–254.

Kass, D. C. L.
1978   *Polycultural cropping systems: Review and analysis.* International Agricultural Bulletin no. 32. Ithaca, New York: Cornell University International Agriculture Program.

King, L. J.
1966   *Weeds of the world: Biology and control.* London: Leonard Hill.

Kimber, C.
1966   Dooryard gardens of Martinique. *Yearbook of the Association of Pacific Coast Geographers* **28:**97–118.

1973   Spatial patterning in dooryard gardens: Implications of a Puerto Rican example. *Geographical Review* **63:**6–26.

1978   A folk context for plant domestication: Or the dooryard garden revisited. *Anthropological Journal of Canada* **16:**2–11.

King, C. E., and W. W. Anderson
1971   Age-specific selection. II: The interaction between *r* and *K* during population growth. *American Naturalist* **105:**137–156.

King, L. J.
1966   *Weeds of the world: Biology and control.* London: Leonard Hill.

Klein, L. L., and D. J. Klein
1975   Social and ecological contrasts between four taxa of neo-tropical primates. In *So-*

References 299

cioecology and psychology of primates, edited by R. H. Tuttle. The Hague: Mouton. Pp. 59–85.

Kohn, D.
1980   Theories to work by: Rejected theories, reproduction, and Darwin's path to natural selection. *Studies in the History of Biology* **4**:67–170.

Kuhn, T. S.
1962   *The structure of scientific revolutions.* Chicago: University of Chicago: University of Chicago Press.

Kurkland, J. A.
1979   Paternity, mother's brother and human sociality. In *Evolutionary biology and human social behavior,* edited by N. A. Chagnon and W. G. Irons. North Scituate, Mass.: Duxbury Press. Pp. 4–39.

Ladizinsky, G.
1979   Seed dispersal in relation to the domestication of Middle East legumes. *Economic Botany* **33**:284–289.

Lancaster, J. B.
1975   *Primate behavior and the emergence of human culture.* New York: Holt.
1976   Sex roles in primate societies. In *Sex differences: Social and biological perspectives,* edited by M. S. Teitelbaum. Garden City: Anchor Press. Pp. 22–62.

Lathrap, D. W.
1977   Our father the cayman, our mother the gourd: Spinden revisited, or a unitary model for the emergence of agriculture in the New World. In *Origins of Agriculture,* edited by Charles A. Reed. The Hague: Mouton. Pp. 713–751.

Laughlin, C. D., and E. G. d'Aquili
1974   *Biogenic structuralism.* New York: Columbia University Press.

Lee, J. A.
1960   A study of plant competition in relation to development. *Evolution* **14**:18–28.

Lee, R. B.
1968   What hunters do for a living, or, how to make out on scarce resources. In *Man the hunter,* edited by R. B. Lee and I. DeVore. Chicago: Aldine. Pp. 30–48.
1972a   Population growth and the beginnings of sedentary life among the !Kung bushmen. In *Population growth: Anthropological implications,* edited by Brian Spooner. Cambridge: MIT Press. Pp. 329–342.
1972b   The intensification of social life among !Kung bushmen. In *Population growth: Anthropological implications,* edited by B. Spooner. Cambridge: MIT Press. Pp. 343–350.

Leeds, A., and V. Dusek
1981   Editors' note. *The Philosophical Forum* 13:i–xxxv.

Leigh, E.
1965   On the relation between productivity, biomass, diversity, and stability of a community. *Proceedings of the National Academy of Science of the United States* **53**:777–783.
1968   The ecological role of Volterra's equation. *Some mathematical problems in biology, (lectures on mathematics in the life sciences)* **I**:1–61.

Lemmon, J. G.
1885   Indigenous potatoes of North America. *Transactions of the American Horticultural Society* **3**:141–150.

Levins, R.
1975   Evolution in communities near equilibrium. In *Ecology and evolution of communities,* edited by M. L. Cody and J. M. Diamond. Cambridge: Harvard University Press. Pp. 16–50.

Lewis, H. T.
1972   The role of fire in the domestication of plants and animals in Southwest Asia: A hypothesis. *Man* **7**(2):195–222.
Lewontin, R.
1958   The adaptations of populations to varying environments. *Cold Spring Harbor Symposium on Quantitative Biology* **22**:395–408.
1962   Interdeme selection controlling a polymorphism in the house mouse. *American Naturalist* **96**:705–722.
1968   *Population biology and evolution*, edited by R. Lewontin. Syracuse: Syracuse University Press.
1969   The meaning of stability. In *Diversity and stability in ecological systems*. Brookhaven Symposium in Biology No. 2. Springfield, Virginia: National Bureau of Standards, U.S. Dept. of Commerce. Pp. 13–24.
1970   The units of selection. *Annual Review of Ecology and Systematics* **11**:1–18.
1974   *The genetic basis of evolutionary change.* New York: Columbia University Press.
1979a   Fitness, survival, and optimality. In *Analysis of ecological systems*, edited by D. J. Horn, G. R. Stairs, and R. D. Mitchell. Columbus: Ohio State University Press. Pp. 3–22.
1979b   Sociobiology as an adaptationist program. *Behavioral Science* **24**:5–14.
Little, M. A., and G. E. B. Morren, Jr.
1976   *Ecology, energetics, and human variability.* Dubuque: William C. Brown.
Lotka, A. J.
1925   *Elements of physical biology.* Baltimore: Williams and Wilkins.
Lovejoy, A. O.
1936   *The great chain of being.* Cambridge: Harvard University Press.
Lovett, J. V., and G. R. Sagar
1978   Influence of bacteria in the phyllosphere of *Camelina sativa* (L.) Crantz on germination of *Linum usitatissimum* (L.) *New Phytologist* **81**:617–625.
Lovett, J. V., and H. F. Jackson
1980   Allelopathic activity of *Camelina sativa* (L.) Crantz in relationship to phyllosphere bacteria. *New Phytologist* **86**:273–277.
Lupton, F. G. H.
1977   The plant breeders' contribution to the origin and solution of pest and disease problems. In *Origins of pest, parasite, disease, and weed problems*, edited by J. M. Cherret and G. R. Sager. Oxford: Blackwell. Pp. 71–82.
Lynch, T. F.
1973   Harvest timing, transhumance, and the process of domestication. *American Anthropologist* **75**:1254–1259.
1980   Guitarrero cave in its Andean context. In *Guitarrero cave: Early man in the Andes*, edited by T. F. Lynch. New York: Academic Press. Pp. 293–320.
MacArthur, R. H.
1955   Fluctuations of animal populations and a measure of community stability. *Ecology* **36**:533–536.
MacArthur, R. H., and J. H. Connell
1966   *The biology of populations.* New York: Wiley.
MacArthur, R. H., and E. O. Wilson
1967   *The theory of island biogeography.* Princeton: Princeton University Press.
MacFayden, A.
1963   *Animal ecology: Aims and Methods* (second edition). London: Pitman and Sons.

MacNeish, R. S.
   1964a   Ancient Mesoamerican civilization. *Science* **143**:531–537.
   1964b   The food-gathering and incipient agricultural stage of prehistoric Middle America. In *Handbook of Middle American Indians* (volume 1), edited by R. C. West. Austin: University of Texas Press. Pp. 413–426.
   1967   A summary of the subsistence. In *Environment and subsistence: The prehistory of the Tehuacan Valley* (volume 1), edited by D. S. Beyers. Austin: University of Texas Press. Pp. 290–309.

Mangelsdorf, P. C.
   1974   *Corn: Its origin, evolution, and improvement.* Cambridge: Harvard University Press.

Manning, A.
   1975   Behavioral genetics and the study of behavioral evolution. In *Function and evolution in behavior,* edited by G. Baerends, C. Beer, and A. Manning. Oxford: Clarendon. Pp. 71–91.

Margalef, R.
   1968   *Perspectives in ecological theory.* Chicago: University of Chicago Press.

Martin, J. H., and W. H. Leonard
   1967   *Principles of field crop production* (second edition). London: Macmillan.

May, R. M.
   1974   *Stability and complexity in model ecosystems* (second edition). Princeton: Princeton University Press.
   1977   Population genetics and cultural inheritance. *Nature* **268**:11–13.

Maynard Smith, J.
   1969   The status of neo-Darwinism. In *Towards a Theoretical Biology,* 2, (an I.U.B.S. Symposium), edited by C. H. Waddington. Chicago: Aldine. Pp. 82–105.
   1974   *Models in ecology.* Cambridge: Cambridge University Press.
   1976   Group selection. *Quarterly Review of Biology* **51**:277–283.
   1978   Optimization theory in evolution. *Annual Review of Ecology and Systematics* **9**:31–56.

Mayr, Ernst
   1961   Cause and effect in biology. *Science* **134**:1501–1506.
   1963   *Animal species and evolution.* Cambridge: Harvard University Press.
   1976   Typological versus population thinking. In *Evolution and the diversity of life,* edited by E. Mayr. Cambridge: Harvard University Press. Pp. 26–29.
   1982   *The growth of biological thought.* Cambridge: Harvard University Press.

Mayr, Ernst, and W. Provine (Editors)
   1980   *The evolutionary synthesis.* Cambridge: Harvard University Press.

McGrew, W. C.
   1975   Patterns of plant food sharing by wild chimpanzees. In *Contemporary primatology,* edited by S. Chevalier-Skolnikoff and E. E. Poirier. New York: Garland. Pp. 261–287.

McKey, Doyle
   1975   The ecology of coevolved seed dispersal systems. In *Coevolution of animals and plants,* edited by L. E. Gilbert and P. H. Raven. Austin: University of Texas Press. Pp. 159–191.

McNaughton, S. J.
   1975   *r*- and *K*-selection in *Typha. American Naturalist* **109**:251–263.

Medawar, P. B.
   1969   *The art of the soluble.* London: Methuen.

1958   A possible case of interdependence between a mammal and a higher plant. *Archives Nearlandaises de Zoologie* **13**:314–318.

Menzel, E.
1975   Natural language of young chimpanzees. *New Scientist* **65**:127–130.

Mills, S. K., and J. H. Beatty
1979   The propensity interpretation of fitness. *Philosophy of Science* **46**:263–286.

Mitchell, R. D.
1975   The evolution of oviposition tactics in the bean weevil, *Callosobruchus maculatus*. *Ecology* **56**:696–702.

Mitchell, R. D., and M. B. Williams
1979   Darwinian analyses: The new natural history. In *Analysis of ecological systems*, edited by D. J. Horn, G. R. Stairs and R. D. Mitchell. Columbus: Ohio State University Press. Pp. 23–50.

Montagu, M. F. Ashley (Editor)
1968   *Culture: Man's adaptive dimension.* London: Oxford University Press.

Morgan, L. H.
1877   *Ancient Society.* New York: Holt.
1964   *Ancient society,* edited by Leslie White. Cambridge: Belknap Press (Harvard University Press).

Munson, R.
1971   Biological adaptation. *Philosophy of Science* **38**:200–215.
1972   Biological adaptation: A reply. *Philosophy of Science* **39**:529–532.

Murdoch, W. W.
1975   Diversity, complexity stability and pest control. *Journal of Applied Ecology* **12**:795–807.

National Academy of Sciences
1975   *Underexploited tropical plants with promising economic value.* Washington: National Academy of Sciences.

Nesom, G.
1981   Ant dispersal in *Wedelia hispida* HBK (Heliantheae: Composititae). *The Southwestern Naturalist* **25**:5–12.

Nichols, C.
1974   Darwinism and the social sciences. *Philosophy of the Social Sciences* **4**:255–277.

Odum, E. P.
1959   *Fundamentals of ecology.* Philadelphia: W. B. Saunders.

Opler, M. E.
1962   Integration, evolution and Morgan. *Current Anthropology* **2**:478–479.
1965   Cultural dynamics and evolutionary theory. In *Social change in developing areas,* edited by Herbert R. Barringer, G. I. Blacksten and Raymond Mack. Cambridge: Schenkman. Pp. 68–96.

Oppenheimer, J. R., and G. E. Lang
1969   Cebus monkeys: Effect on tracking of *Gustavia* trees. *Science* **165**:187–188.

Orians, G. H., and N. E. Pearson
1979   On the theory of central place foraging. In *Analysis of ecological systems,* edited by D. F. Hern. Columbus: Ohio State University Press. Pp. 155–177.

Oster, G. F.
1975   Review of *Ecological stability* (1974, edited by M. B. Usher and M. H. Williamson). *Ecology* **56**:1462.

Owadally, A. W.
1979   The dodo and the tambalacoque tree. *Science* **203**:1363–1364.

Papendick, R. I., P. A. Sanchez, and G. B. Triplett (Editors)
    1976   *Multiple cropping.* Madison: American Society of Agronomy.
Parker, C.
    1977   Prediction of new weed problems, especially in the developing world. In *Origins of pest, parasite, disease, and weed problems,* edited by J. M. Cherret and G. R. Sager. Oxford: Blackwell. Pp. 249–264.
Pearson, N. E.
    1976   *Optimal foraging theory.* Quantitative Science Paper No. 39. Seattle: University of Washington Press.
Perrin, R. M.
    1977   Pest management in multiple cropping systems. *Agro-Ecosystems* 3:93–118.
Peters, C. R., and E. M. O'Brien
    1981   The early hominid plant-food niche: Insights from an analysis of plant exploitation by *Homo, Pan* and *Papio* in eastern and southern Africa. *Current Anthropology* 22:127–140.
Pfeiffer, J. E.
    1976   A note on the problem of basic causes. In *Origins of African plant domestication,* edited by J. R. Harlan, J. M. J. de Wet, and A. B. L. Stempler. The Hague: Mouton. Pp. 23–28.
Pianka, Eric R.
    1970   On *r*- and *K*-selection. *American Naturalist* 104:592–597.
    1972   *r*- and *K*-selection or *b*- and *d*-selection. *American Naturalist* 106:581–588.
    1978   *Evolutionary ecology.* New York: Harper and Row.
Pielou, E. C.
    1966   Species diversity and pattern diversity in the study of ecological succession. *Journal of Theoretical Biology* 10:370–383.
    1977   *Mathematical ecology.* New York: Wiley.
Pimentel, D.
    1961   Species diversity and insect population outbreaks. *Annals of the Entomological Society of America* 54:76–86.
    1977   The ecological basis of insect, pest, pathogen and weed problems. In *Origins of pest, parasite, disease, and weed problems,* edited by J. M. Cherret and G. R. Sager. Oxford: Blackwell. Pp. 3–34.
Polgar, S.
    1975   Population, evolution and theoretical paradigms. In *Population, ecology, and social evolution,* edited by S. Polgar. Chicago: Aldine. Pp. 1–26.
Provine, W. B.
    1971   *The origins of theoretical population genetics.* Chicago: University of Chicago Press.
Pulliam, H. R.
    1974   On the theory of optimal diets. *American Naturalist* 108:59–74.
Pulliam, H. R. and C. Dunford
    1980   *Programmed to learn: An essay on the evolution of culture.* New York: Columbia University Press.
Pumpelly, R.
    1908   *Explorations in Turkestan. Expedition of 1904: Prehistoric civilization of Anau.* Pittsburgh: Carnegie Institute, Publication No. 73.
Pyke, G. H., H. R. Pulliam, and E. L. Charnov
    1977   Optimal foraging: A selective review of theory and test. *Quarterly Review of Biology* 52:137–154.

Redman, C.
    1977   Man, domestication, and culture in southeastern Asia. In *Origins of agriculture*,
              edited by C. A. Reed. The Hague: Mouton. Pp. 523–542.
Reed, C. A.
    1977a   Introduction: prologue. In *Origins of agriculture*, edited by C. A. Reed. The Hague:
              Mouton. Pp. 1–5, 9–21.
    1977b   A model for the origin of agriculture in the Near East. In *Origins of agriculture*, edited
              by C. A. Reed. The Hague: Mouton. Pp. 543–568.
    1977c   Origins of agriculture: Discussion and some conclusions. In *Origins of agriculture*,
              edited by C. A. Reed. The Hague: Mouton. Pp. 879–953.
Rehr, S. S., P. P. Feeney, and D. H. Janzen
    1973   Chemical defense in Central American non-ant acacias. *Journal of Animal Ecology*
              42:405–416.
Richerson, P. J.
    1976   Review of M. Sahlins' *Culture and practical reason*. *Human Ecology* **6**:117–121.
    1977   Ecology and human ecology: A comparison of theories in the biological and social
              sciences. *American Ethnologist* **4**:1–26.
Richerson, P. J., and R. Boyd
    1978   A dual inheritance model of the human evolutionary process I: Basic postulates and a
              simple model. *Journal of Social and Biological Structure* **1**:127–154.
Rick, C. M.
    1950   Pollination relations of *Lycopersicon esculentum* in native and foreign regions. *Evolu-
              tion* **4**:110–112.
    1958   The role of natural hybridization in the derivation of cultivated tomatoes of western
              South America. *Economic Botany* **12**:346–367.
    1978   The tomato. *Scientific American* **239**:76–87.
Ridley, N. H.
    1930   *The dispersal of plants throughout the world*. Ashford: Reeves.
Rindos, David
    1980   Symbiosis, instability, and the origins and spread of agriculture: A new model.
              Current Anthropology **21**:751–772.
Risch, S.
    1980   The population dynamics of several herbivorous beetles in a tropical agroecosystem:
              The effect of intercropping corn, beans and squash in Costa Rica. *Journal of Applied
              Ecology* **17**:593–612.
Ritterbush, P. C.
    1964   *Overtures to biology: The speculations of Eighteenth century naturalists*. New Haven:
              Yale University Press.
Rockwood, L.
    1976   Plant selection and foraging in two species of leaf-cutting ants (*Atta*). *Ecology*
              57:48–61.
Rogers, D. J.
    1965   Some botanical and ethnological consideration of *Manihot esculenta*. *Economic Bota-
              ny* **19** (4):369–377.
Rogers, D. J., and S. G. Appan
    1973   *Manihot*. Flora Neotropica Monograph No. 13. New York: Hafner Press.
Root, R. B.
    1973   Organization of a plant-arthropod association in simple and diverse habitats: The
              fauna of collards (*Brassica oleracea*). *Ecological Monographs* **43**:94–124.
Rosenberg, A.
    1982   On the propensity definition of fitness. *Philosophy of Science* **49**:268–273.

Ruse, M.
    1971   Functional statements in biology. *Philosophy of Science* **38**:87–95.
    1972   Biological adaptation. *Philosophy of Science* **39**:525–528.
    1973   *The philosophy of biology*. London: Hutchinson.
    1979   *The Darwinian revolution*. Chicago: University of Chicago Press.
Ruthenberg, H.
    1976   *Farming systems in the tropics* (second edition). London: Clarendon.
Ruyle, E. E.
    1973   Genetic and cultural pools: some suggestions for a unified theory of biocultural evolution. *Human Ecology* **1**:201–215.
Ryder, E. J.
    1976   *Latuca*. In *Evolution of crop plants*, edited by N. W. Simmonds. London: Longmans. Pp. 29–31.
Sahlins, M. D.
    1960   Evolution: Specific and general. In *Evolution and culture*, edited by M. D. Sahlins and E. R. Service. Ann Arbor: University of Michigan Press. Pp. 12–44.
Sahlins, M. D., and E. R. Service (Editors)
    1960   *Evolution and culture*. Ann Arbor: University of Michigan Press.
Salisbury, E.
    1964   *Weeds and aliens* (second edition). London: Collins.
Sarton, G.
    1970   *A history of science* (vol. 1). New York: The Norton Library.
Sauer, C. O.
    1936   American agricultural origins: A consideration of nature and culture. In *Essays in anthropology in honor of A. L. Kroeber*, edited by Robert H. Lowie. Berkeley: University of California Press. Pp. 279–298.
    1947   Early relations of man to plants. *Geographical Review* **37**:1–25.
    1952   *Agricultural origins and dispersals*. New York: American Grographical Society.
    1956   The agency of man on the earth. In *Man's role in changing the face of the earth* (volume 1), edited by Sol Tax. Chicago: University of Chicago Press. Pp. 46–69.
    1969   *Seeds, spades, hearths, and herds* (second edition). Cambridge: MIT Press.
Schieman, E.
    1939   Gedanken zur Genzentrentheorie Vavilovs. *Naturwissenshaften* **27**:377–383, 494–401.
Schleiden, M. J.
    1853a   *Die Pflanze und ihr Leben* (second edition). Leibzig: Wilhelm Engelman.
    1853b   *The plant: A biography*. Translated by Arthur Henfrey. London: H. Bailliere.
Schoener, T. W.
    1971   Theory of feeding strategies. *Annual Review of Ecology and Systematics* **2**:369–404.
Schwanitz, F.
    1966   *The origin of cultivated plants*. Cambridge: Harvard University Press.
Schweber, S.
    1977   The origin of the *Origin* revisited. *Journal of the History of Biology* **10**:229–316.
Service, E. R.
    1975   *Origins of the state and civilization: The process of cultural evolution*. New York: Norton.
    1981   The mind of Lewis H. Morgan. *Current Anthropology* **22**:25–43.
Shukovskii, P. M.
    1962   *Cultivated plants and their wild relatives*, translated by P. S. Hudson. Farnham Royal, Bucks, Enlgand: Commonwealth Agricultural Bureaus.

Silk, J. B.
    1978   Patterns of food sharing among mother and infant chimpanzees at Bombe National Park, Tanzania. *Folia Primatologica* **29**:129–141.

Simon, J.
    1978   An integration of the invention-pull and population-push theories of economic-demographic history. *Research in Population Economics* **1**:165–187.

Simmonds, N. W.
    1976   *Quinoa and relatives.* In *Evolution of crop plants,* edited by N. W. Simmonds. London: Longmans. Pp. 29–30.

Simpson, G. G.
    1953   *The major features of evolution.* New York: Simon and Schuster.
    1961   *Principles of animal taxonomy.* New York: Columbia University Press.
    1967   *The meaning of evolution* (revised edition). New Haven: Yale University Press.

Slobodkin, L. B.
    1964   The strategy of evolution. *American Scientist* **52**:342–357.

Smartt, J.
    1978   The evolution of pulse crops. *Economic Botany* **32**:185–198.

Smith, C. E., Jr.
    1969   From Vavilov to the present—A review. *Economic Botany* **23**:2–19.

Smith, C. C.
    1970   The coevolution of pine squirrels (*Tamiasi uris*) and conifers. *Ecological Monographs* **40**:349–371.
    1975   The coevolution of plants and seed predators. In *Coevolution of animals and plants,* edited by L. E. Gilbert and P. H. Raven. Austin and London: University of Texas Press. Pp. 53–77.

Smith, P. E. L.
    1972a   Changes in population pressure in archaeological explanation. *World Archaeology* **4**:5–18.
    1972b   Land-use, settlement patterns, and subsistence agriculture: A demographic perspective. In *Man, settlement, and urbanism,* edited by P. J. Ucko, R. Tringham, and G. W. Dimbleby. London: Duckworth. Pp. 409–425.

Smythe, N.
    1970   Relationships between fruiting season and seed dispersal methods in a neotropical forest. *American Naturalist* **104**:25–35.

Sober, E.
    1980   Evolution, population thinking and essentialism. *Philosophy of Science* **47**:368–386.
    1981   Holism, individualism, and the units of selection. *PSA 1980,* **2**:93–121.

South, S.
    1955   Evolutionary theory in archaeology. *Southern Indian Studies* **7**:10–32.

Spooner, B. (Editor)
    1972   *Population growth: Anthropological implications.* Cambridge: MIT Press.

Stanley, S. M.
    1975   A theory of evolution above the species level. *Proceedings of the National Academy of Science* **72**:4604–4607.
    1979   *Macroevolution: Patterns and process.* San Francisco: W. H. Freeman.

Stearns, S. C.
    1976   Life-history tactics: a review of the ideas. *Quarterly Review of Biology* **51**:3–47.

Stebbins, G. L.
    1971   Adaptive radiation of reproductive characteristics in angiosperms: seeds and seedlings. *Annual Review of Ecology and Systematics* **2**:237–260.

1974   *Flowering plants: Evolution above the species level.* Cambridge: Harvard University Press.

Stern, J. T., Jr.
1970   The meaning of "adaptation" and its relation to the phenomenon of natural selection. *Evolutionary Biology* **4:**39–66.

Steward, J.
1955   *Theory of cultural change.* Urbana: University of Illinois Press.

Stocking, G. W., Jr.
1968   *Race, culture, and evolution: Essays in the history of anthropology.* New York: The Free Press.
1974   Some problems in the understanding of nineteenth century cultural evolution. In *Readings in the history of anthropology,* edited by R. Darnell. New York: Harper and Row. Pp. 407–425.

Struever, S.
1962   Implications of vegetal remains from an Illinois Hopewell site. *American Antiquity* **27:**584–586.

Struhsaker, T. T.
1967   Ecology of vervet monkeys (*Cercopithecus aethiops*) in the Masai-Amboseli Game Reserve, Kenya. *Ecology* **48:**891–904.

Suneson, C. A.
1949   Survival of four barley varieties in mixture. *Agronomy Journal* **41:**459–461.

Suzuki, Akira
1975   The origin of hominid hunting: A primatological perspective. In *Socioecology and psychology of primates,* edited by R. H. Tuttle. The Hague: Mouton. Pp. 259–278.

Tahvanainen, J. O., and R. B. Root
1972   The influence of vegetational diversity on the population ecology of a specialized herbivore, *Phyllotketa cruciferae. Oecologia* 10:321–346.

Tanaka, Jiro
1976   Subsistence ecology of Central Kalahari San. In *Kalahari hunter-gatherers,* edited by R. B. Lee and I. De Vore. Cambridge: Harvard University Press. Pp. 99–119.

Tanner, N., and A. Zihlman
1976   Women in evolution. Part 1: Innovation and selection in human origins! *Signs* **1:**585–608.

Teleki, G.
1973   The omnivorous chimpanzee. *Scientific American* **228:**32–42.
1975   Primate subsistence patterns: Collector-predators and gatherer-hunters. *Journal of Human Evolution* **4:**125–184.

Temple, S. A.
1977   Plant–animal mutualism: Coevolution with dodo leads to near extinction of plant. *Science* **197:**885–886.

Thoday, J. M.
1958a   Effects of disruptive selection: experimental production of a polymorphic population. *Nature* **181:**1124–1125.
1958b   Natural selection and biological processes. In *A century of Darwin,* edited by S. A. Barnett. London: Heinemann. Pp. 313–333.

Thomson, G. M.
1922   *The naturalization of animals and plants in New Zealand.* Cambridge: Cambridge University Press.

Thompson, K. F.
1976   *Brassica oleracea.* In *Evolution of crop plants,* edited by N. W. Simmonds. London: Longman's. Pp. 49–52.

Tiger, L., and R. Fox
   1966   The zoological perspective in social science. *Man* (NS) **1**:75–81.
   1971   *The imperial animal.* New York: Holt.
Tindale, N. B.
   1977   Adaptive significance of the Panara or grass seed culture of Australia. In *Stone tools as cultural markers: Change, evolution, and complexity,* edited by R. V. S. Wright. Canberra: Australian Institute of Aboriginal Studies. Pp. 345–349.
Toynbee, A. J.
   1935   *A study of history* (Volume 1, second edition). London: Oxford University Press.
Trager, W.
   1970   *Symbiosis.* New York: Van Nostrand Reinhold.
Trenbath, B. R.
   1974   Biomass productivity of mixtures. *Advances in Agronomy* **26**:177–210.
Trivers, R. L.
   1971   The evolution of reciprocal altruism. *Quarterly Review of Biology* **46**:35–57.
   1972   Parental investment and sexual selection. In *Sexual selection and the descent of man,* edited by B. Campbell. Chicago: Aldine. Pp. 136–179.
   1974   Parent–offspring conflict. *American Zoologist* 14:249–263.
Ucko, Peter J., Ruth Tringham, and G. W. Dimbleby (Editors)
   1972   *Man, settlement and urbanism.* London: Duckworth.
Van den Berghe, P.
   1975   *Man in society: A biosocial view.* New York: Elsevier.
Van den Berghe, P. L., and D. P. Barash
   1977   Inclusive fitness and human family structure. *American Anthropologist* **79**:809–823.
van der Pijl, L.
   1972   *Principles of dispersal in higher plants* (second edition). New York: Springer.
Vander Wall, S. B., and R. P. Balda
   1977   Coadaptations of the Clark's nutcracker and the pinon pine for efficient seed harvest and dispersal. *Ecological Monographs* **47**:89–111.
Van Valen, L.
   1973   A new evolutionary law. *Evolutionary Theory* **1**:1–30.
   1975   Group selection, sex, and fossils. *Evolution* **29**:87–94.
Vavilov, N. I.
   1926   Studies on the origins of cultivated plants. *Bulletin of Applied Botany and Plant Breeding* **16**:1–245.
   1951   *The origin, variation, immunity, and breeding of cultivated plants,* translated by K. S. Chester. New York: The Ronald Press.
Vayda, A. P., and B. J. McCay
   1975   New directions in ecology and ecological anthropology. *Annual Review of Anthropology* **4**:293–306.
Vorzimmer, P.
   1970   *Charles Darwin: The years of controversy.* Philadelphia: Temple University Press.
Wade, M. J.
   1978   A critical review of the models of group selection. *Quarterly Review of Biology* **53**:101–114.
Wagener, P. L.
   1977   The concept of environmental determinism in cultural evolution. In *Origins of agriculture,* edited by C. A. Reed. The Hague: Mouton. Pp. 49–74.
Wallace, J. A.
   1972   Tooth chipping in the Australopithicines. *Nature* **244**:117–118.

1975   Dietary adaptations of *Australopithicus* and early *Homo*. In *Paleoanthropology, morphology, and paleoecology,* edited by R. H. Tuttle. The Hague: Mouton. Pp. 203–223.

Watson, R. A., and P. J. Watson
1969a   Early cereal cultivation in China. In *The domestication and exploitation of plants and animals,* edited by P. J. Ucko and and G. W. Dimbleby. London: Duckworth. Pp. 397–405.
1969b   *Man and nature.* New York: Harcourt.

Weber, N. A.
1966   Fungus-growing ants. *Science* **153**:587–604.
1972   Gardening ants, the attines. *Members of the American Philosophical Society* **92**:i–xvii, 1–146.

Werren, J. H., and H. R. Pulliam
1981   An intergenerational transmission model for the cultural evolution of helping behavior. *Human Ecology* **9**:465–483.

Wheeler, W. M.
1973   *The fungus growing ants of North America.* New York: Dover.

Whitaker, T. W., and W. P. Bemis
1975   Origin and evolution of the cultivated *Cucurbita*. *Bulletin of the Torrey Botany Club* **102**(6):362–368.

White, L. A.
1949   *The science of culture.* New York: Farrer-Straus.
1959a   The concept of evolution in anthropology. In *Evolution and anthropology: A centennial appraisal,* edited by B. J. Meggers. Washington, D.C.: The Anthropological Society of Washington. Pp. 106–124.
1959b   *The evolution of culture.* New York: McGraw-Hill.

Wickler, W.
1968   *Mimicry in plants and animals.* New York: McGraw-Hill.

Williams, G. C.
1966   *Adaptation and natural selection.* Princeton: Princeton University Press.
1971   *Groups selection,* edited by G. C. Williams. Chicago: Aldine.

Williams, M.
1973   The logical status of the theory of natural selection and other evolutionary controversies. In the *Methodological unity of science,* edited by M. Bunge. Dordrecht: Reidel. Pp. 84–102.

Wilson, D. S.
1980   *The natural selection of populations and communities.* Menlo Park: Benjamin/Cummings.

Wilson, E. O.
1975   *Sociobiology: The new synthesis.* Cambridge: Harvard University Press.
1978   *On human nature.* Cambridge: Harvard University Press.

Wimsatt, W.
1980   Reductionistic research strategies and their biases in the units of selection controversy. In *Scientific discovery: Case studies,* edited by T. Nickles. Dordrecht: D. Reidel. Pp. 213–259.
1981   Units of selection and the structure of the multi-level genome. *PSA 1980* **2**:122–183.

Winterhalder, B., and E. A. Smith (Editors)
1981   *Hunter–gatherer foraging strategies: Ethnographic and archaeological analyses.* Chicago: University of Chicago Press.

Wolsky, M., and A. Wolsky
1976   *The mechanism of evolution: A new look at old ideas.* Basel: S. Karger.
Wolpoff, M. H.
1973   Posterior tooth size, body size, and diet in South African Cracile Australopithicines. *American Journal of Physical Anthropology* **39**:375–394.
Woodburn, J.
1968   An introduction to Hazda ecology. In *Man the hunter*, edited by R. B. Lee and I. DeVore. Chicago: Aldine. Pp. 49–55.
Woodmause, R. G.
1977   Clusters of lumber pine trees: A hypothesis of plant–animal coaction. *Southwest Naturalist* **161**:334–339.
Wright, S.
1945   Tempo and mode in evolution: A critical review. *Ecology* **26**:415–419.
Wright, H. E., Jr.
1968   Natural environment and early food production north of Mesopotamia. *Science* **161**:334–339.
1970   Environmental changes and the origin of agriculture in the Near East. *Bioscience* **20**:210–213.
1977   Environmental change and the origin of agriculture in the Old and New Worlds. In *Origins of agriculture*, edited by C. A. Reed. The Hague: Mouton. Pp. 281–318.
Wynne-Edwards, V. C.
1962   *Animal dispersion in relation to social behavior.* New York: Hafner.
Yen, D. E.
1980   Pacific production systems. In *South Pacific agriculture: Choices and constraints*, edited by R. G. Ward and A. Proctor. Manila: Asian Development Bank. Pp. 73–106.
Young, R. M.
1969   Malthus and the evolutionists: The common context of biological and social theory. *Past and Present* **43**:109–141.
1971   Darwin's metaphor: Does nature select? *The Monist* **55**:442–503.
1980   *Mind, brain and adaptation in the nineteenth century.* Oxford: Clarendon.
Zihlman, A.
1978   Women and evolution. Part II: Subsistence and social organization among early hominids. *Signs* **4**:4–20.
Zihlman, A., and N. Tanner
1978   Gathering and hominid adaptation. In *Female hierarchies*, edited by L. Tiger and H. Fowler. Chicago: Beresford. Pp. 163–194.
Zohary, D.
1969   The progenitors of wheat and barley in relation to domestication and agricultural dispersal in the Old World. In *The domestication and exploitation of plants and animals*, edited by P. J. Ucko and G. W. Dimbleby. London: Duckworth. Pp. 47–66.
1970   Centers of diversity and centers of origin. In *Genetic resources in plants—their exploration and conservation*, edited by O. H. Frankel and E. Bennet. Oxford: Blackwell Scientific Publications Pp. 33–42.
Zubrow, E. B. W.
1975   *Prehsitoric carrying capacity: A model.* Menlo Park: Cummings.

Index